CHEMICAL MODELING
FROM ATOMS TO LIQUIDS

CHEMICAL MODELING FROM ATOMS TO LIQUIDS

Alan Hinchliffe

Reader in Chemistry
UMIST
(University of Manchester Institute of Science and Technology)
Manchester M60 1QD, United Kingdom

JOHN WILEY & SONS, LTD
Chichester · New York · Weinheim · Brisbane · Singapore · Toronto

Other Wiley Editorial Offices

John Wiley & Sons, Inc., 605 Third Avenue,
New York, NY 10158–0012, USA

WILEY-VCH Verlag GmbH, Pappelallee 3,
D-69469, Weinheim, Germany

Jacaranda Wiley Ltd, 33 Park Road, Milton,
Queensland 4064, Australia

John Wiley & Sons (Asia) Pte Ltd, 2 Clement Loop #02–01
Jin Xing Distripark, Singapore 129809

John Wiley & Sons (Canada) Ltd,
22 Worcester Road, Rexdale,
Ontario. M9W 1L1, Canada

British Library Cataloguing in Publication Data

A catalogue record for this book is available from the British Library

ISBN 0-471-99903-2
ISBN 0-471-99904-0

Typeset in 10/12pt Times by C.K.M. Typesetting, Salisbury, Wiltshire.
Printed and bound in Great Britain by Bookcraft, Midsomer Norton, Somerset.
This book is printed on acid-free paper responsibly manufactured from sustainable forestry,
for which at least two trees are planted for each one used for paper production.

Contents

Introduction

I am sure you know that gases expand when they are heated, that glass is brittle, that elastic bands can be stretched almost to infinity before they break and that metals are good conductors of electricity, that your home PC contains 'silicon chips', and that your chance of winning the UK national lottery with a single ticket is about 1 in 14 million, but do you know *why*?

And if I turn to the more esoteric properties of materials

- Why should we hope to describe the thermodynamic properties of 6×10^{23} gaseous molecules in terms of a simple equation of state such as $pV = nRT$?
- Why does a rubber band *contract* on heating, and give out heat on stretching?
- Why does liquid ^{4}He show totally different experimental behavior to ^{3}He, and what exactly is 'zero-point energy'?
- Why do electrons exhibit antisocial behavior, yet protons love to crowd together?
- Why are amino acid chains almost invariantly bent?

The word 'model' has a special meaning in science. In the old days, chemists would sit down with bits of plastic tubing of precise length together with plastic 'atoms', and construct mechanical models (which promptly fell apart) of complicated molecules. The pioneers of liquid modeling also constructed mechanical models comprising steel balls in a gelatin-filled tray, which they then shook and photographed. But the computer has now come of age in modeling, and the more streetwise (in an information technology sense) scientists reach for the keyboard in order to construct their models.

'Modeling' means having a set of mathematical equations which are capable of representing accurately the phenomenon under study. Thus, we can have a model of the UK economy just as we can have a model of the diesel engine in a GM estate car, a hydrogen molecule-ion, a polymer chain, a liquid detergent or a superconducting fluid.

You have probably read in the scientific press about chemical models with names such as Molecular Mechanics (MM), Monte Carlo (MC) and Molecular Dynamics (MD), and how chemists, physicists and biologists routinely turn to these models in order to understand and predict the properties of molecules in solution and molecular liquids. But what is the mathematical basis for these models, and what is involved in the calculations?

In my book *Modeling Molecular Structures* (a volume in the John Wiley Theoretical Chemistry Series), I concentrated on the properties of individual atoms and molecules and explained in some depth the procedures by which we can model these properties. The plan was to follow that text with one devoted to models of materials, aimed at a similar expert audience. The book proposal went out to referees, and it became apparent that there was a need for a more elementary text dealing with the modeling of materials, and one that would include a great deal of background material.

So here we are, and I make no apology for including substantial chapters on introductory classical mechanics, quantum mechanics and electrostatics. Perhaps I should have added a chapter on relativistic mechanics, and perhaps the electrostatics chapter should have been extended to include a fuller treatment of Maxwell's equations, but a line has to be drawn somewhere. It was decided as a matter of principle that no punches would be pulled with mathematical notation, particularly vectors and partial differentiation. To sweeten the pill, a Mathematical Toolkit has been included as an appendix.

The emphasis of our study is on the modeling of the material itself—the way the molecules interact with each other, the quantum mechanical consequences of having a collection of identical particles rather than a single one, and so on. A key concept for modeling materials is the intermolecular interaction potential, which describes the way the basic building blocks of atoms and molecules interact with each other when they are combined into a material. If you learn nothing else from the text, you will learn about the intermolecular and the intramolecular potentials.

Next come the statistical concepts, and in particular we need to address the problem as to how energy is divided out amongst an assembly of particles. For classical particles, this gives us the Boltzmann distribution but the interesting thing is that electrons in metals and photons emitted from black bodies don't conform to the Boltzmann distribution. They each have their own distribution.

I have taken time to give a quantum mechanical treatment of atoms and molecules. There are many commercial computer packages available if this is what you want, and I have illustrated those chapters with typical output.

Metals are completely different 'animals' from individual atoms, and we can visualize a metal as islands of cations surrounded by a sea of electrons. The important concept here is the quantum mechanical treatment of the total system of electrons. I will teach you about the strange world of the quantum, and the distinguishability (or otherwise) of seemingly identical particles.

We also have to take account of macroscopic energies, and so I have included an introductory chapter on thermodynamics.

It is easier to understand the structures and properties of gases and solids compared to the structure and properties of liquids. Solids are relatively easy to model because, to a first-approximation, they have a regular infinite lattice. Similarly, models of gases generally exploit the fact that the molecules are on average so far apart that their mean interactions are extremely weak. Although the *ideal gas model* is widely used in science, it is a non-starter in real applications, as any chemical engineer will tell you.

The advent of computing has led to great advances in our understanding of liquid structures. Computer modeling based on MC and MD techniques have stimulated the development of analytical theories, and have also challenged experimentalists to provide detailed structural and dynamical data. Modeling is a 'hot topic' in science and technology research. Many undergraduate science and technology courses now have modules that deal with modeling. Unfortunately, physicists, chemists and biologists come away from modeling courses with very different views of the subject. I hope that this book will show that we are all speaking the same language.

Acknowledgments

It is no mean task to read and critically comment on a new book manuscript. I want to put on record my thanks to my colleague Dr Joe Lee from UMIST. Joe read every word that I wrote in the first draft, and put me right on very many topics. All remaining errors, and I am sure that there are many, are entirely mine.

Dr Andrew Slade from John Wiley & Sons, Ltd gave me excellent advice about the market for this text, and agonized with me about the title.

Software Packages

I have used several specialist software packages to prepare the material for this book. The easiest thing is to give their Internet home page and a brief description of the product(s).

TableCurve3D

TableCurve3D is the first and only program that combines a powerful surface fitter with the ability to find the ideal equation to describe three-dimensional empirical data. TableCurve3D uses a selective subset procedure to fit 36 000 of its 453 697 387 built-in-equations from all disciplines to find the one that provides the ideal fit.

http://www.spss.com/

Mathcad 7 Professional

The professionals' choice for solving technical problems, with the most complete set of numeric, symbolic and visualization tools. Perform your calculations in Mathcad's highly usable free-form interface using real math notation. Supplement your calculations with text, graphs and objects.

http://www.mathsoft.com/

Gaussian94/Gaussian98

The Gaussian series of electronic structure programs are designed to model a broad range of molecular systems under a variety of conditions, performing computations starting from the basic laws of quantum mechanics. Theoretical chemists can use Gaussian to perform basic research in established and emerging areas of chemical interest. Experimental chemists can use it to study molecules and reactions of definite or potential interest, including both stable species and those compounds which are difficult or impossible to observe experimentally (short-lived intermediates, transition structures, and so on).

http://www.gaussian.com/

HyperChem

According to the *Journal of Chemical Education*, 'The new release 5 of HyperChem, an outstanding computational chemistry program, offers many significant enhancements in visualization tools such as plotting electrostatic potentials on isoelectron density surfaces. By including ... Monte Carlo simulations, the new release of HyperChem slightly extends the already significant array of quantum mechanical (Ab Initio and semi-empirical but not density functional) and classical mechanical (molecular mechanics and molecular dynamics) methods that were already featured in version 4.5.'

http://www.hyper.com

ChemDraw and Chem3D

According to the information posted on their webpage in January 1999, 'CambridgeSoft is the only company offering an integrated suite of desktop chemical applications running under Windows and Macintosh, and available in English and Japanese. CS ChemOffice combines the chemical drawing standard CS ChemDraw, the 3D modeling and simulation program CSChem3D, and the chemical structure-based information storage and retrieval facilities of CS ChemFinder. Versions of these programs with different functional capabilities are packaged with add-ons, other chemistry programs, and chemical databases to create a range of products serving the needs of all chemists, from students to experienced professionals. ...'

http://www.cambridgesoft.com

DTMM3

Desktop Molecular Modeller (DTMM) Version 3 is a simple-to-use molecular modeling program that enables you to perform powerful molecular synthesis, editing, energy minimization and display.

http://www.oup.co.uk/

Gateway2000

Unless otherwise stated, I ran all of the calculations on my office PC, a lowly Gateway2000 P5-166 with 64 MByte of RAM and 5 GByte of disk space.

http://www.gateway.com/

1 Describing Macroscopic Systems

Most chemical systems consist of very large numbers of atoms and molecules; a mole of any substance contains 6.022×10^{23} particles, a number of breathtaking and unimaginable magnitude. Nevertheless, it is found that many such collections of particles satisfy very simple laws. For example, it is found experimentally that only three of the four variables, amount of substance (n), pressure (p), volume (V) and temperature (T), are independently variable for a pure substance. The four variables are related somehow by an *equation of state* and so where better then to start this text than with the equation of state for an *ideal* (or *perfect*) gas

$$pV = nRT \qquad (1.1)$$

This simplistic equation encapsulates the work of Boyle (who showed that pV for an ideal gas was a constant, provided the temperature and the amount of substance were kept constant) and Gay-Lussac (who showed that the ratio of the volume to the temperature was a constant, provided that the pressure and the amount of substance were kept constant). The gas constant R is a fundamental constant, independent of the nature of the gas. A gas that obeys the ideal equation of state is called ideal (or perfect) gas, and all gases approach ideal behavior as their densities are reduced.

Once the range and accuracy of experimental measurements were extended, it became apparent that the product pV_m was not a constant at a given temperature, rather that it varied from gas to gas, and that it varied markedly with pressure (I will denote molar quantities such as V/n as V_m throughout this text; the term 'molar' means 'divided by the amount of substance'). A convenient quantity often used to discuss deviations from ideal behavior is the compression factor z

$$z = \frac{pV}{nRT} \qquad (1.2)$$

For an ideal gas, $z = 1$ at all temperatures and pressures and so deviations from $z = 1$ indicate deviations from ideality. Typical data are shown in Figure 1.1. The solid line refers to dihydrogen at 273 K, the dotted line refers to dinitrogen at 273 K and the dashed line refers

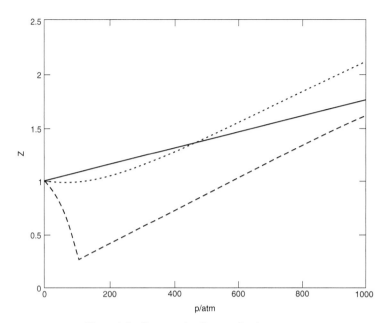

Figure 1.1   Compression factor z for three gases

to CO_2 at 313 K. The CO_2 data refer to a higher temperature because liquefaction occurs below 31°C.

Over the years, very many equations of state have been suggested. A few equations of state have some kind of a basis in theory, whilst others are just a mathematical exercise in fitting pVT data.

1.1   THE VAN DER WAALS EQUATION OF STATE

Starting from the ideal gas equation of state

$$pV = nRT$$

we make an allowance for the finite volume of particles by subtracting a quantity nb (where b is a characteristic of the particular gas) from the volume. It turns out that nb can be regarded as three times the volume of the particles present.

In a real gas there are both attractive and repulsive forces between the individual particles. The repulsive forces are most important at close separations so that at average separations of gaseous molecules the attractive forces dominate.

Near the walls of a container, single particles are not surrounded by other particles on the wall side, and this means that the pressure exerted by molecules on the walls of a container is less than that we would otherwise expect. Van der Waals suggested that we should subtract a quantity $n^2 a/V^2$ that depends on the square of the number density from the right-hand side of this equation:

$$p = \frac{nRT}{(V - nb)}$$

in order to compensate. The Van der Waals equation of state is

$$\left(p + \frac{n^2 a}{V^2}\right)(V - nb) = nRT \tag{1.3}$$

It is shown in every elementary physical chemistry textbook that the critical constants are related to a and b as follows

$$T_c = \frac{8a}{27bR}$$

$$V_v = 3bn$$

$$p_c = \frac{a}{27b^2}$$

At first sight the Van der Waals parameters a and b can be evaluated from a knowledge of the critical constants, but it turns out that the critical volume is almost impossible to measure experimentally. Instead, they are normally evaluated by numerically fitting experimental pVT data.

Using the three values of the critical constants we predict a critical compression factor

$$z_c = \frac{p_c V_c}{n T_c} \tag{1.4}$$

which works out as $3/8 = 0.375$ for the Van der Waals equation of state.

Critical constants for a few simple substances are shown in Table 1.1, together with the critical value of z (z_c). It turns out that z_c is indeed roughly constant across a wide range of substances, but it does not equal the prediction of $3/8$ from the Van der Waals equation of state.

It is an interesting fact that the Van der Waals equation of state can be rewritten in terms of the so-called reduced variables (which are dimensionless)

$$p_r = \frac{p}{p_c} \qquad T_r = \frac{T}{T_c} \qquad V_r = \frac{V}{V_c}$$

to give

$$\left(p_r + \frac{3}{V_r^2}\right)(3V_r - 1) = 8T_r \tag{1.5}$$

Table 1.1   Critical constants

	p (atm)	$V_{\mathrm{m,c}}$ (cm^3 mol^{-1})	T_c (K)	z_c
He	2.26	58.0	5.2	0.307
Ar	48.0	75.2	150.7	0.292
H$_2$	12.77	65.5	32.99	0.309
CO$_2$	72.8	94.0	304.2	0.274
CH$_4$	45.6	98.7	190.6	0.288

This dimensionless equation has the same form as the original equation, but the parameters a and b seem to have disappeared. In fact they have done no such thing, they are contained in the critical constants and hence in the reduced variables.

Van der Waals is credited for investigating the idea that substances should show the same behavior at the same values of the reduced pressure, volume and temperature. This idea turned out to be remarkably accurate, and it is called the *Principle of Corresponding States* (PCS).

The Van der Waals equation is often referred to as a *two-parameter equation of state*, because it contains the two parameters a and b that have to be fixed for each substance by appeal to experiment. The agreement with experiment is modest, and over the years more accurate equations of state have been proposed. We can crudely classify their complexity and accuracy by the number of parameters they contain; two examples (much used by chemical engineers) are given below.

1.2   BEATTIE–BRIDGEMAN EQUATION OF STATE

This widely quoted equation of state contains five parameters

$$pV_m^2 = RT\left[V_m + B_0\left(1 - \frac{b}{V_m}\right)\right]\left(1 - \frac{c}{V_m T^3}\right) - A_0\left(1 - \frac{a}{V_m}\right) \tag{1.6}$$

The five parameters a, b, c, A_0 and B_0 which are again characteristic of a given gas are found by fitting experimental pVT data. The equation ordinarily fits experimental data to within 0.15% for low densities, but is seriously in error near the critical point.

As far as I am aware, there is no particular theoretical 'justification' for this equation of state in terms of molecular interactions.

1.3   BENEDICT–WEBB–RUBIN EQUATION OF STATE

An equation of state with eight parameters has been formulated by Benedict, Webb and Rubin for the lighter hydrocarbons. This equation of state is normally written in terms of the molar density ρ_m $(1/V_m)$ as follows

$$p = RT\rho_m + \left(B_0 RT - A_0 - \frac{C_0}{T^2}\right)\rho_m^2 + (bRT - a)\rho_m^3 + a\alpha\rho_m^6 + c\rho_m^3\left(1 + \frac{\gamma\rho_m^2}{T^2}\right)\exp(-\gamma\rho_m^2) \tag{1.7}$$

The eight parameters have to be determined from experimental pVT data.

1.4   THE VIRIAL EQUATION OF STATE

In order to reproduce the experimental behavior of gases at high pressures or near their condensation temperatures, it is necessary to have equations of state that contain many

Table 1.2 Boyle temperatures

Substance	Boyle temperature T_B (K)
He	22.64
Ar	411.52
H2	110.04
CO_2	714.81
CH_4	509.66

parameters. The commonest form of the virial equation of state can be written

$$z = 1 + \frac{B(T)n}{V} + \frac{C(T)n^2}{V^2} + \cdots \tag{1.8}$$

where the coefficients $B(T)$, $C(T)$, and so on depend on the temperature and on the molecular species present. $B(T)$ is usually called the *second virial coefficient*, $C(T)$ is the *third virial coefficient*, and so on. The virial coefficients are usually determined by fitting experimental pVT data to the virial equation of state. The idea is that successive terms in the virial expansion will be of successively smaller importance.

The virial equation of state can also be written as a polynomial in p rather than V in which case we have

$$z = 1 + B'(T)p + C'(T)p^2 + \cdots \tag{1.9}$$

The virial equation of state has a special significance in that the virial coefficients can be related directly to molecular properties. The second virial coefficient $B(T)$ depends on the interaction potential between pairs of particles (to be discussed).

It sometimes happens that there is a temperature for which $B(T) = 0$ and in this case z is approximately 1. At this temperature the gas usually behaves ideally over an extended range of pressures, since $B(T)$ is the largest correction term in the virial expansion. This temperature is called the *Boyle temperature*, and it is denoted T_B. A selection of Boyle temperatures is shown in Table 1.2.

A point often missed in introductory chemistry textbooks is that equations of state should relate to all phases, not just the gas phase. Also, they should apply to mixtures as well as pure substances. Chemical engineers are well aware of these points because they have to deal with mixtures of very many components in all phases, but chemists tend not to be. The equations of state given above are reasonable for pure substances in the gaseous phase, poor for the liquid state and useless for the solid state.

1.5 THE PRINCIPLE OF CORRESPONDING STATES (PCS)

In 1873 Van der Waals first introduced the term *reduced condition* and presented the *theorem of corresponding states* that all pure substances have the same compression factor when measured at the same reduced conditions of temperature and pressure. There is ample experimental evidence for this assumption, and the concept was extended to liquids by Young in 1899.

According to the PCS theorem, the compression factor z should depend on only the reduced pressure and the reduced temperature

$$z = f(p_r, T_r)$$

Based on this Principle, generalized charts and tables were prepared for the compression factors of pure gases and fluids.

According to the Van der Waals equation the critical compression factor z_c should have the same value of 3/8 for all substances. In fact the values range from 0.20 to 0.30, and it became apparent that the charts would have to be extended to include values of z_c as a third variable.

Table 1.3 is a sample extracted from the literature.

The values of z are recorded for substances whose $z_c = 0.27$. At other values of z_c we use the correction formula

$$z' = z + D(z_c - 0.27) \qquad (1.10)$$

where D is D_b if z is <0.27, and D is D_a if z is >0.27.

The line under the value of $z = 0.0929$ indicates that condensation takes place below the corresponding reduced temperature (here 0.92). The values above the line in the z column refer to the liquid state and the values below the line refer to the gaseous state.

For example, methane has critical constants of 45.6 atm, 98.7 cm^3 mol^{-1} and 190.6 K. These give

$$z_c = \frac{45.6 \times 1.101325 \times 10^5 \,\text{N m}^{-2} \times 98.7 \,\text{cm}^3 \,\text{mol}^{-1}}{8.31451 \,\text{J K}^{-1} \,\text{mol}^{-1} \times 190.6 \,\text{K}}$$

$$= 0.287$$

z is found from Table 1.3 using this z_c and the calculated reduced temperature and pressure of interest. Methane has a reduced critical compression factor of 0.287 and so we use the value of D_a. Thus, at a reduced pressure of 0.60 and reduced temperature of 0.96, the compression factor for methane is $0.700 + 0.67(0.287 - 0.270) = 0.711$. Given this, we can calculate V_m for a particular T and p.

Table 1.3   Compression factors

$p_r = 0.60$ T_r	D_b	z	D_a
0.90	0.55	0.0908	0.60
0.92	0.55	0.0929	0.60
0.94	0.77	0.660	0.73
0.96	0.65	0.700	0.67
0.98	0.54	0.729	0.62

The $z = 0.0929$ entry is underlined to indicate a phase change from gas to liquid.

2 Thermodynamics

Thermodynamics is a study of energy changes. It leads to precise relationships between the bulk properties of systems in equilibrium and it permits answers to be given to questions like

- What is the effect of the change in temperature on the vapour pressure of a pure liquid?
- What is the position of equilibrium in a certain chemical reaction?

There are two approaches to thermodynamics; *classical thermodynamics* deals with the macroscopic properties of matter and does not rely on the fact that matter is molecular, whilst *statistical thermodynamics* aims to relate the properties of collections of large numbers of particles to their bulk behavior. Classical thermodynamics is correct in the sense that it gives demonstrably correct answers to problems such as the two posed above. It has nothing whatever to say about atoms or molecules, and indeed it was developed before the time of a reliable atomic/molecular theory. Statistical thermodynamics is an active subject because our knowledge of molecules and intermolecular forces is increasing. No evidence has ever been produced to suggest that the laws of classical thermodynamics are incorrect, and indeed statistical thermodynamics has given us increasing confidence in them.

2.1 WORK, HEAT AND HEAT CAPACITY

The concept of work will appear at several points in the text. An easy place to start is with the work done in the expansion or compression of a perfect gas. Consider a container such as the cylinder shown in Figure 2.1, filled with an amount n of a perfect gas. The temperature is constant, and the piston is free to move in order to keep the pressure of the gas p equal to the pressure of the outside world, p_{ext}.

Thermodynamics is a precise subject and we need to learn the language and the technical terms. That part of the universe under study (the ideal gas in the cylinder, in this example) is called the *system*. The remainder of the universe is called the *surroundings*.

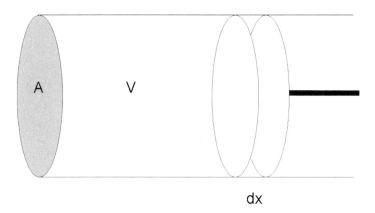

dx

Figure 2.1   Ideal gas in a cylinder

The force exerted by the surroundings on the system is the external pressure p_{ext} times the cross-sectional area of the piston, A. If, by applying extra pressure, I push in the piston by a differential distance dx and then release the extra pressure, the piston will return to its original position. We say that the system is *in equilibrium*. Systems are in equilibrium if they are stable with respect to an infinitesimal change.

Suppose that I now again gently push the piston into the cylinder. If the piston moves by distance dx then the volume will decrease by $dV = A\,dx$ (or more formally, the volume will increase by $-A\,dx$). If at any instant the external pressure is p_{ext}, then the force on the system is $A\,p_{ext}$ and so work is done in the compression

Work done = force × distance moved in direction of force

which gives

$$dw = -p_{ext}\,dV$$

In thermodynamics, it is usual to keep a note of the work done (by the surroundings) *on* the system. I therefore write

$$dw_{on} = -p_{ext}\,dV \tag{2.1}$$

The negative sign is included here because dw_{on} and p_{ext} are both positive quantities and dV (the volume increase) actually corresponds to a volume decrease and so it is negative.

If we now consider a compression from V_i to V_f then we have to integrate the above expression in order to find the total work done. A change that takes place through a series of equilibrium stages is referred to as a *reversible* change; in this case, the pressure of the system is always equal to the external pressure and so

$$w_{on} = -\int_{V_i}^{V_f} p_{gas}\,dV$$

which gives on integration, for an ideal gas being compressed at constant temperature,

$$w_{on} = -nRT \ln\left(\frac{V_f}{V_i}\right) \tag{2.2}$$

If on the other hand the change takes place very quickly, we often speak about an *irreversible change*. If the system is compressed by a constant pressure p_{ext} the work done is

$$w_{on} = -\int_{V_i}^{V_f} p_{ext}\, dV$$

which now gives on integration

$$w_{on} = -p_{ext}(V_f - V_i) \tag{2.3}$$

Mathematicians say that we need to specify the path of the integration; the work done is seen from my two examples to depend on the path taken.

Work and heat are just forms of a more general quantity called energy. The law of conservation of energy teaches us that energy cannot be created or destroyed.

We have to be careful in thermodynamics (just like in the rest of the world) to keep a profit and loss account for energy. If we compress the gas in the cylinder, then we (the surroundings) expend energy and the gas (the system) gains energy. If we let the gas expand then the gas loses energy. In both cases, I am assuming that no heat passes through the cylinder walls. As I explained above, it is usual to write w_{on} for the energy gained by the system.

Instead of doing mechanical work on the gas in the piston, we could have applied heat dq. In this case, we would say that the gas had gained energy dq.

2.2   ISOTHERMAL AND ADIABATIC CHANGES

There are two particularly important types of change that a system can undergo. An *isothermal* change occurs at constant temperature; isothermal changes usually occur when there is thermal contact between the system and the surroundings.

Adiabatic changes involve no exchange of heat with the surroundings; they usually occur in thermally isolated systems.

Heat changes are measured by calorimetry, when a temperature change dT is measured. The *heat capacity* of the system C is defined as

$$C = \frac{dq}{dT} \tag{2.4}$$

and so is the ratio of the energy gained divided by the temperature rise. C depends on the amount of substance present, and the *specific heat capacity* c is defined as the ratio of C to mass. The molar heat capacity C_m is defined likewise

$$C_m = \frac{1}{n}\frac{dq}{dT} \tag{2.5}$$

Molar heat capacities generally depend on temperature, volume and pressure (only two of which are independent). It is helpful to define two further heat capacities which relate to

Table 2.1   Molar heat capacities at 1 atm $C_{p,m}$ (J K^{-1} mol^{-1})

	300 K	800 K
C_2H_2	44.06	62.47
C_2H_4	43.72	84.52
C_2H_6	52.93	108.1
C_2F_6	106.7	160.3
C_2Cl_6	136.8	172.8
C_6H_6 (benzene)	82.22	188.5
$C_{12}H_{18}$ (hexamethylbenzene)	249.9	474.9

measurements made at constant pressure and at constant volume

$$C_{V,m} = \frac{1}{n}\left(\frac{dq}{dT}\right)_V \tag{2.6}$$

$$C_{p,m} = \frac{1}{n}\left(\frac{dq}{dT}\right)_p \tag{2.7}$$

which are the molar heat capacities at constant volume and pressure respectively. They are generally functions of temperature, and they generally increase with molecular size as Table 2.1 shows.

Heat capacities of pure substances are obviously strongly temperature dependent, and they are often reported in the literature by expressions such as

$$C_{p,m} \, (\text{J K}^{-1}\,\text{mol}^{-1}) = a + b\left(\frac{T}{K}\right) + c\left(\frac{T}{K}\right)^2 + \cdots \tag{2.8}$$

where a, b and c are pure numerical constants that have been derived by fitting experimental values. These constants $a, b, c, \ldots$ can be positive or negative, but these heat capacities always increase with temperature. To look ahead a little, I will tell you that they reach a different limiting value at high temperatures depending on the complexity of the molecule.

2.3   SYSTEMS AND STATES

At this point, it is convenient to record some further definitions. The most restrictive type of thermodynamic system is an *isolated* one, where neither material nor energy can be exchanged with the surroundings. This does not preclude chemical reactions within the system. In a *closed* system, material cannot be exchanged with the surroundings. An *open* system is a much more general construct, for both energy and matter can flow into and out of such a system.

The thermodynamic *state* of a system can be described by a set of observables. The state of a homogeneous system can be described by stating the amount of each substance present together with two additional quantities such as the temperature and pressure. Only two of the three properties temperature, pressure and volume are independent; if any two are specified then the other is fixed. The equation of state gives a relationship between these properties.

For a heterogeneous system, we must specify the amounts of each substance present in each phase, in addition to the temperature and the pressure.

Classical thermodynamics is a very precise and succinct subject for it consists of just four laws together with a handful of useful definitions.

2.4 THE ZEROTH LAW

The zeroth law states that if system A is in thermal equilibrium with system B, and system B is in thermal equilibrium with system C, then system A must be in thermal equilibrium with system C.

This law may seem obvious or even trivial, but it indicates that systems have a property that is independent of their nature and which tells us about the existence of thermal equilibrium. This property is called the *temperature*. According to popular legend, the zeroth law is so named because it was discovered after the first, second and third laws but logically it precedes them in concept.

2.5 THE FIRST LAW

The first law is to do with heat and work, and rests on the experimental fact that both are just different forms of energy that are freely interconvertible. It deals with the internal energy U of a system, which can be thought of as the sum of the kinetic and potential energies of the molecules comprising the system. The first law does not specify U fully, rather it is a statement about the changes in U when a closed system gains or loses energy. Thus, if dq is added to such a system and dw_{on} is performed on the system, then the system gains internal energy dU

$$dU = dq + dw_{on}$$

The law can also be written in terms of the initial and final states of the system

$$U_f - U_i = q + w_{on}$$

where q is the total heat added to, and w_{on} the total work done on the system A convenient shorthand is the Δ symbol; ΔU means the change in U or in other words $U_f - U_i$. Note that ΔU could be positive or negative. Thus the first law can be written

$$\Delta U = q + w_{on} \tag{2.9}$$

It was shown earlier that the work done in a change generally depended on how the change was made. The same is true of the heat absorbed. ΔU is quite different; it depends only on the initial and final states of the system and not on how we made the change from the initial state to the final state.

Functions like U (as distinct from q and w_{on}) play an important role in thermodynamics, and they are given a special name. They are called *functions of state*.

A mathematical digression is in order here. In the Appendix, exact and inexact differentials are discussed. A necessary and sufficient condition for the general differential

$$dz = M\,dx + N\,dy$$

to be an exact differential is that

$$\left(\frac{\partial M}{\partial y}\right)_x = \left(\frac{\partial N}{\partial x}\right)_y$$

If dz is an exact differential, then

$$M = \left(\frac{\partial z}{\partial x}\right)_y \quad \text{and} \quad N = \left(\frac{\partial z}{\partial y}\right)_x$$

and when this condition is satisfied, it is possible to integrate dz to give

$$\int_i^f dz = z_f - z_i$$

which depends only on the initial and final states.

2.6 THE ENTHALPY H

It should be clear from my statement of the first law that $\Delta U = q$ when no work is done on the system. Commonly, we only concern ourselves with the work done by a reversibly expanding system and so changes in U are equal to the heat put into the system provided that the volume is constant (since $dw_{on} = -p_{ext} \, dV$).

In many cases we are interested in reversible changes that take place at constant pressure rather than constant volume, usually with the pressure equal to atmospheric pressure. If the pressure is indeed constant and no form of work other than that incurred by a volume change is involved, then

$$U_f - U_i = q - p(V_f - V_i)$$

and this can be conveniently rewritten as

$$(U_f + pV_f) - (U_i - pV_i) = q \tag{2.10}$$

This suggests that the quantity $U + pV$ will be a useful one under conditions of constant pressure, since changes in $(U + pV)$ are related to the heat change.

The *enthalpy H* is therefore defined as

$$H = U + pV \tag{2.11}$$

H, like U, is a function of state; changes in H depend only on the final and initial state of the system. At constant pressure, $\Delta H = q$.

If we measure a value of ΔU or ΔH for a pure substance, then this value will depend on the initial and final values of three of the four quantities, temperature, the amount of substance present, the pressure and the volume (three rather than four since the four are related by an equation of state). Of particular interest are changes done at constant volume or at constant pressure.

We can record the temperature and pressure in parentheses ΔH (298 K, 1 atm) or we can add pressure information as a superscript as in ΔH^0. The presence of the superscript by itself does not define this so-called *standard state*, the value of the pressure (or volume, etc.) has to be explicitly stated. The subscript *m* as in ΔH_m means a molar quantity.

The heat capacities of pure substances at constant volume and constant pressure can be easily related to U and H as follows

$$C_V = \left(\frac{\partial U}{\partial T}\right)_V \tag{2.12}$$

$$C_p = \left(\frac{\partial H}{\partial T}\right)_p \tag{2.13}$$

2.6.1 Bond Energies and Enthalpies

Information about the strengths of chemical bonds can be deduced by measuring the enthalpy change (or the internal energy change) for a 'reaction' such as

$$AB(g) \rightarrow A(g) + B(g)$$

The molar enthalpy change for this reaction is called the *bond dissociation enthalpy*. This quantity is usually discussed in terms of diatomic molecules, and some typical values are shown in Table 2.2.

This bond dissociation enthalpy has to be distinguished from the *mean* bond dissociation enthalpy for a polyatomic molecule. Consider the simple case of CH_4. The molar enthalpy changes for the four reactions

$$CH_4 \rightarrow CH_3 + H$$
$$CH_3 \rightarrow CH_2 + H$$
$$CH_2 \rightarrow CH + H$$
$$CH \rightarrow C + H$$

ought to be slightly different, on chemical grounds.

If we take an average of the values for the four reactions given above, then we arrive at the *mean C–H bond dissociation enthalpy*. This is usually referred to as (simply) the C–H *bond enthalpy*. Bond enthalpies are important because they allow us to predict the enthalpy change in a general chemical reaction.

Table 2.2 Bond dissociation enthalpies at 298 K

	Value (kJ mol^{-1})
H–H	436
H–F	565
H–Cl	431
H–Br	366
O=O	497

2.7 ENTROPY AND THE SECOND LAW

The first law was derived from experiments concerned with the interconversion of various forms of energy, and it is a succinct statement of the fact that energy cannot be created or destroyed.

Experience of mechanics suggests that systems when left to themselves will tend to a state of minimum energy. This turns out to be only half of the story. Consider for example the simple experiment shown in Figure 2.2. The left-hand bulb contains Ne gas and the right-hand bulb contains Ar gas, and the apparatus is surrounded by a constant temperature bath.

If we open the stopcock and measure the composition of the gases in the bulbs at a later time, then chemical experience tells us that the gases will be equally distributed between the left- and right-hand bulbs. I chose a pair of inert gases deliberately to divert you from thinking about energy changes. The energy change of mixing can be measured experimentally, and it is essentially zero. There is clearly a second driving force for chemical and physical equilibria.

The second law of thermodynamics is based on the experimental observation that it is impossible to achieve perpetual motion, and it is concerned with a function of state called the *entropy S*. The second law determines which way a change will proceed if left to itself. It can be stated in many different (but equivalent) ways, and simple mathematical statement as is

$$dS = \frac{dq_{rev}}{T} \tag{2.14}$$

which relates the gain in entropy dS by the system when it absorbs heat dq_{rev} in a reversible infinitesimal change.

Many texts have been written about entropy. It is often said that entropy is 'time's arrow'. To understand such statements we have to consider the system, the surroundings, reversible and irreversible changes. The entropy of a system can stay constant, increase or decrease when a change is made. There is however no 'law of conservation of entropy'; roughly speaking, the entropy of the system plus the surroundings is constant for a reversible change, but the entropy of the system plus the surroundings always increases for an irreversible change. Natural changes are irreversible, and the entropy of an isolated (constant energy) system tends to increase until it reaches a state of maximum entropy. The total entropy of the system plus surroundings is not conserved, it tends to a maximum.

Entropy S is a function of state, just like U and H. For a pure substance it is determined exactly by three of the four variables pressure, volume, amount of substance and temperature.

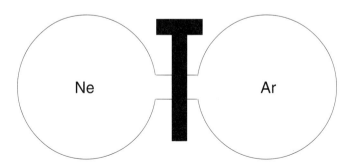

Figure 2.2   Mixing two inert gases

The definition given above does not determine entropy absolutely since there is an unknown constant of integration, S_0 in the equation

$$S_T = S_0 + \int_{0\,K}^{T} \frac{\mathrm{d}q_{\mathrm{rev}}}{T} \tag{2.15}$$

The definition does allow changes in entropy to be found. For example, consider the phase change

$$\text{ice} \rightarrow \text{water at 273 K and constant pressure of 1 atm}$$

These are equilibrium conditions for the ice/water system.

The entropy change is $\Delta S = S_{\mathrm{ice}} - S_{\mathrm{water}}$ which is given, in such a reversible situation, by the enthalpy of melting ΔH_f divided by the temperature of melting T_f

$$\Delta S = \frac{\Delta H_f}{T_f}$$

The entropy change when a substance is heated from T_i to T_f at constant pressure can also be found quite simply to be

$$\Delta S = \int_{T_i}^{T_f} \frac{C_p}{T} \, \mathrm{d}T \tag{2.16}$$

In order to find this change in entropy, we have to measure the heat capacity of the substance as a function of temperature between T_i and T_f. Integration yields the entropy change.

2.8   THE THIRD LAW

The third law fixes the constant of integration S_0 and so permits us to determine absolute values of the entropy. A statement of the law is:

The entropy of a perfectly crystalline substance is zero at 0 K

This statement probably raises a question as to what is a 'perfectly crystalline substance', but we will let that pass for the minute. The phrase should be intuitively obvious at this point.

Standard molar entropies (usually given at 298 K and under conditions of either 10^5 Pa or 1 atm pressure) are widely reported for pure substances. They are found from experimental measurements of heat capacity with temperature, and from measurements of the enthalpy changes that accompany phase changes (including mesomorphic and allotropic solid → solid, solid → liquid, and so on). These quantities are usually given the prefix *thermodynamic* because they rely on measurements of heat capacity and enthalpies of transition, but most importantly on the third law for their absolute values.

2.8.1   Molar Enthalpies of Formation and Molar Entropies

Consider the formation of ethane from the reaction

$$2C \, (\text{graphite}) + 3H_2(g) \rightarrow C_2H_6(g)$$

Internal energies, enthalpies and entropies are all functions of state and so the values of ΔU, ΔH and ΔS for this reaction only depend on the initial and final states of the reactants and products and not on the way the change gets carried out. We make use of this idea by tabulating the so-called standard molar internal energies and enthalpies of formation, and standard entropies, for molecules. This is usually done for a temperature of 298 K. These values are just the experimental values of ΔU_m (1 atm or 10^5 Pa) for the reaction that forms the molecule of interest starting from the constituent elements in their *standard state*, that is their most stable form at the specified temperature and pressure. The physical state of the substance is that which is stable at the specified temperature and pressure.

All elements in their standard state have zero molar enthalpy of formation, by definition.

Ions in solution pose a problem because we cannot imagine a reaction for the formation of H^+ that would not also involve a negative ion. We therefore fix, arbitrarily, the standard enthalpy of formation of aqueous H^+ at zero at all temperatures.

2.8.2   Atom and Bond Contributions

Atom and bond concepts dominate chemistry. Many molecular properties such as the molar volume, the diamagnetic susceptibility and several thermodynamic functions have been shown to obey additivity rules where the molecular property can be calculated as a sum of atom and/ or bond contributions. Benson's definitive work gives the atom contributions to the standard (at 1 atm pressure) molar heat capacity at 298 K (Table 2.3).

Thus for example the molar heat capacity of CH_3CHO would be evaluated as $(2 \times 15.7 + 4 \times 3.56 + 14.2)$ J K^{-1} mol^{-1} = 59.840 J K^{-1} mol^{-1}, which compares moderately well with the experimental value at 298 K of 54.8 J K^{-1} mol^{-1}.

The next higher level of approximation is to consider the additivity of bond properties. Tables of bond properties have been developed, and they can be used to predict thermodynamic properties such as the heat capacities and enthalpies of formation.

For example, Table 2.4 gives C_p^0 values at 298 K which gives a value of $(4 \times 7.28 + 1 \times 8.28 + 1 \times 17.6)$ J K^{-1} mol^{-1} = 55.0 J K^{-1} mol^{-1} for the standard molar heat capacity of CH_3CHO at 298 K, in better agreement with experiment than the 'atom' value.

Such calculations do not distinguish between isomers, and the next order of approximation is to consider molecular properties as the sum of contributions due to groups. The word 'group' here has a different meaning than in organic chemistry; a group is defined as an

Table 2.3   Atom contributions to molar heat capacities

Atom	$C_{p,m}^0$(J K^{-1}mol^{-1})
H	3.56
D	5.02
C	15.7
N	14.2
O	14.2
F	10.0
Cl	15.5
Br	17.8
Si	24.7
S	19.7

Table 2.4 Bond contribution to molar heat capacities

Bond	C_p^0 ($J K^{-1} mol^{-1}$)
CH	7.28
CC	8.28
CO as in CHO	17.6

atom in a molecule that is bound to at least two other atoms. Benson gives parameters for the estimation of heat capacities, standard entropies and enthalpies of formation.

2.9 DIFFERENTIAL RELATIONSHIPS

This is a good point to summarize the three laws and a useful function. I will restrict myself to systems where the amounts of substance are kept constant.

1. $dU = dq + dw_{on}$

2. $dS = \dfrac{dq_{rev}}{T}$

3. The entropy of a perfectly crystalline substance is zero at 0 K

A useful function

$$H = U + pV$$

Since U is a function of state dU is an exact differential which means that it can be written in the form

$$dU = \left(\frac{\partial U}{\partial T}\right)_V dT + \left(\frac{\partial U}{\partial V}\right)_T dV \qquad (2.17)$$

if T and V are taken to be the independent variables, or

$$dU = \left(\frac{\partial U}{\partial T}\right)_p dT + \left(\frac{\partial U}{\partial p}\right)_T dp \qquad (2.18)$$

if T and p are taken to be the independent variables. The first equation is preferred on experimental grounds.

Just to remind you, both these equations assume that the amount of substance is constant. If we were to consider a pure, single phase system where n could vary, then we would have to consider the variation of U with n and we would write (for example)

$$dU = \left(\frac{\partial U}{\partial T}\right)_{V,n} dT + \left(\frac{\partial U}{\partial V}\right)_{T,n} dV + \left(\frac{\partial U}{\partial n}\right)_{V,T} dn \qquad (2.19)$$

Likewise we write for the enthalpy

$$dH = \left(\frac{\partial H}{\partial T}\right)_{p,n} dT + \left(\frac{\partial H}{\partial p}\right)_{T,n} dp + \left(\frac{\partial H}{\partial n}\right)_{p,T} dn \qquad (2.20)$$

Consider now a closed system that undergoes a reversible pressure–volume change. Combination of the first and second laws gives, after a little manipulation

$$dU = T\,dS - p\,dV \tag{2.21}$$

with a corresponding expression for dH

$$dH = T\,dS + V\,dp \tag{2.22}$$

Comparison between equations (2.17) and (2.21), and between equations (2.20) and (2.22) gives

$$\left(\frac{\partial U}{\partial S}\right)_V = T$$

$$\left(\frac{\partial U}{\partial V}\right)_S = -p$$

$$\left(\frac{\partial H}{\partial S}\right)_p = T \tag{2.23}$$

$$\left(\frac{\partial H}{\partial p}\right)_S = -V$$

which are a set of useful thermodynamic relations.

If we relax the conditions to cover the case of irreversible (i.e. spontaneous) pressure-volume changes, then my previous statement of the second law needs modification to read

$$dS \geq \frac{dq}{T}$$

and we find

$$dU \leq T\,dS - p\,dV$$

$$dH \leq T\,dS + V\,dp$$

Internal energy/enthalpy and entropy are useful basic concepts for discussing the direction of natural changes. Under conditions of constant pressure, there are two competing effects. The enthalpy change ΔH and the entropy change ΔS.

Systems move to equilibrium to minimize H (if S is constant)

Systems move to equilibrium to maximize S (if H is constant)

Under conditions of constant volume, there are also two competing effects, in this case the internal energy change ΔU and the entropy change ΔS.

Systems move to equilibrium to minimize U (if S is constant)

Systems move to equilibrium to maximize S (if U is constant)

2.10 THE GIBBS AND THE HELMHOLTZ ENERGIES

It proves useful to introduce two new state functions, the Gibbs energy

$$G = H - TS \qquad (2.24)$$

and the Helmholtz energy

$$A = U - TS \qquad (2.25)$$

G and A are the arbiters between energy and entropy; they take account of the different demands of energy and entropy, depending on whether the pressure is constant (in which case G is the driving force) or whether the volume is constant (in which case A is the driving force).

Assume once more the same restrictive conditions as above, that is to say we have a closed system that experiences a reversible pressure–volume change. Starting from the first and second laws, and the definitions of G and A, you will easily demonstrate that

$$dG = -S\,dT + V\,dp$$

$$dA = -S\,dT - p\,dV$$

If we now allow for irreversible changes then

$$dG \leq -S\,dT + V\,dp$$

$$dA \leq -S\,dT - p\,dV$$

Finally, we should consider the possibility that a number of substances are present. If n_i is the amount of substance i then (for example) the Gibbs energy will depend on some complicated and unknown way on p, T, n_1, n_2 and so on,

$$G = G(p, T, n_1, n_2, \ldots)$$

and so a small change dG can be

$$dG = \left(\frac{\partial G}{\partial p}\right)_{T,n_i} dp + \left(\frac{\partial G}{\partial T}\right)_{p,n_i} dT + \sum_i \left(\frac{\partial G}{\partial n_i}\right)_{p,T,n_j \neq n_i} dn_i \qquad (2.26)$$

The summation in the final term runs over all the substances present, and all the n_j except for the one under consideration, n_i, are left constant during the differentiation of G in $\partial G/\partial n_i$.

2.11 THE CHEMICAL POTENTIAL

The chemical potential of species i can be defined by

$$\mu_i = \left(\frac{\partial G}{\partial n_i}\right)_{p,T,n_j \neq n_i} \qquad (2.27)$$

The following equivalent formulae for μ, which are useful when different quantities are kept constant, can be easily derived

$$\mu_i = \left(\frac{\partial U}{\partial n_i}\right)_{S,V,n_j \neq n_i}$$

$$\mu_i = \left(\frac{\partial H}{\partial n_i}\right)_{S,p,n_j \neq n_i}$$

$$\mu_i = \left(\frac{\partial G}{\partial n_i}\right)_{p,T,n_j \neq n_i}$$ (2.28)

$$\mu_i = \left(\frac{\partial A}{\partial n_i}\right)_{V,T,n \neq n_i}$$

It is easily demonstrated that

$$\sum_i \mu_i \, dn_i = 0$$

$$\left(\frac{\partial G}{\partial p}\right)_{T,n_i} = V$$

$$\left(\frac{\partial G}{\partial T}\right)_{p,n_i} = -S$$

so that

$$dG = V \, dp - S \, dT + \sum_i \mu_i \, dn_i$$ (2.29)

and this essential equation is sometimes called the fundamental equation of chemical thermodynamics. It is useful for treating physical and chemical equilibria.

2.12   APPLICATIONS TO EQUILIBRIUM

The variation of the chemical potential with pressure for a pure ideal gas at a constant temperature can be found by integrating $(\partial G/\partial p)_{T,n_i} = V$; if the temperature is constant then $dT = 0$, and for an ideal gas $V = nRT/p$ so that

$$G = G^0 + \int_{p_0}^{p} nRT \, \frac{dp}{p}$$

which gives

$$G = G^0 + nRT \, \log_e\left(\frac{p}{p^0}\right)$$ (2.30)

If we take the standard pressure to be (for example) 1 atm, then we can also write

$$G = G^0(1 \, \text{atm}) + nRT \, \log_e\left(\frac{p}{1 \, \text{atm}}\right)$$

Since the chemical potential for a pure substance can be expressed as

$$\mu = \left(\frac{\partial G}{\partial n}\right)_{p,T}$$

we see that

$$\mu = \mu^0 (1 \, atm) + RT \, \log_e \left(\frac{p}{1 \, atm}\right) \tag{2.31}$$

I have proved this only for a pure ideal gas. It is also true for any ideal gas mixture. Consider now a simple chemical reaction

$$A \ \rightarrow \ B$$

and that an infinitesimal amount $d\xi$ of A turns into B. Then

- change in amount of A is $-d\xi$
- change in amount of B is $+d\xi$

At constant temperature and pressure

$$dG = \mu_A dn_A + \mu_B \, db_B$$

and therefore

$$dG = (\mu_B - \mu_A) \, d\xi$$

Now the chemical potentials depend on the composition of the system and their values will change as the reaction proceeds.

- So long as $\mu_A > \mu_B$, dG is negative and the reaction proceeds from A to B.
- So long as $\mu_B > \mu_A$, dG is positive and the reaction proceeds from B to A.
- When $\mu_A = \mu_B$, the system is in equilibrium.

For a standard state of 1 atm, we have shown above that

$$\mu = \mu^0(1 \, atm) + RT \, \log_e \left(\frac{p}{1 \, atm}\right)$$

so if we write p_A and p_B for the equilibrium gas pressures we have

$$\mu_A^0(1 \, atm) + RT \, \log_e \left(\frac{p_A}{1 \, atm}\right) = \mu_B^0(1 \, atm) + RT \, \log_e \left(\frac{p_B}{1 \, atm}\right) \tag{2.32}$$

which rearranges to give

$$RT \, \log_e \left(\frac{p_B}{p_A}\right) = -[\mu_B^0(1 \, atm) - \mu_A^0(1 \, atm)]$$

The standard molar Gibbs energy change for the reaction is

$$\Delta G_m^0(1 \, atm) = \mu_B^0(1 \, atm) - \mu_A^0(1 \, atm)$$

and the ratio of the equilibrium pressures is just the equilibrium constant K_p. Thus

$$RT \, \log_e(K_p) = -\Delta G_m^0 \tag{2.33}$$

Although this equation was derived for the simple reaction A $\rightarrow$ B, the result is perfectly general for a reaction involving ideal gases.

3   Résumé of Classical Mechanics

Mechanics is a branch of math that deals with two kinds of phenomena; the forces acting on bodies at rest (statics) and the study of why things move (dynamics). Many authors just refer to 'Mechanics' for the subject as a whole, and do not make any distinction between things in motion and things at rest.

I have to make a start with this topic at some convenient place, and I therefore give a little of the historical background before launching directly into the three laws of mechanics.

One of the most obvious facts about the world around us is that it contains matter in motion. You should not be surprised to learn that mechanics has been a central theme in the development of science. The ancient Greeks were obviously spectacularly wrong in their attempts to explain why and how things move. For example, Aristotle claimed that if one body were twice as heavy as another, it would fall to earth twice as quickly. This idea held sway for almost 2000 years until Galileo (1564–1642) actually did the experiments in Pisa, Italy.

Galileo's great achievement was to realize the importance of observation and measurement in science.

In the seventeenth century, the Danish astronomer Tycho Brahe made detailed and accurate measurements of planetary motions. The German mathematician Johannes Kepler was able to show that the planets moved in elliptic paths, thus disproving Aristotle's assertion of circular orbits.

3.1   NEWTON'S LAWS

Isaac Newton (1642–1727) presented mechanics to the world as a scientific subject with his publication of the *Principia Mathematica*. Newton gave us three laws, which are valid in all but the most extreme circumstances. He brought order and simplicity to dynamical processes, and showed that the entire universe obeyed the same set of three laws.

1. **Any body remains in a state of rest or of uniform motion unless an unbalanced force acts upon it.**

This is just a statement that accelerations are caused by something. The name given to that something is the force. I have used the term 'unbalanced force' because there may be many forces acting on a body at any one time. Only a non-zero resultant produces a change from the state of rest or uniform motion.

2. **In order to make a body of mass *m* undergo an acceleration a, a force F is required that is equal to the product of the mass times the acceleration. In symbols F = *ma*.**

Newton's second law is a vector statement, and I have therefore cheated a little. Vector analysis was not discovered until the time of J. Willard Gibbs (about 1880).

This particular statement of the second law is valid only if the mass is constant. Think of a space probe moving towards a distant star. The probe will have a certain mass on blast-off, but will gradually use its fuel and so its mass will decrease. We obviously have to allow for this change in the mass in order to calculate the path of the probe.

The correct statement of the law should be that force is the rate of change of linear momentum (which we have yet to meet).

Also we should note that force is a vector quantity, and occasionally the forces acting on a body will add up vectorially to zero.

Now consider two bodies, one labelled A and the other one labelled B. If we denote the force exerted by body A on body B as $F_{A\ on\ B}$ then Newton's third law tells us that

3. **When two bodies A and B interact with each other, the force exerted by body A on body B, $F_{A\ on\ B}$ is equal and opposite to the force exerted by body B on body A $F_{B\ on\ A}$.**

3.2   MOMENTUM

Although Newton's laws enable us to solve problems involving motion and forces, we are often faced with problems where we do not know the forces, or where the forces vary rapidly over the timescale of the experiment. Consider an experiment where two atoms collide. As I will show in later chapters, the force between the two atoms changes dramatically with the distance between the atoms. We can certainly calculate the force at every instant of time during the collision of the two atoms, but we would have to work out the accelerations at every instant of time in order to use Newton's law in order to study the problem.

The way round this problem is to introduce a new concept, the *linear momentum*. This is defined as

$$\mathbf{p} = m\mathbf{v} \tag{3.1}$$

for a body of mass *m* moving with velocity vector **v**.

I will demonstrate shortly that the momentum is conserved, and this simple fact enables us to say everything there is to say about the motion of two bodies before and after a collision.

The statement of Newton's second law $\mathbf{F} = m\mathbf{a}$ is clearly consistent with the statement that $\mathbf{F} = d\mathbf{p}/dt$, provided the mass is constant. In fact, Newton's second law should be written in terms of the linear momentum as $\mathbf{F} = d\mathbf{p}/dt$ rather than the statement I gave above linking force to acceleration.

To demonstrate that linear momentum is conserved, consider my statement of the third law given above.

$$\mathbf{F}_{A\ on\ B} = -\mathbf{F}_{B\ on\ A}$$

I now identify the definitions of the forces with the rates of changes of the momenta, using Newton's second law; $\mathbf{F}_{A \text{ on } B}$ is the force exerted by body A on body B, and so it produces a change in the linear momentum $\mathbf{p}_B$ of body B. Likewise $\mathbf{F}_{B \text{ on } A}$ is the force exerted by body B on body A, and so it produces a change in the linear momentum $\mathbf{p}_A$ of body A.

Hence

$$\frac{d\mathbf{p}_B}{dt} = -\frac{d\mathbf{p}_A}{dt}$$

or

$$\frac{d\mathbf{p}_B}{dt} + \frac{d\mathbf{p}_A}{dt} = 0$$

or

$$\frac{d}{dt}(\mathbf{p}_A + \mathbf{p}_B) = 0$$

Thus the rate of change of linear momentum in body A is always balanced by the rate of change of the linear momentum in body B, and this is true at every instant of time.

Not only that, but the total linear momentum $\mathbf{p}_A + \mathbf{p}_B$ is constant throughout the collision of two bodies, even though each momentum might be separately varying.

It turns out that this is a fundamental law of physics, and we refer to it as *the law of conservation of momentum*.

The total linear momentum of an isolated system is conserved.

3.3   ENERGY

I now need to discuss two kinds of energy; the energy that a body has by virtue of its motion (the *kinetic energy*), and the energy that a body has by virtue of its shape or position in space (the *potential energy*). I should also remind you that these two forms of energy can be inter-converted in such a way that the total amount of energy in a system is constant. This is the *law of conservation of energy*, one of the most important laws of nature.

The easiest place to start the discussion is to consider the motion of a body of mass m, moving with a constant acceleration. The speed acquired and the distance travelled can be calculated by using a simple set of equations known as the *constant acceleration equations*.

These equations are most easily solved in one dimension, so I am going to consider motion along the x axis. If I denote the time as t, then the speed is dx/dt and the (constant) acceleration is d^2x/dt^2.

(i) If d^2x/dt^2 is a constant (call it a) then

$$\frac{d^2x}{dt^2} = a$$

$$\frac{dx}{dt} = at + b$$

where b is a constant of integration. If the speed $dx/dt = u$ when $t = 0$ and the speed $dx/dt = v$ at time t then

$$v = u + at$$

which equation relates the final (v) and initial (u) speeds of a body that undergoes a constant acceleration a.

In vector notation the equation becomes

$$\mathbf{v} = \mathbf{u} + \mathbf{a}t \qquad (3.2)$$

(ii) The second equation relates the distance travelled by a moving body subject to a constant acceleration. We start from (i) above

$$v = u + at$$

$$\frac{dx}{dt} = u + at$$

where u is the initial speed (a constant). Integrating again we find

$$x = ut + \frac{1}{2}at^2 + c$$

where c is a constant of integration. If the body starts from position x_0 along the x axis then $x = x_0$ when $t = 0$ and so

$$x = ut + \frac{1}{2}at^2 + x_0$$

In vector notation, the equation relates the position vector $\mathbf{r}$ to the velocity $\mathbf{v}$ and acceleration $\mathbf{a}$ by

$$\mathbf{r} = \mathbf{u}t + \frac{1}{2}\mathbf{a}t^2 + \mathbf{r}_0 \qquad (3.3)$$

(iii) Starting from the definition of constant acceleration

$$\frac{d^2x}{dt^2} = a$$

which we can rewrite as

$$\frac{d}{dt}\left(\frac{dx}{dt}\right) = a$$

or

$$\frac{1}{2}\frac{d}{dt}(v^2) = a$$

$$\frac{1}{2}\frac{dx}{dt}\frac{d}{dx}(v^2) = a$$

If we now integrate with respect to distance x, and substitute that the initial speed is u, we find

$$v^2 = u^2 + 2as \qquad (3.4)$$

where s is the distance travelled in the time t. This is our third equation for constant acceleration; it tells us the relationship between the final velocity v, the initial velocity u, the acceleration a and the distance travelled s.

This final equation is an interesting one, especially if we rearrange it as follows;

$$v^2 = u^2 + 2as$$

$$\frac{m}{2}v^2 = \frac{m}{2}u^2 + ams$$

$$\frac{m}{2}v^2 = \frac{m}{2}u^2 + Fs$$

or, rearranging the final line we get

$$\frac{m}{2}v^2 - \frac{m}{2}u^2 = Fs \tag{3.5}$$

The left-hand side seems to correspond to a change in a certain quantity given by $\frac{1}{2}$ mass times the square of the speed. The right-hand side of the equation depends only on the force and the distance moved in the direction of the force.

The quantity $\frac{1}{2}mv^2$ is called the *kinetic energy* of the body of mass m, and a physical statement of the above equation is that the body has gained kinetic energy by virtue of having moved through a distance s under the influence of a constant force F. In the case where the constant force $\mathbf{F}$ and the displacement vector $\mathbf{s}$ are not parallel, the equation is

$$\frac{m}{2}v^2 - \frac{m}{2}u^2 = \mathbf{F} \cdot \mathbf{s} \tag{3.6}$$

3.3.1 Work

There is a useful phrase in physical science that is to do with the energy transferred, and it is 'work'. Work measures the energy transferred in any change, and is generally calculated from the change in energy of a body when it moves through a distance under the influence of a force.

Work done by force = change in energy of body

3.4 THE LAW OF CONSERVATION OF ENERGY

One of the most important ideas in physics is contained in this law. When a force acts to make something move, energy is transferred. There are many kinds of energy besides kinetic energy, and whenever one form of energy is converted into another it is always found that the total amount of energy is fixed.

Consider for example a body thrown vertically upward with a certain kinetic energy. Eventually the body will come to rest and then fall back again to earth, as everyday experience teaches us. At its highest point, the kinetic energy of the body is zero and so according to the law of conservation of energy, this kinetic energy must have been converted into another form of energy which has been stored by the body.

The energy that a body has by virtue of its height above the earth's surface is called the gravitational potential energy, and I know from everyday observation that it is given by the formula

Increase in gravitational potential energy when a body of mass m is raised through height h is mgh where $g = 9.81\,\mathrm{m\,s^{-2}}$ is the acceleration due to gravity.

Provided one is concerned only with the motion of bodies which take place in a small region of space, then it is reasonable to say that the effect of gravity is that it results in a uniform downward acceleration which is given the symbol g. If one considers the motion of a body like a space probe, this approximation is no longer satisfactory; in fact, the magnitude of the acceleration is inversely proportional to the square of the distance of the point from the center of the earth. I want to formalize our study of these topics by considering the force of gravity in more detail.

3.4.1   The Gravitational Field

Figure 3.1 shows two bodies A and B, with masses M_A and M_B. Body A is at the centre of coordinates, and the distance between the centers of the bodies is R. It is known experimentally that there is a gravitational attractive force between these two bodies of magnitude

$$F_{AB} = G\frac{M_A M_B}{R^2} \tag{3.7}$$

where G is the gravitational constant equal to $6.673 \times 10^{-11}\,\mathrm{N\,m^2\,kg^{-2}}$. The gravitational force always acts along the line of centers of the masses A and B, and it is always attractive.

Notice that the force is not a constant, for it depends on the distance between the two bodies. In the case where one body is the earth and the other body (say) an object on the surface of the earth, then the force is almost constant provided the object does not move far from the earth's surface.

Force is a vector quantity, and it is sometimes profitable to write the above equation as

$$\mathbf{F}_{AB} = -G\frac{M_A M_B}{R^2}\,\hat{\mathbf{R}} \tag{3.8}$$

$\mathbf{F}_{AB}$ is the force exerted by body A on body B, and the unit vector points from A (the coordinate origin) to B. The minus sign simply indicates that the force is attractive. This

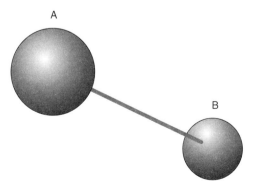

Figure 3.1   Force of gravity between two bodies

statement is consistent with Newton's third law, since

$$\mathbf{F}_{BA} = -G\,\frac{M_A M_B}{R^2}\,(-\hat{\mathbf{R}})$$

the force exerted by body B on body A is minus $\mathbf{F}_{AB}$ by virtue of the change in sign of the unit vector. The equation is more usually written in terms of the vector $\mathbf{R}$ as

$$\mathbf{F}_{AB} = -G\,\frac{M_A M_B}{R^3}\,\mathbf{R} \qquad (3.9)$$

Suppose we now calculate the gravitational attraction between M_A and a number of other masses M_B, M_C,... keeping the distance fixed. The ratios F_{AB}/M_B, F_{AC}/M_C,... will all be equal, and I say that body A has generated a *gravitational field* at that (and every other) point in space. This field is an example of a *vector field* and I will write it as

$$\mathbf{E}_{A,\,grav} = -G\,\frac{M_A}{R^3}\,\mathbf{R} \qquad (3.10)$$

The field exists at all points in space irrespective of whether the second mass is present or not. Vector fields are often represented as arrows drawn in space in directions indicating the direction of the field (Figure 3.2).

The interpretation is that, if we place a second body B (with mass M_B) in the field due to A, then body B experiences a force caused by the interaction of B with the field due to A. The force is M_B times the field at the point of B.

Let me now calculate the work done when body B moves from one position (I) to another position (II) in the field of A. We have to distinguish between the work done on B and the work done by B. One is just the negative of the other, and I am going to focus on the former, w_{on}.

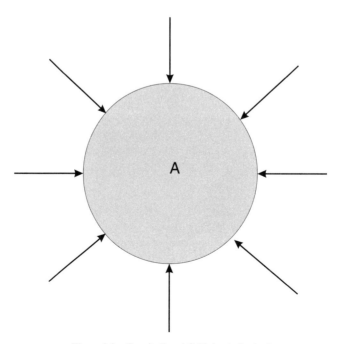

Figure 3.2 Gravitational field due to body A

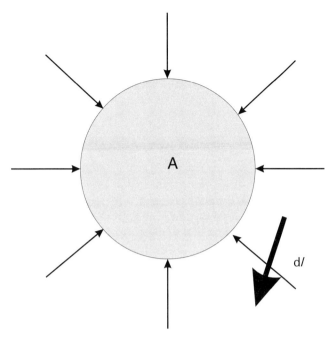

Figure 3.3 Body B moves by d*l*

Suppose we move body B through a differential element d*l*, which need not be along a radius (see Figure 3.3). The work done on B is $-\mathbf{F} \cdot \mathrm{d}l$ which is $-\mathbf{F}\,\mathrm{d}l\cos(\theta)$ where θ is the angle between $\mathbf{F}$ and d*l*. It can be seen from the Figure 3.3 $\mathrm{d}R = \mathrm{d}l\cos(\theta)$ and so

$$\mathrm{d}w_{\mathrm{on}} = GM_{\mathrm{A}}M_{\mathrm{B}}\frac{\mathrm{d}R}{R^2}$$

Integrating and inserting limits we find

$$w_{\mathrm{on}} = \int_{\mathrm{I}}^{\mathrm{II}} GM_{\mathrm{A}}M_{\mathrm{B}}\frac{\mathrm{d}R}{R^2}$$

$$= GM_{\mathrm{A}}M_{\mathrm{B}}\left(\frac{1}{R_{\mathrm{I}}} - \frac{1}{R_{\mathrm{II}}}\right)$$

Notice that the difference is independent of the path taken in moving from point I to point II. Gravitational fields depend only on the distance from the body causing the field to the point in space of interest, there is no angular dependence. Such fields are called central fields, and the result derived above is true for all central fields.

Of particular interest is the work done when we bring up body B from infinity (point I) to the point R (point II). According to our calculation above, we have

$$w_{\mathrm{on}} = -G\frac{M_{\mathrm{A}}M_{\mathrm{B}}}{R} \tag{3.11}$$

This quantity represents the energy change when body B is brought from infinity to point R away from body A. Alternatively it represents the energy change when body A is brought from infinity to a point distant R from body B. It is called the *mutual gravitational potential* energy of the bodies A and B.

3.5 THE GRAVITATIONAL POTENTIAL ENERGY

The equation

$$w_{\mathrm{on}} = -GM_{\mathrm{A}}M_{\mathrm{B}}\left(\frac{1}{R_{\mathrm{II}}} - \frac{1}{R_{\mathrm{I}}}\right)$$

relates the energy change when body B moves from point I to point II in the gravitational field due to body A. As I noted above, the change in energy does not depend on the path taken in the gravitational field, it only depends on the distance from the body A. It is useful to define a quantity called the gravitational potential energy ϕ (of body A) which takes account of these ideas.

The gravitational potential ϕ at position $\mathbf{R}$, due to body A at the coordinate origin, is given by

$$\phi(R) = -G\,\frac{M_{\mathrm{A}}}{R} \tag{3.12}$$

$\phi(R)$ is a useful quantity in the same way that the gravitational field is a useful quantity but it is a scalar quantity rather than a vector quantity.

3.5.1 Relationship Between Force and Mutual Potential Energy

I am going to derive a general relationship between force and mutual potential energy. To do this, I will consider a body of mass m moving with velocity $\mathbf{v}$ under the influence of some potential U.

The kinetic energy of a body of mass m moving with velocity $\mathbf{v} = \mathrm{d}\mathbf{r}/\mathrm{d}t$ can be written in vector notation as

$$\frac{1}{2}mv^2 = \frac{1}{2}m\,\frac{\mathrm{d}\mathbf{r}}{\mathrm{d}t}\cdot\frac{\mathrm{d}\mathbf{r}}{\mathrm{d}t}$$

According to the law of conservation of energy the total energy is a constant that I will call ε. We want to find a U so that the ε is a constant.

$$\varepsilon = \frac{1}{2}m\,\frac{\mathrm{d}\mathbf{r}}{\mathrm{d}t}\cdot\frac{\mathrm{d}\mathbf{r}}{\mathrm{d}t} + U \tag{3.13}$$

First of all, if ε is a constant then $\mathrm{d}\varepsilon/\mathrm{d}t$ is zero. Differentiation of the energy expression with respect to time gives

$$\frac{\mathrm{d}\varepsilon}{\mathrm{d}t} = \frac{\mathrm{d}}{\mathrm{d}t}\left(\frac{1}{2}m\,\frac{\mathrm{d}\mathbf{r}}{\mathrm{d}t}\cdot\frac{\mathrm{d}\mathbf{r}}{\mathrm{d}t} + U\right)$$

$$= m\,\frac{\mathrm{d}\mathbf{r}}{\mathrm{d}t}\cdot\frac{\mathrm{d}^2\mathbf{r}}{\mathrm{d}t^2} + \frac{\mathrm{d}U}{\mathrm{d}t}$$

which is zero because ε is constant.

By the chain rule, dU/dt is given by

$$\frac{dU}{dt} = \frac{\partial U}{\partial x}\frac{dx}{dt} + \frac{\partial U}{\partial y}\frac{dy}{dt} + \frac{\partial U}{\partial z}\frac{dz}{dt}$$

which is

$$\text{grad } U \cdot \frac{d\mathbf{r}}{dt}$$

Hence on rearrangement

$$\frac{d\mathbf{r}}{dt} \cdot \left(m\frac{d^2\mathbf{r}}{dt^2} + \text{grad } U \right) = 0$$

According to Newton's second law,

$$\mathbf{F} = m\frac{d^2\mathbf{r}}{dt^2}$$

and so we find

$$\mathbf{F} = -\text{grad } U \qquad\qquad (3.14)$$

which defines the potential energy U.

In one dimension (say the x dimension) the equation can be rewritten as

$$F_x = -\frac{dU}{dx}$$

which shows the relationship between force and mutual potential energy in one dimension.

3.6 VIBRATIONAL MOTION

To get us started in our study of the classical mechanics of vibrational motion, consider a particle of mass m lying on a frictionless horizontal table, and attached to a spring as shown in Figure 3.4.

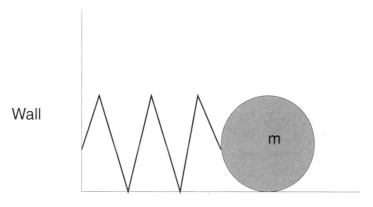

Figure 3.4 Hooke's law

The particle is initially at rest, and we call this the equilibrium position. The length of the spring is then R_e. If we stretch the spring, it exerts a restoring force on the particle. Likewise, if we compress the spring, it acts to restore the particle to the equilibrium position. I will denote the length of the spring R, and the extension is then $R - R_e$. If we denote the restoring force F_s, experimental measurements show that many springs obey Hooke's law (after Robert Hooke, a contemporary of Newton)

$$F_s = -k_s(R - R_e)$$

where the constant of proportionality k_s is called the force constant. I have included the negative sign because the restoring force acts in the opposite direction to that of the displacement from equilibrium, and I would like k_s to be a positive quantity. Hooke's law usually breaks down when we stretch springs by large amounts. A spring that obeys Hooke's law is sometimes called a perfect spring.

We now somehow set the particle in motion and let it oscillate about the equilibrium position. In order to find the position R of the particle at time t, we can make use of Newton's second law which relates force to acceleration

$$m \frac{d^2 R}{dt^2} = -k_s(R - R_e)$$

Such an equation is called an *equation of motion*. In this particular case, the equation of motion is a second order differential equation, and it has the general solution

$$R = R_e + A \sin\left(\sqrt{\frac{k_s}{m}}\, t\right) + B \cos\left(\sqrt{\frac{k_s}{m}}\, t\right) \tag{3.15}$$

where A and B are constants of integration. We can also write this in the alternative form

$$R = R_e + C \cos\left(\sqrt{\frac{k_s}{m}}\, t + \phi\right)$$

where C and ϕ are also constants of integration.

The quantity $\sqrt{k_s/m}$ occurs repeatedly in the study of such vibrational motion, and so we give it a special symbol.

$$\omega = \sqrt{\frac{k_s}{m}} \tag{3.16}$$

Recall that the arguments of sines and cosines are dimensionless, and so the SI units of ω are s^{-1} (or Hz). You might have been expecting me to write radians s^{-1}, but I should remind you that plane angles are dimensionless quantities, since they are defined as a ratio of arc length to radius.

Sines and cosines repeat every 2π (radians), and a little thought shows that we can also write the solution

$$R = R_e + A \sin(\omega t) + B \cos(\omega t)$$

as

$$R = R_e + A \sin\left[\omega\left(t + \frac{2\pi}{\omega}\right)\right] + B \cos\left[\omega\left(t + \frac{2\pi}{\omega}\right)\right]$$

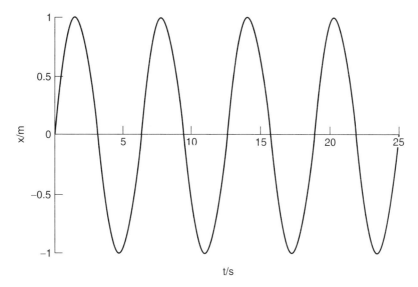

Figure 3.5   Simple harmonic motion

where I have also substituted ω for $\sqrt{k_s/m}$. The addition of $2\pi/\omega$ to t has no effect on the solution; the value of R repeats itself every $2\pi/\omega$. For this reason, ω is called the angular frequency and $2\pi/\omega$ the period.

The solutions given above are general solutions; they contain two constants of integration. To find the particular solution for any given problem, I have to take account of the initial conditions. For the sake of argument, let us start the motion at $t = 0$ with the particle at the equilibrium position ($R = R_e$) and with an initial speed of v_0 (which is dR/dt or $d(R - R_e)/dt$, evaluated at $t = 0$). Substituting $R = R_e$ at $t = 0$ gives $B = 0$. If we then differentiate

$$\frac{dR}{dt} = A\omega \, \cos(\omega t)$$

and put $dR/dt = v_0$ when $t = 0$, we see that $A = v_0/\omega$ and so

$$R - R_e = \frac{v_0}{\omega} \sin(\omega t + 2\pi)$$

The graph of a typical solution is shown in Figure 3.5. Motion of this type, in which the position is related to time by sine function, is known as simple harmonic motion.

(I put $A = 1\,\text{m}$, $B = 0$, $m = 1\,\text{kg}$ and $k_s = 1\,\text{N m}^{-1}$ into the solution given above, in order to generate Figure 3.5.).

3.6.1   Average Values

At any given time, the displacement of the particle from its equilibrium position may be non-zero. Sometimes the displacement will be positive, sometimes it will be negative, but on average it will be zero, as can be seen from Figure 3.5. If we denote the average value of R by $\langle R \rangle$, then it should be clear by inspection that $\langle R \rangle = R_e$ and that the mean value of the displacement $\langle R - R_e \rangle = 0$.

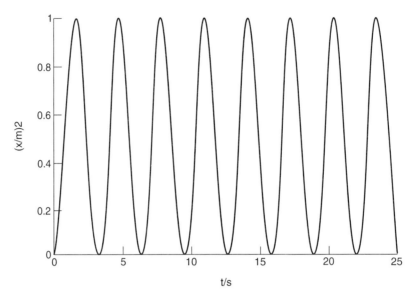

Figure 3.6   Square of position *versus* time for simple harmonic motion

More useful quantities for our present purpose are the root mean square displacements such as the root mean square displacement of $R - R_e$

$$(R - R_e)_{rms} = \langle (R - R_e)^2 \rangle^{1/2} \tag{3.17}$$

which is the average of the square of $(R - R_e)^2$. Do not confuse this with the square of the average value of $(R - R_e)$; they are not the same thing (Figure 3.6).

3.7   THE POTENTIAL ENERGY

In Section 3.5, I showed you how to interrelate the force and the mutual potential energy. In the special case of one-dimensional motion, we have

$$U = - \int F(x) \, dx$$

For a Hooke's law spring, integration gives

$$U(R - R_e) = U_0 + \frac{1}{2} k_s (R - R_e)^2 \tag{3.18}$$

where U_0 is a constant of integration. We can eliminate U_0, if we wish, by defining the zero of potential to correspond to the equilibrium position $R = R_e$.

The law of conservation of energy states that the sum of the kinetic and potential energies is a constant throughout the motion of the particle. You might like to see how the kinetic energy T and the potential energy U vary with time, for the particular values of A, B, etc., used above (Figures 3.7 and 3.8).

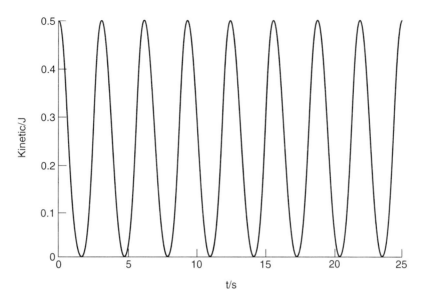

Figure 3.7   Kinetic energy *versus* time for simple harmonic motion

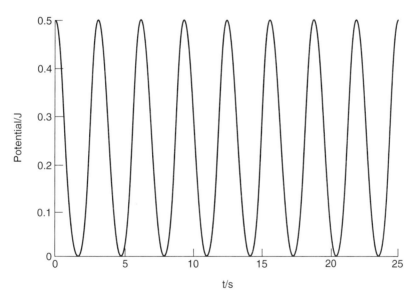

Figure 3.8   Potential energy *versus* time for simple harmonic motion

The kinetic energy and the potential energy both vary with time, but if we add them together then you will find that they come to a constant.

3.8   TWO PARTICLES, ONE SPRING

Consider next a simple classical model for the vibrational motion of a diatomic molecule. We assume that the molecule can be modeled as two particles joined by a perfect spring. You may

Figure 3.9   Hooke's law with two balls on one spring

think it unrealistic in trying to model the vibrations of a molecule using classical mechanics rather than quantum mechanics, but it is surprising how accurate the model can be for many applications.

Consider two atoms of masses m_1 and m_2 connected by a perfect spring with force constant k_s and natural length R_e, as shown in the Figure 3.9. I will consider motion along the horizontal axis, and I will call this the x axis. The x coordinates of the two particles are denoted x_1 and x_2.

We now pull the particles away from each other so that the length of the spring is R (which is given by $x_2 - x_1$). The spring extension is $R - R_e$, which is $(x_2 - x_1 - R_e)$. The spring exerts restoring forces on the two atoms. Considering atom 1, the spring exerts a restoring force of magnitude $k_s(x_2 - x_1 - R_e)$ in the direction of increasing x_1 on atom 1, and so, using Newton's second law

$$m_1 \frac{\mathrm{d}^2 x_1}{\mathrm{d}t^2} = k_s(x_2 - x_1 - R_e) \tag{3.19}$$

The extended spring exerts a force of magnitude $k_s(x_2 - x_1 - R_e)$ in the direction of decreasing x_2 on atom 2, and so

$$m_2 \frac{\mathrm{d}^2 x_2}{\mathrm{d}t^2} = -k_s(x_2 - x_1 - R_e) \tag{3.20}$$

At first sight, these two differential equations look like they are going to be difficult to solve because they contain mixtures of x_1 and x_2. If I consider a simpler question, how the bond length varies with time, it turns out that the equations of motion have a very simple solution.

$$\frac{\mathrm{d}^2 R}{\mathrm{d}t^2} = \frac{\mathrm{d}^2 x_2}{\mathrm{d}t^2} - \frac{\mathrm{d}^2 x_1}{\mathrm{d}t^2}$$

$$= -\frac{k_s}{m_2}(R - R_e) - \frac{k_s}{m_1}(R - R_e)$$

$$= -k_s\left(\frac{1}{m_1} + \frac{1}{m_2}\right)(R - R_e)$$

We define a quantity μ called the reduced mass of the system, satisfying

$$\frac{1}{\mu} = \frac{1}{m_1} + \frac{1}{m_2} \tag{3.21}$$

and we can then write the equation of motion

$$\mu \frac{d^2 R}{dt^2} = -k_s (R - R_e)$$

which looks very similar to the equation of motion for the single particle of mass m. The only difference is that m is replaced by μ. The general solution is

$$R = R_e + A \sin\left(\sqrt{\frac{k_s}{\mu}} t\right) + B \cos\left(\sqrt{\frac{k_s}{\mu}} t\right) \qquad (3.22)$$

and so the diatomic molecule undergoes simple harmonic motion with angular frequency $\sqrt{k_s/\mu}$, which is exactly what we would expect for a *single* particle of mass μ attached to a fixed point by a perfect spring with force constant k_s.

How does all this square with our ideas about energy conservation, and what is the potential U in this case?

It is easy to demonstrate (using the link between force and potential) that the potential energy is

$$U = \frac{1}{2} k_s (x_2 - x_1 - R_e)^2$$

and so the total energy of the system is

$$\varepsilon_{\text{vib}} = \frac{1}{2} m_1 \left(\frac{dx_1}{dt}\right)^2 + \frac{1}{2} m_2 \left(\frac{dx_2}{dt}\right)^2 + \frac{1}{2} k_s (x_2 - x_1 - R_e)^2 \qquad (3.23)$$

You might like to check by direct differentiation of this expression that ε_{vib} truly is a constant.

3.8.1 Summary

We have modeled a vibrating diatomic molecule as two masses joined together with a Hooke's law spring. The classical treatment has equation of motion

$$\mu \frac{d^2 R}{dt^2} = -k_s (R - R_e)$$

and the molecule vibrates with a fundamental vibration frequency of $\sqrt{k_s/\mu}$. There are no restrictions on the values of the vibrational energies.

3.9 NORMAL MODES OF VIBRATION

Let me now make the classical problem a bit harder. I will take two particles of mass m_1 and m_2 as before, but I am going to take two springs, and join the system up as in Figure 3.10.

As usual, we assume that the springs are perfect springs. I will call the force constant of the left-hand spring k_1 and the force constant of the right-hand spring k_2. The equilibrium position corresponds to the two masses with x coordinates $R_{1,e}$ and $R_{2,e}$. When we stretch

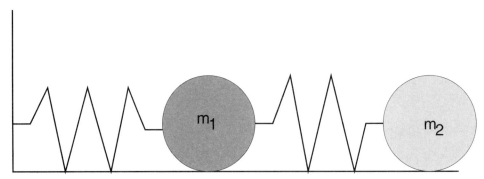

Figure 3.10   Two particles and two springs

the system, the two springs extend and I will call the instantaneous positions x_1 and x_2. The first thing to do is to write down the equations of motion for the two particles, and I will do this using Newton's second law.

Look at the left-hand spring, which has spring force constant k_1; the spring exerts a restoring force on particle 1 of $-k_1(x_1 - R_{1e})$.

Now look at the right-hand spring. This spring is stretched by an amount $(x_2 - x_1)$, and so it exerts a force of $k_2(x_2 - R_{2e} - x_1 + R_{1e})$; this force acts to the left on particle 2 and to the right on particle 1.

For the sake of neatness, I will write

$$X_1 = x_1 - R_{1e}$$

$$X_2 = x_2 - R_{2e}$$

and so

$$k_2(X_2 - X_1) - k_1 X_1 = m_1 \frac{d^2 X_1}{dt^2}$$

$$-k_2(X_2 - X_1) = m_2 \frac{d^2 X_2}{dt^2} \tag{3.24}$$

There are many different solutions to these two equations, but it proves possible to find two particularly simple ones called *normal modes of vibration*. These have the property that both particles execute simple harmonic motion at the same angular frequency. Not only that, every possible vibrational frequency of the two particles can be described in terms of the normal modes, so they are obviously very important.

Having said that it proves possible to find such solutions where both particles vibrate with the same frequency, let me assume that there exist such solutions to the equations of motion such that

$$X_1(t) = A \sin(\omega t + \phi_1)$$
$$X_2(t) = B \sin(\omega t + \phi_2) \tag{3.25}$$

where A, B, ϕ_1 and ϕ_2 are constants that have to be determined from the boundary conditions.

Differentiating these two equations with respect to time gives

$$\frac{d^2 X_1(t)}{dt^2} = \omega^2 A \sin(\omega t + \phi_1)$$

$$\frac{d^2 X_2(t)}{dt^2} = \omega^2 B \sin(\omega t + \phi_2)$$

and substituting these expressions into the equations of motion gives

$$-\frac{(k_1 + k_2)}{m_1} X_1 + \frac{k_2}{m_1} X_2 = -\omega^2 X_1$$

$$\frac{k_2}{m_2} X_1 - \frac{k_2}{m_2} X_2 = -\omega^2 X_2$$

It turns out that these two equations are valid only when ω has one of two possible values called the normal mode angular frequencies. In either case, both particles oscillate with the same angular frequency.

In order to find the normal modes of vibration, I am going to write the above equations in matrix form, and then find the eigenvalues and eigenvectors of a certain matrix. If you are not *au fait* with matrices, then go straight to the results.

In matrix form, we write

$$\begin{pmatrix} -\dfrac{(k_1 + k_2)}{m_A} & \dfrac{k_2}{m_1} \\ \dfrac{k_2}{m_2} & -\dfrac{k_2}{m_2} \end{pmatrix} \begin{pmatrix} X_1 \\ X_2 \end{pmatrix} = -\omega^2 \begin{pmatrix} X_1 \\ X_2 \end{pmatrix} \tag{3.26}$$

which is obviously a matrix eigenvalue problem. We have to find the values of $-\omega^2$ for which these equations hold, and then for each value of $-\omega^2$ we need to find the relevant combinations of the coordinates.

To make the algebra simple, I will work through a numerical example with $m_1 = 1\,\text{kg}$, $m_2 = 5\,\text{kg}$, $k_1 = 30\,\text{N}\,\text{m}^{-1}$ and $k_2 = 45\,\text{N}\,\text{m}^{-1}$.

Substituting these values into the matrix equation gives

$$\begin{pmatrix} -75s^{-2} & 45s^{-2} \\ 9s^{-2} & -9s^{-2} \end{pmatrix} \begin{pmatrix} X_1 \\ X_2 \end{pmatrix} = -\omega^2 \begin{pmatrix} X_1 \\ X_2 \end{pmatrix} \tag{3.27}$$

since the matrix equation is only 2×2, we do not need any complicated numerical analysis, we can just expand the determinant

$$\begin{vmatrix} -75s^{-2} + \omega^2 & 45s^{-2} \\ 9s^{-2} & -9s^{-2} + \omega^2 \end{vmatrix} = 0$$

which gives $\omega^2 = 3.3\,\text{s}^{-2}$ or $\omega^2 = 80.7\,\text{s}^{-2}$.

There are thus two frequencies at which the two particles will show simple harmonic motion at the same frequency. The normal mode with angular frequency $\omega_1 = 1.8\,\text{s}^{-1}$ is called the first normal mode and the one with angular frequency $\omega_2 = 9.0\,\text{s}^{-1}$ is called the second normal mode.

To a certain extent, the amplitudes and phases of the normal modes are arbitrary; they are determined by the starting positions and velocities of the particles. The two eigenvectors of the matrix, which work out as $(0.8468 \quad 0.5319)^{\mathrm{T}}$ and $(0.1246 \quad -0.9922)^{\mathrm{T}}$ are however of some interest. They define the *normal coordinates* for the motion.

3.10 ANGULAR MOMENTUM

In order to discuss the concepts that are needed for a study of rotating objects, it is helpful to begin with a very simple case in which a single particle of mass m moves along a curve in space. At some time t, the particle has position vector $\mathbf{r}(t)$ relative to the origin. The linear momentum vector $\mathbf{p}$ is thus

$$\mathbf{p} = m\,\frac{\mathrm{d}\mathbf{r}}{\mathrm{d}t} \tag{3.28}$$

The *angular momentum* vector $\mathbf{l}$ is defined by the vector cross product

$$\mathbf{l} = m\mathbf{r} \times \frac{\mathrm{d}\mathbf{r}}{\mathrm{d}t} \tag{3.29}$$

which can also be written in terms of the linear momentum as

$$\mathbf{l} = \mathbf{r} \times \mathbf{p} \tag{3.30}$$

The definition of the angular momentum vector depends on the choice of coordinate origin, for if we were to choose a new origin such that the position vector of the particle were

$$\mathbf{r}' = \mathbf{r} + \mathbf{R}$$

where $\mathbf{R}$ is a constant vector, then a little manipulation shows that

$$\mathbf{l}' = \mathbf{l} + m\mathbf{R} \times \frac{\mathrm{d}\mathbf{r}}{\mathrm{d}t} \tag{3.31}$$

According to Newton's second law, the rate of change of linear momentum of a single particle of mass m acted upon by a force $\mathbf{F}$ (Figure 3.11) is given by

$$\mathbf{F} = m\,\frac{\mathrm{d}^2\mathbf{r}}{\mathrm{d}t^2}$$

which can also be written in terms of the linear momentum as

$$\mathbf{F} = m\,\frac{\mathrm{d}\mathbf{p}}{\mathrm{d}t}$$

The linear momentum of such a particle is therefore constant in the absence of an applied force. It turns out that a rather similar equation applies to the angular momentum. To see this, we differentiate both sides of the defining equation with respect to time, giving

$$\frac{\mathrm{d}\mathbf{l}}{\mathrm{d}t} = m\,\frac{\mathrm{d}}{\mathrm{d}t}\left(\mathbf{r} \times \frac{\mathrm{d}\mathbf{r}}{\mathrm{d}t}\right)$$

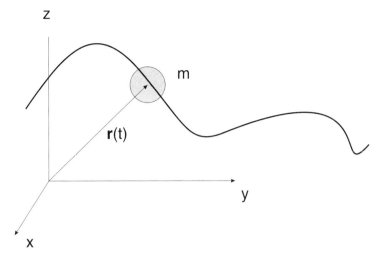

Figure 3.11 Particle of mass *m* moving along path in space

which gives, after a little manipulation

$$\frac{\mathrm{d}\mathbf{l}}{\mathrm{d}t} = \mathbf{r} \times \mathbf{F} \tag{3.32}$$

The vector $\mathbf{r} \times \mathbf{F}$ is known as the *torque* applied to the particle relative to the given coordinate origin. Thus we have proved that

In the absence of an applied torque, a particle's angular momentum remains constant in time.

One example of this occurs when there are no applied forces acting on the particle. But even if the total force acting on the particle is non-zero, it is possible that the torque is zero and hence the angular momentum is constant. An important example is where the force vector is permanently parallel to the position vector, as in planetary motion. In this case, the vector cross product $\mathbf{r} \times \mathbf{F} = 0$.

3.11 CIRCULAR MOTION

The torque law is especially useful when studying the motion of a particle around a circle of constant radius. Suppose then that a particle of mass *m* with position vector

$$\mathbf{r} = x\mathbf{e}_x + y\mathbf{e}_y + z\mathbf{e}_z$$

executes circular motion around the *z* axis, with a constant radius *a*. Thus

$$\mathbf{r} = a \sin \vartheta \mathbf{e}_x + a \cos \vartheta \mathbf{e}_y + d\mathbf{e}_z$$

where *d* is the constant distance along the *z* axis, and θ an angle which measures the rotation about the *z* axis. Differentiation gives

$$\frac{\mathrm{d}\mathbf{r}}{\mathrm{d}t} = -a \frac{\mathrm{d}\vartheta}{\mathrm{d}t} \cos \vartheta \mathbf{e}_x + a \frac{\mathrm{d}\vartheta}{\mathrm{d}t} \sin \vartheta \mathbf{e}_y$$

which can also be written (by making use of the properties of the unit vectors)

$$\frac{d\mathbf{r}}{dt} = \frac{d\vartheta}{dt}\mathbf{e_z} \times \mathbf{r}$$

This compact equation is often written as

$$\frac{d\mathbf{r}}{dt} = \omega \times \mathbf{r} \qquad (3.33)$$

where ω is known as the angular velocity vector of the particle. The magnitude of the angular velocity vector is known as the angular speed.

The angular momentum vector can also be written in terms of the coordinates of the particle; we have from above

$$\mathbf{l} = m\mathbf{r}x\frac{d\mathbf{r}}{dt} \quad \text{and} \quad \frac{d\mathbf{r}}{dt} = \frac{d\vartheta}{dt}\,\mathbf{e_z} \times \mathbf{r}$$

which can be combined to give, after a little manipulation

$$\mathbf{l} = m\frac{d\vartheta}{dt}[-zx\mathbf{e_x} - yz\mathbf{e_y} + (x^2+y^2)\mathbf{e_z}]$$

The angular momentum vector is not directed along the z axis, but has x and y components. This is because of our choice of 'origin' along the z axis. For rotation about the coordinate origin then $z = 0$ and we have

$$\mathbf{l} = m\frac{d\vartheta}{dt}(x^2+y^2)\mathbf{e_z}$$

Remembering that we are dealing with circular motion and so $x^2 + y^2 = a^2$, we see that the magnitude of $\mathbf{l}$ is

$$ma^2\left(\frac{d\vartheta}{dt}\right) \qquad (3.34)$$

The quantity ma^2 turns out to have a special significance for a particle moving in a circle. It is called the *moment of inertia* and we give it the symbol I.

I should tell you that the moment of inertia is a key concept in the study of all rotational motions, not just motion in a circle. Without going into detail, let me sketch some of the results. Consider a collection of particles of masses $m_1, m_2, \ldots, m_n$ with position vectors $\mathbf{R}_1$, $\mathbf{R}_2, \ldots, \mathbf{R}_n$. The moment of inertia is a tensor property which can be represented as a real symmetric 3×3 matrix

$$\mathbf{I} = \begin{pmatrix} I_{xx} & I_{xy} & I_{xz} \\ I_{yx} & I_{yy} & I_{yz} \\ I_{zx} & I_{zy} & I_{zz} \end{pmatrix} \qquad (3.35)$$

where for example

$$I_{xy} = \sum_{i=1}^{n} m_i x_i y_i \qquad (3.36)$$

For a certain choice of axes, the tensor will be diagonal and we speak about the *principal axes of inertia*.

4   Modeling Simple Solids (i)

4.1   THE LAWS OF ELECTROSTATICS

As far as we know, there are four fundamental types of interactions between bodies. There is the weak nuclear force that governs beta decay, the strong nuclear force that is responsible for binding together the particles in a nucleus; the gravitational force that holds the earth very firmly in its orbit round the sun and finally the electromagnetic force that is responsible for binding atomic electrons to nuclei and for holding atoms together when they form molecules and aggregates.

These four forces all depend on the distance between the bodies, but other things being equal they can be ranked as follows in order of magnitude:

strong nuclear > electromagnetic > weak nuclear > gravitational.

You may be surprised to find that I have ranked the gravitational force as the weakest; after all, it is the only one we come across every day in our macroscopic world. The reason why we are constantly aware of the gravitational force is that gravitational forces always add, because there appears to be only one type of matter and so if a large number of particles are aggregated together (a mole of particles comprises 6.023×10^{23} units) into a large mass then the gravitational force of that mass is correspondingly large.

It turns out that there are two kinds of electrical charges which we might choose to call types X and Y. The experimental evidence is that the force between X type charges is repulsive, as is the force between Y type charges. The force between an X and a Y type charge is attractive. For this reason, the early experimentalists decided to classify charges as positive or negative, because a positive quantity times a positive quantity gives a positive quantity, a negative quantity times a negative quantity gives a positive quantity but a negative quantity times a positive quantity gives a negative quantity.

But which charged fundamental particle correspond to X, and which to Y? I will simply tell you that we decide quite arbitrarily to take

electrons as the negative charge, and protons as the positive charge

Molecules normally consist of equal numbers of electrons and protons whose charges are exactly equal and opposite, which explains why we rarely see bodies carrying an excess of either charge. An electrolyte such as sodium chloride generates an equal number of positive and negative ions when made into a solution, but rarely are we aware of an imbalance of electric charge.

As far as we know, electric charge is rather like matter in that it cannot be created or destroyed. We say that

electric charge is conserved

A thunderstorm results when nature separates out positive and negative charges on the macroscopic scale. It is thought that friction between moving masses of air and water vapor detaches electrons from some molecules and attaches them to others. This results in a cloud (which is of course a macroscopic object) being left with an excess charge, with spectacular results.

It was the investigations into such atmospheric phenomena that first gave clues about the nature of the electromagnetic force.

The force between two charged particles is described by Coulomb's law, which is a fundamental law of electromagnetism. At its simplest, Coulomb's law relates to the mutual force between a pair of point charges Q_A and Q_B separated by distance r_{AB}, as shown in Figure 4.1. The position vector of charge Q_A is $\mathbf{r}_A$ and the position vector of Q_B is $\mathbf{r}_B$.

From the laws of vector analysis, the vector

$$\mathbf{r}_{AB} = \mathbf{r}_B - \mathbf{r}_A$$

therefore points from Q_A to Q_B.

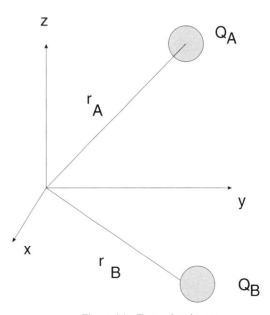

Figure 4.1   Two point charges

Coulomb (1785) was the first person to give a mathematical form to the interaction. He measured the force between two very small charged bodies (which allowed him to assume that he was dealing with point charges). Coulomb was able to show that the force exerted by Q_A on Q_B, $\mathbf{F}_{A \text{ on } B}$ was

- proportional to the inverse square of the distance between Q_A and Q_B
- proportional to Q_A when Q_B and r_{AB} were fixed
- proportional to Q_B when Q_A and r_{AB} were fixed.

Coulomb also noted that the force always acted along the line joining the centres of the two point charges, and that the force was either attractive or repulsive depending on whether A or B carried different or similar types of charge. The direction of the force is therefore determined by the product $Q_A Q_B$. How do we incorporate all this information into the basic electrostatic force law?

A mathematical result of these observations can be written in scalar terms as

$$F_{A \text{ on } B} \propto \frac{Q_A Q_B}{r_{AB}^2}.$$

Forces are of course vector quantities. In vector notation we find

$$\mathbf{F}_{A \text{ on } B} \propto \frac{Q_A Q_B}{r_{AB}^3} \mathbf{r}_{AB}$$

The law can also be written in terms of a unit vector pointing from Q_A to point charge Q_B as

$$\mathbf{F}_{A \text{ on } B} \propto \frac{Q_A Q_B}{r_{AB}^2} \hat{\mathbf{r}}_{AB}$$

In order to change this proportionality into an equality, we need a constant. This proportionality constant is usually written as $1/(4\pi\epsilon_0)$, where ϵ_0 is an experimentally determined quantity that we refer to as the permittivity of free space. It has a value

$$\epsilon_0 = 8.854 \times 10^{-12} \text{ C}^2 \text{ N}^{-1} \text{ m}^{-2}.$$

Finally then we can then write Coulomb's law

$$\mathbf{F}_{A \text{ on } B} = \frac{1}{4\pi\epsilon_0} \frac{Q_A Q_B}{r_{AB}^3} \mathbf{r}_{AB} \qquad (4.1)$$

All forces must satisfy Newton's third law. We can demonstrate that this force law does indeed satisfy Newton's law by reversing the roles of Q_A and Q_B. The vector $\mathbf{r}_B$ now points from Q_B to Q_A and so is the negative of the one in the definition above, and so $\mathbf{F}_{B \text{ on } A} = -\mathbf{F}_{A \text{ on } B}$.

4.2   THE ELECTROSTATIC FIELD

Once we know the force exerted by one electric charge on another, we can generalise and define the electric field $\mathbf{E}$ at all points in space. Let me start from the force law

$$\mathbf{F}_{A \text{ on } B} = \frac{1}{4\pi\epsilon_0} \frac{Q_A Q_B}{r_{AB}^3} \mathbf{r}_{AB}$$

and focus attention on the *ratio*

$$\frac{\mathbf{F}_{\text{A on B}}}{Q_B} = \frac{1}{4\pi\epsilon_0} \frac{Q_A}{r_{AB}^3} \mathbf{r}_{AB}$$

I have marked the point in space previously occupied by Q_B by an asterisk in Figure 4.2.

The ratio $\mathbf{F}_{\text{A on B}}/Q_B$ is independent of Q_B, and we say that the charge Q_A generates an electrostatic field at the field point denoted by an asterisk. In order to measure this electrostatic field, we could place a test charge such as Q_B at that point in space and measure the force acting on Q_B, but the field exists whether or not the test charge is present.

I am going to follow a certain convention in labeling field points; I will generally write them as upper case vectors such as $\mathbf{R}$ to distinguish them from the position vectors associated with point charges. This is why I have replaced $\mathbf{r}_B$ by $\mathbf{R}$ in Figure 4.2. Electric fields are usually given the symbol $\mathbf{E}$. We say that the charge Q_A generates an electrostatic field $\mathbf{E}$ at all points in space. The electrostatic field at $\mathbf{R}$ due to point charge Q_A with position vector $\mathbf{r}_A$ is therefore

$$\mathbf{E}(\mathbf{R}) = \frac{1}{4\pi\epsilon_0} \frac{Q}{|\mathbf{R} - \mathbf{r}_A|^3} (\mathbf{R} - \mathbf{r}_A) \tag{4.2}$$

4.3 THE MUTUAL ELECTROSTATIC POTENTIAL ENERGY U_{AB} OF Q_A AND Q_B

Consider now an arrangement of two point charges, Q_A at the origin and Q_B at infinity. We now bring up charge Q_B from infinity to the position denoted by an asterisk in Figure 4.2 (distance $\mathbf{r}_{AB}$ from Q_A). There is a mutual force between the two charges Q_A and Q_B, and so energy is transferred in the process because we have to either overcome the repulsion between

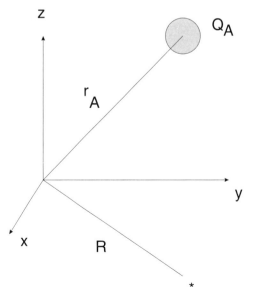

Figure 4.2 Construct needed to define the electrostatic field

two charges of the same sign, or allow for the attraction felt between two charges, one of either sign.

We often refer to the system of electrostatic charges under study as the 'electrostatic system'. In the case of two like charges, energy has to be expended on the electrostatic system in order to bring up the second charge from infinity and so the system energy increases. In the case of two unlike charges, they attract each other and so the electrostatic system loses energy.

Some texts focus attention on the work done *by* the system; be very careful to distinguish one from the other; they are exactly equal and opposite. Again, most texts write w_{on} for work done *on* the system and w_{by} for work done by the system. If the system gains energy, then w_{on} is positive and w_{by} is exactly equal and opposite to this. If w_{on} is negative, then the system has lost energy in the change under discussion.

By analogy with our discussion of gravitational fields discussed in Chapter 3, the work done in bringing up charge Q_B from infinity to the position denoted by an asterisk is given by

$$w_{on} = \frac{1}{4\pi\epsilon_0} \frac{Q_A Q_B}{r_{AB}} \tag{4.3}$$

and this is called the mutual potential energy of the pair of charges

$$U = \frac{1}{4\pi\epsilon_0} \frac{Q_A Q_B}{r_{AB}} \tag{4.4}$$

4.4 THE ELECTROSTATIC POTENTIAL $\phi(\mathbf{R})$

In my discussion of gravitational fields in Chapter 3, I calculated the energy change when body B moves from point I to point II in the gravitational field due to body A as

$$U(\mathrm{II}) - U(\mathrm{I}) = -GM_A M_B \left(\frac{1}{R_{\mathrm{II}}} - \frac{1}{R_{\mathrm{I}}} \right)$$

The change in energy does not depend on the path taken in the gravitational field, it only depends on the distance from the body A.

Electrostatic fields show the same behavior; the energy change moving Q_B from point I to point II in the field due to Q_A is independent of the path taken, and it is useful to define a scalar quantity called the electrostatic potential energy ϕ which takes account of these ideas. This idea is discussed at length in all the electromagnetism textbooks, I do not have the space to go into detail here. I will just quote a few results.

The electrostatic potential $\phi(\mathbf{R})$ at a field point with vector position $\mathbf{R}$, caused by charge Q_A at position vector $\mathbf{r}_A$, is given by

$$\phi(\mathbf{R}) = \frac{1}{4\pi\epsilon_0} \frac{Q_A}{|\mathbf{R} - \mathbf{r}_A|} \tag{4.5}$$

$\phi(\mathbf{R})$ is a useful quantity in electrostatic calculations in the same way that the gravitational potential is a useful quantity in mechanics calculations.

All electrostatic fields are conservative and so the potential energy change in moving Q_B through any electrostatic field depends only on the starting point in the field and the finishing point. It does not depend on the route taken through the field. In other words, the electrostatic potential $\phi(\mathbf{R})$ only depends on the position in space $\mathbf{R}$.

4.5 RELATIONSHIPS BETWEEN E, F, ϕ AND U

The following equations should be read with reference to Figures 4.1 and 4.2. Figure 4.1 has two point electric charges Q_A and Q_B separated by distance $\mathbf{r}_{AB}$. The vector $\mathbf{r}_{AB}$ points from Q_A to Q_B and the vector $\mathbf{R}$ in Figure 4.2 describes an arbitrary field point.

Force law

$$\mathbf{F}_{A \text{ on } B} = \frac{1}{4\pi\epsilon_0} \frac{Q_A Q_B}{r_{AB}^3} \mathbf{r}_{AB}$$

Mutual electrostatic potential energy U of Q_A and Q_B

$$U = \frac{1}{4\pi\epsilon_0} \frac{Q_A Q_B}{r_{AB}}$$

Relationship between $\mathbf{F}$ and U

$$\mathbf{F}(\mathbf{R}) = -\text{grad } U(\mathbf{R}) \tag{4.6}$$

Electrostatic field

$$\mathbf{E}(\mathbf{R}) = \frac{1}{4\pi\epsilon_0} \frac{Q_A}{|\mathbf{R} - \mathbf{r}_A|^3} (\mathbf{R} - \mathbf{r}_A)$$

Electrostatic potential

$$\phi(\mathbf{R}) = \frac{1}{4\pi\epsilon_0} \frac{Q_A}{|\mathbf{R} - \mathbf{r}_A|}$$

Relationship between $\mathbf{E}$ and ϕ

$$\mathbf{E}(\mathbf{R}) = -\text{grad } \phi(\mathbf{R}) \tag{4.7}$$

4.6 MUTUAL POTENTIAL ENERGY OF AN ARRAY OF POINT CHARGES

Consider now an array of point charges $Q_1, Q_2, \ldots, Q_n$ such as shown in Figure 4.3. I am going to calculate the mutual potential energy U of this array of charges. U gives the work done on the array (or the energy change) in building up the array of charges, starting with them all at infinity, and so it is an important quantity in physical science.

I am going to build up U from the contributions made by one point charge at a time.

First of all, I bring up Q_1 from infinity and place it at the coordinate origin. There is no electrostatic field (because there are no other charges present) and so this change makes no contribution to U.

Next I bring up charge Q_2 from infinity and place it distance r_{12} from Q_1. The energy change is given as the mutual potential energy of this pair of point charges

$$U_{12} = \frac{1}{4\pi\epsilon_0} \frac{Q_1 Q_2}{r_{12}}$$

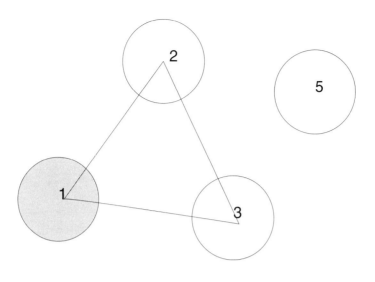

Figure 4.3   Array of point charges

Next I bring up charge Q_3 to position $\mathbf{r}_3$ which is distant r_{13} from charge Q_1 and r_{23} from charge Q_2. The extra energy needed is $U_{13} + U_{23}$ where

$$U_{13} = \frac{1}{4\pi\epsilon_0} \frac{Q_1 Q_3}{r_{13}}$$

and

$$U_{23} = \frac{1}{4\pi\epsilon_0} \frac{Q_2 Q_3}{r_{23}}$$

and the total so far is

$$U_{12} + (U_{13} + U_{23})$$

Next I bring up charge Q_4 to position $\mathbf{r}_4$ which is distant r_{14} from charge Q_1, r_{24} from charge Q_2 and r_{34} from charge Q_3. The extra energy needed is $U_{14} + U_{24} + U_{34}$ where

$$U_{14} = \frac{1}{4\pi\epsilon_0} \frac{Q_1 Q_3}{r_{14}}$$

$$U_{24} = \frac{1}{4\pi\epsilon_0} \frac{Q_2 Q_4}{r_{24}}$$

and

$$U_{34} = \frac{1}{4\pi\epsilon_0} \frac{Q_3 Q_4}{r_{34}}$$

and the total so far is

$$U_{12} + (U_{13} + U_{23}) + (U_{14} + U_{24} + U_{34})$$

By building up an entire array of n point charges, we see that the total mutual potential energy is

$$U_{12} +$$

$$(U_{13} + U_{23}) +$$

$$(U_{14} + U_{24} + U_{34}) +$$

$$\cdots +$$

$$(U_{1n} + U_{2n} + \cdots + U_{n-1,n})$$

Notice that we do not 'double count' terms like U_{12} and U_{21} (which are equal).

The total is

$$U = \sum_{i=1}^{n-1} \sum_{j=i+1}^{n} U_{ij} \tag{4.8}$$

In order to write the sum more symmetrically by including all terms like U_{12} and U_{21} (but not terms such as U_{11}), we have to write

$$U = \frac{1}{2} \sum_{i=1}^{n} \sum_{j\neq i=1}^{n} U_{ij} \tag{4.9}$$

Finally, if I substitute the expressions for each U_{ij}

$$U_{ij} = \frac{1}{4\pi\epsilon_0} \frac{Q_i Q_j}{r_{ij}}$$

then

$$4\pi\epsilon_0 U = \frac{1}{2} \sum_{i=1}^{n} \sum_{j\neq i=1}^{n} \frac{Q_i Q_j}{r_{ij}}$$

or

$$4\pi\epsilon_0 U = \frac{1}{2} \sum_{i=1}^{n} Q_i \sum_{j\neq i=1}^{n} \frac{Q_j}{r_{ij}} \tag{4.10}$$

Using the definition of the potential due to a point charge and the obvious fact that such potentials are additive, the final equation is seen to be

$$U = \frac{1}{2} \sum_{i=1}^{n} Q_i \phi(\mathbf{r}_i) \tag{4.11}$$

That is to say, we calculate the electrostatic potential at the position $\mathbf{r}_i$ of each point charge Q_i due to all the other point charges, and then sum all of these terms in order to obtain the total mutual potential energy of the point charge array.

You should note that I have limited my discussion to point charges. The expression given above for U turns out to be exact for any charge distribution.

4.7   THE BINDING ENERGY OF A CRYSTAL

Solids may be classified according to the kind of bonding within them. In order of decreasing bond strength the classes are *ionic, covalent, metallic* and *van der Waals.*

The bonding in an ionic solid such as sodium chloride derives primarily from the coulomb interaction between the ions. So, for example, in sodium chloride an electron can be thought of as being formally transferred from a Na atom to a Cl atom giving Na^+ and Cl^- ions which both have the stable electronic configuration of inert gas atoms. Ionic compounds always contain positive and negative ions, but the assumption that an electron is completely transferred is only part of the story, as we will see shortly.

The bonding in covalent solids results from atoms with partly filled outer shells that share their valence electrons. Elements in groups III, IV, V, VI and VII of the periodic table normally participate in covalent bonds. In diamond each carbon atom is tetrahedrally arranged and shares its valence electrons with four surrounding carbons.

The metals form solids in which the outer electrons are essentially ionized and can be visualized as being delocalized over the metal. Bonding in such solids involves a Coulomb interaction between the positively charged metal ions and the 'sea' of electrons in which they are embedded.

At low temperatures, even the so-called inert gases can be liquefied and solidified, demonstrating that there must be weak attractive and repulsive forces between the atoms. A great deal of research effort has been expended into a study of so-called 'van der Waals molecules' such as HCl...Ar, and van der Waals forces are responsible for the bonding in many molecular solids such as polymers and liquid crystals.

Ultimately it is the Coulomb inverse square law that is responsible for holding things together, or making them repel. Once we start to aggregate fundamental particles such as electrons and protons into atoms and molecules, then the overall effective pair potential is no longer given by a simple expression such as the inverse square law. It is necessary to try and quantify the residual forces and give them algebraic expressions.

To make quantitative calculations of the physical properties of solids we need quantitative expressions for the pair potential U_{ij} between pairs of particles making up the solid. In the case of ionic solids, the particles involved are ions and not point charges.

Chemists often visualize an Na^+Cl^- ion pair as two point charges (the nuclei) surrounded by the electron clouds (Figure 4.4). The NaCl internuclear distance is 282 pm in the crystal, and we often take the radii of the Na^+ and the Cl^- ions as 102 and 180 pm. As far as calculations of electrostatic fields outside these spheres are concerned, we can think of the electrons as being distributed through the sphere, situated at the nuclei or resident on the sphere surfaces.

The two ions are composite particles, and whilst it is easy to write an expression for the pair potential of two point charges Q_A and Q_B distance r_{AB} (from Coulomb's law) as

$$U_{AB}(R) = \frac{Q_A Q_B}{4\pi\epsilon_0} \frac{1}{r_{AB}}$$

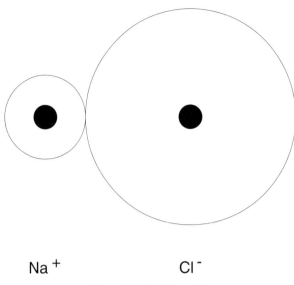

$$Na^+ \qquad\qquad Cl^-$$

Figure 4.4 NaCl ion pair

such a simple pair potential ignores the fact that each individual ion consists of nuclei and electrons and as the two ions approach, they eventually repel each other because the nuclei are no longer fully screened by the electron clouds. The magnitude of this repulsive force also depends on the distance between the nuclei of A and B and so we can write

$$U_{AB}^{\text{repulsive}} = \frac{C}{r_{AB}^m} \tag{4.12}$$

where C is a constant and m a high power, typically 12 or 13.

In general then the pair potential for two ions A and B is best written as

$$U_{AB} = \frac{Q_A Q_B}{4\pi\epsilon_0} \frac{1}{r_{AB}} + \frac{C}{r_{AB}^m} \tag{4.13}$$

If the two ions have the same sign then the first term on the right-hand side of the above equation leads to a repulsive force. If Q_A and Q_B have opposite signs then it represents an attractive interaction.

The other type of solid for which a reliable pair potential can be written down is the van der Waals solid, for which

$$U_{AB} = -\frac{C_6}{r_{AB}^6} + \frac{C_{12}}{r_{AB}^{12}} \tag{4.14}$$

These equations give quantitative expressions for the dependency of the pair potential on the interatomic separation between particles A and B. It is not possible to write down such expressions for covalent and metallic species. In such solids, the interactions cannot be described as simple pair potentials. For example, covalent solids consist of molecules that have directional bonds, and the strength of the interaction depends on the number of bonds formed and the angles between the bonds.

In a metal the atoms do not interact in a pairwise manner; it is the interaction of the whole sea of the electrons with the positive cations that is responsible for the bonding. In a covalent solid we have to take account of the directional properties of the bonds.

In the remainder of this chapter I will show how to model ionic and Van der Waals solids. In later chapters I will show how to take account of the properties of ionic and metallic solids.

4.8 A SIMPLE IONIC SOLID

In this section I am going to show how to estimate the binding energy of a very simple crystal such as NaCl.

We will think for the minute of NaCl as a purely ionic crystal made up from Na^+ and Cl^- ions localized at the appropriate points of the crystal lattice. Each ion has six nearest neighbors of the opposite kind. We describe this by saying that the co-ordination number z is 6 for this particular crystal.

In fact, the NaCl crystal was one of the first ones to be studied using X-ray diffraction. It is one of the simplest crystal types known, with alternate Na and Cl species at the corners of a cube. Crystallographers refer to the NaCl lattice as a simple cubic lattice. The distance between adjacent ions in the three-dimensional lattice is 282 pm.

I want to investigate the binding energy for a typical ion buried deep within a crystal and to do this I will first calculate the potential energy ε^{pot} due to the interaction of this ion with the other ions in the crystal. The total potential energy of the crystal is then one half the sum of all the ε^{pot}'s, and the binding energy is the negative of this quantity per ion. The binding energy gives a measure of the energy needed to separate the crystal into its constituent ions.

I can illustrate the principles involved by considering first the one-dimensional crystal shown in Figure 4.5. This consists of an infinite array of alternately charged ions $+Q$, $-Q$, $+Q$, and so on, as shown, with a constant spacing a. In order to calculate an average quantity I will focus attention on the 'black' ion. All my calculations refer to $0\,K$, and the ions have no vibrational motion.

From our considerations above, the mutual potential energy of the black ion (which I will take to be number 0 in the array) and the remaining ions in the crystal is

$$\varepsilon_0^{pot} = \frac{+Q}{4\pi\varepsilon_0} \sum_{j\neq 0}^{\infty} \frac{Q_j}{r_{0j}} \tag{4.15}$$

I have written ε^{pot} rather than U for the mutual potential energy to emphasize that I am dealing with a single ion rather than all ions; the quantity ε^{pot} gives the work done in bringing the black ion from infinity to the point in the crystal lattice as shown. I will now sum the series for the very simple linear ionic crystal shown above.

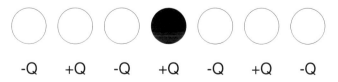

-Q   +Q   -Q   +Q   -Q   +Q   -Q

Figure 4.5   A one-dimensional ionic crystal

I can exploit the symmetry of the crystal by summing contributions to ε^{pot} in pairs, one contribution from an ion to the right of the black ion, one contribution from the equivalent ion on the left hand side.

The nearest neighbor ions each give contributions of

$$-\frac{Q^2}{4\pi\varepsilon_0}\frac{1}{a}$$

The second nearest neighbor ions each give contributions of

$$+\frac{Q^2}{4\pi\varepsilon_0}\frac{1}{2a}$$

The third nearest neighbor ions each give contributions of

$$-\frac{Q^2}{4\pi\varepsilon_0}\frac{1}{3a}$$

and so on.

The total is

$$\varepsilon_0^{pot} = -\frac{2Q^2}{4\pi\epsilon_0}\frac{1}{a}\left(1 - \frac{1}{2} + \frac{1}{3} - \frac{1}{4} + \cdots\right)$$

This is usually written

$$\varepsilon_0^{pot} = -\frac{Q^2}{4\pi\varepsilon_0}\frac{\alpha}{a} \tag{4.16}$$

where α is called the Madelung constant, to emphasize the connection with the mutual potential energy of a pair of point charges $+Q$ and $-Q$ distance a apart. The Madelung constant is a characteristic for crystals of a given symmetry, and can be evaluated by summing the infinite series in the expression above. For a single ion pair, $\alpha = 1$ and in the case of our infinite one-dimensional crystal,

$$\alpha = 2\left(1 - \frac{1}{2} + \frac{1}{3} - \cdots\right)$$

which happens to be 2 log$_e$ 2. The Madelung constant is therefore 2 log$_e$ 2 = 1.38629 in this case.

Direct summation shows that the series is very slowly convergent (Table 4.1) and over the years, mathematicians have calculated Madelung constants for all known three-dimensional crystals.

In order to find the total mutual potential energy U of the entire crystal, I have to sum over all the ion pairs and multiply by $\frac{1}{2}$ (in order to avoid double counting of the interactions). I chose a typical ion in an essentially infinite lattice, and so just need to multiply by $\frac{1}{2}N$ (where N is the number of ions) to give

$$U = -\frac{1}{2}N\frac{Q^2}{4\pi\varepsilon_0}\frac{\alpha}{a} \tag{4.17}$$

where there are N ions present in the crystal. The factor of $\frac{1}{2}$ appears in order to avoid double counting between pairs of ions.

Table 4.1 Direct summation of the series for α

Number of ion pairs	α
10	1.2913
100	1.3763
1000	1.3853
10 000	1.3862
1 000 000	1.3863

Finally I can estimate the average binding energy ε_b per ion by dividing $-U$ by N;

$$\varepsilon_b = -U/N$$

$$= \frac{1}{2} \frac{Q^2}{4\pi\varepsilon_0} \frac{\alpha}{a} \tag{4.18}$$

4.9   BETTER PAIR POTENTIALS

A more realistic pair potential for Na and Cl is the Mie potential discussed above

$$U_{\text{Na..Cl}} = -\frac{e^2}{4\pi\varepsilon_0} \frac{1}{r} + \frac{C}{r^{12}} \tag{4.19}$$

where C is a constant, which depends on the nature of the anion and cation. The first term is the purely attractive Coulomb term, whilst the second term caters for repulsion between the nuclei once they approach closely enough. This repulsive term falls off very quickly with interionic separation, so much so that we only need to consider nearest neighbor contributions.

If we repeat the calculation given above for the potential energy ε^{pot} of a single ion in the one-dimensional crystal then we find

$$\varepsilon^{\text{pot}} = -\frac{Q^2}{4\pi\varepsilon_0} \frac{\alpha}{a} + 2 \frac{C}{a^{12}} \tag{4.20}$$

(where we have only allowed for the short range repulsion between nearest neighbor pairs of ions).

The revised estimate of the binding energy of the one-dimensional crystal is therefore

$$\varepsilon_b = \frac{1}{2} \left(\frac{Q^2}{4\pi\varepsilon_0} \frac{\alpha}{a} - \frac{2C}{a^{12}} \right) \tag{4.21}$$

This calculation can also be extended to three dimensional ionic crystals. In the NaCl example, we find

$$\varepsilon^{\text{pot}} = -\frac{Q^2}{4\pi\varepsilon_0} \frac{\alpha}{a} + z \frac{C}{a^{12}} \tag{4.22}$$

where $z = 6$ and $\alpha = 1.7476$.

Table 4.2 Lennard–Jones parameters

	$(\varepsilon/k_B)\,(\text{K})$	$\sigma\,(\text{pm})$
He	10.22	258
Ne	35.7	279
Ar	124	342
H_2	33.3	297
C_6H_6	440	527

4.10 THE BINDING ENERGY OF A VAN DER WAALS SOLID

I mentioned earlier the Van der Waals solid; this is a solid comprised of species whose pair potentials can be written in Lennard–Jones form as

$$U_{ij}(R) = \frac{C_{12}}{r_{ij}^{12}} - \frac{C_6}{r_{ij}^6} \tag{4.23}$$

where C_{12} and C_6 are constants that depend on the nature of species i and j.

This pair potential can also be written in terms of the parameters ε and σ as

$$U_{ij}(r) = 4\varepsilon\left[\left(\frac{\sigma}{r}\right)^{12} - \left(\frac{\sigma}{r}\right)^6\right] \tag{4.24}$$

where ε is the depth of the potential well and σ is the distance of closest approach.

Typical values of the Lennard–Jones parameters are shown in Table 4.2. Workers in this field normally divide the energy by the Boltzmann constant k_B to give a temperature.

Calculation of the binding energy is quite straightforward, given the pair potential and a knowledge of the crystal structure. Consider the one-dimensional crystal shown in Figure 4.6, and I will focus attention on the central atom that I have drawn in black and labelled 0. If we call the interatomic distance a, then atoms 1 and $1'$ each make contributions of

$$U_{01} = \frac{C_{12}}{a^{12}} - \frac{C_6}{a^6}$$

Atoms 2 and $2'$ each make contributions of

$$U_{02} = \frac{C_{12}}{(2a)^{12}} - \frac{C_6}{(2a)^6}$$

Figure 4.6 A one-dimensional Van der Waals crystals

Table 4.3   Direct summation of the attractive and repulsive terms

Number of terms in sum	Attractive term	Repulsive term
1	1	1
2	1.0156	1.0002
3	1.0700	1.0002
10	1.0173	1.0002
20	1.0173	1.0002

and so on. The total $\varepsilon^{\text{atom}}$ for atom 0 is therefore

$$\varepsilon^{\text{atom}} = \frac{2C_{12}}{a^{12}}\left(1 + \frac{1}{2^{12}} + \frac{1}{3^{12}} + \cdots\right) - \frac{2C_6}{a^6}\left(1 + \frac{1}{2^6} + \frac{1}{3^6} + \cdots\right)$$

The two series

$$\left(1 + \frac{1}{2^{12}} + \frac{1}{3^{12}} + \cdots\right) \quad \text{and} \quad \left(1 + \frac{1}{2^6} + \frac{1}{3^6} + \cdots\right)$$

converge very quickly so that

$$\varepsilon^{\text{atom}} = -\frac{2.0347C_6}{a^6} + \frac{2.0005C_{12}}{a^{12}} \tag{4.25}$$

The total mutual potential energy of a system of N such atoms is therefore

$$U = \frac{1}{2}N\left(-\frac{2.0347C_6}{a^6} + \frac{2.0005C_{12}}{a^{12}}\right) \tag{4.26}$$

and so the binding energy for a given atom is

$$\varepsilon_b = \frac{1}{2}\left(-\frac{2.0347C_6}{a^6} + \frac{2.0005C_{12}}{a^{12}}\right) \tag{4.27}$$

I have used the term 'atom', but the same derivation is valid for all particles that satisfy the Van der Waals potential (Table 4.3). Notice that the binding energy is almost exactly that due to the two nearest neighbor terms. With this in mind, you will not be surprised to find that the corresponding results for a two-dimensional and a three-dimensional cubic crystal come out as

$$\varepsilon_b = \frac{1}{2}\left(-\frac{4.6589C_6}{a^6} + \frac{4.0640C_{12}}{a^{12}}\right) \tag{4.28}$$

and

$$\varepsilon_b = \frac{1}{2}\left(-\frac{8.4006C_6}{a^6} + \frac{6.2021C_{12}}{a^{12}}\right) \tag{4.29}$$

5 Introduction to Quantum Mechanics

5.1 PARTICLES AND WAVES

Quantum mechanics is the theory of nuclear, atomic and molecular systems. It has been developed from the two main branches of classical physics, namely Newtonian (i.e. classical) mechanics and Maxwell's electromagnetic theory. I mentioned classical mechanics when we discussed vibrational motion in Chapter 3, and I gave you some of the basic ideas of classical electromagnetism when I discussed the stability of an ionic crystal in Chapter 4.

Electromagnetic theory is concerned with electric and magnetic phenomena, and these are best described in terms of certain electric and magnetic fields $\mathbf{E}$ and $\mathbf{B}$. These fields are related to charge and current densities through a set of four equations called Maxwell's equations. Maxwell's equations are beautifully concise statements of many electromagnetic phenomena, and they play the same role in electromagnetism that Newton's equations play in classical mechanics. I am not going to state them here because they can be found in any standard textbook on electromagnetism, and are best written in terms of certain vector operators that we have not met.

The key point for our discussion is that they lead to the conclusion that, in free space, the electric and magnetic fields satisfy the wave equations

$$\frac{\partial^2 \mathbf{E}}{\partial x^2} + \frac{\partial^2 \mathbf{E}}{\partial y^2} + \frac{\partial^2 \mathbf{E}}{\partial z^2} = \frac{1}{c_0^2} \frac{\partial^2 \mathbf{E}}{\partial t^2} \tag{5.1}$$

$$\frac{\partial^2 \mathbf{B}}{\partial x^2} + \frac{\partial^2 \mathbf{B}}{\partial y^2} + \frac{\partial^2 \mathbf{B}}{\partial z^2} = \frac{1}{c_0^2} \frac{\partial^2 \mathbf{B}}{\partial t^2} \tag{5.2}$$

These are classical wave equations, and they state that the fields $\mathbf{E}$ and $\mathbf{B}$ propagate through free space at the speed of light. For motion along the x axis we can write solutions for these wave equations as

$$\mathbf{E} = \mathbf{E}_0 \exp[j(\omega t - kx)] \tag{5.3}$$

$$\mathbf{B} = \mathbf{B}_0 \exp[j(\omega t - kx)] \tag{5.4}$$

where j is the square root of -1 (i.e. $j^2 = -1$), $\mathbf{E}_0$ and $\mathbf{B}_0$ are maximum wave amplitudes, ω is the angular wave frequency and k the so-called wave vector related to the wavelength λ by $k = 2\pi/\lambda$.

Maxwell made the inspired guess that these waves could be identified with visible light and of course we have now become familiar with other forms of such radiation whose wavelengths range from γ-rays of nuclear fallout through X-rays, ultraviolet and visible radiation, infrared, microwave and radio frequencies.

Interference phenomena, of which diffraction is typical, relies on the wave picture for a satisfactory explanation.

Classical mechanics and electromagnetism are coupled by the Lorentz law

$$\mathbf{F} = Q(\mathbf{E} + \mathbf{v} \times \mathbf{B}) \tag{5.5}$$

which gives us the force $\mathbf{F}$ on a particle of charge Q, moving with velocity vector $\mathbf{v}$ in an electric field $\mathbf{E}$ and a magnetic field $\mathbf{B}$.

By the end of the nineteenth century, this classical picture of the world, with matter consisting of point particles and radiation consisting of waves, seemed to provide explanations for almost all known physical phenomena. There were a few loose ends, and attempts to tidy up the loose ends led to the birth of quantum theory.

5.2 PARTICLE ASPECTS OF RADIATION

The first indication of a breakdown in the classical concepts occurred in the study of 'blackbody radiation', a phenomenon which is concerned with the exchange of energy between matter and radiation. According to the classical theories, this exchange is continuous in the sense that radiation of frequency ν can give up any amount of energy on absorption. Planck showed that the correct thermodynamic formula is obtained only if it is assumed that the energy exchange is discrete. Planck postulated that radiation of frequency ν can only exchange energy with matter in units of $h\nu$, where h is a physical constant called Planck's constant ($h = 6.626 \times 10^{-34}$ J s).

Planck's hypothesis may be stated by saying that the radiation of frequency ν behaves like a stream of particles (called photons) of energy

$$\varepsilon = h\nu \tag{5.6}$$

which may be emitted or absorbed by matter. Photons are subject to the laws of relativistic mechanics rather than classical mechanics and you will have to take my word for it when I tell you that the relativistic equation relating energy ε and momentum p is

$$\frac{\varepsilon^2}{c_0^2} = p^2 + m^2 c_0^2 \tag{5.7}$$

Photons travel at the speed of light and so must have a zero rest mass, giving

$$p = \frac{\varepsilon}{c_0}$$

and so we find

$$p = \frac{h}{\lambda} \qquad (5.8)$$

which relates the momentum to the wavelength.

A much simpler example of the particle nature of radiation is shown by the photoelectric effect. A beam of monochromatic radiation of frequency ν is directed onto the surface of a metal, and electrons may or may not be emitted. If $h\nu$ is less than some limiting energy Φ, which depends on the particular metal under study, then no electrons are emitted. If $h\nu$ is greater than Φ then electrons are emitted with kinetic energy T given by

$$T = h\nu - \Phi \qquad (5.9)$$

Note that even when electrons are emitted, their kinetic energy does not depend on the intensity of the electromagnetic radiation, only on its frequency. Such behavior is completely unintelligible from the point of view of classical electromagnetic theory, but it is quite easy to understand on the basis of Planck's hypothesis.

The photoelectric effect, and studies of black body radiation, demonstrate that energy exchange takes place in units of quanta $h\nu$. The particle nature of the radiation itself is shown most clearly in the scattering of X-rays by electrons, the Compton effect (Figure 5.1).

Consider a photon with momentum $\mathbf{p}_1$ which collides with a stationary electron. After the collision the photon has momentum $\mathbf{p}_2$ and the electron has momentum $\mathbf{p}_e$. According to the law of conservation of momentum $\mathbf{p}_1 = \mathbf{p}_2 + \mathbf{p}_e$ and therefore $p_e^2 = p_1^2 + p_2^2 - 2p_1p_2\cos(\theta)$ where θ is the angle shown in Figure 5.1.

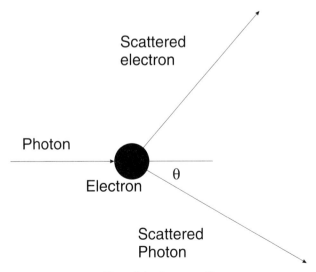

Figure 5.1 Compton effect

According to the (relativistic) principle of energy conservation we have

$$p_1 + mc_0 = p_2 + \sqrt{(p_e^2 + mc_0^2)}$$

and eliminating p_e^2 from these equations gives

$$mc_0(p_1 - p_2) = 2p_1p_2 \sin^2\left(\frac{\theta}{2}\right)$$

Dividing by p_1p_2 and expressing the result in terms of the wavelength $\lambda = h/p$ we obtain

$$\lambda_2 - \lambda_1 = 2\lambda_e \sin^2\left(\frac{\theta}{2}\right) \tag{5.10}$$

where λ_e is the Compton wavelength of the electron, h/mc_0.

The change in wavelength of the incoming photon therefore depends only on the angle of scatter of the radiation, and not on the original frequency, and this is exactly in accord with experiment. This explanation follows directly from treating the colliding photon and electron as particles, but it cannot be explained in terms of the electromagnetic wave picture.

5.3 WAVE ASPECTS OF MATTER AND THE de BROGLIE HYPOTHESIS

Davisson and Germer's 1927 experiments showed that a beam of electrons reflected from the surface of a nickel crystal formed a diffraction pattern, just like the diffraction of light by a grating. Diffraction is essentially a wave phenomenon, and so under these circumstances a single electron must in some way be associated with a wave.

Even before the experiments of Davisson and Germer, de Broglie had suggested that the formula $p = h/\lambda$ which relates the particle and wave aspects of radiation should also apply to electrons. Thus an electron with a given energy and momentum should be associated in some mysterious and undefined way with a de Broglie wave of the type

$$\mathbf{A}_0 \exp[j(kt - \omega x)]$$

which is just a wave of the form discussed above.

The procedure whereby atomic phenomena could be 'explained' in terms of classical mechanics was actually carried very much further than I have intimated. There is little point in going down that path, unless you happen to be interested in the History of Science, and I do not intend to do so. It is obviously unsatisfactory to have both radiation and matter being treated sometimes as waves and sometimes as matter in an apparently arbitrary manner. What was needed was a complete reformulation of the theory in such a way that the classical concepts remained intact, but that the Planck and de Broglie rules should appear as natural consequences of some coherent theory. This theory is the quantum theory.

5.4 THE PROBABILISTIC NATURE OF SOME EXPERIMENTS

Experimental evidence indicating the wave nature of light had been around for some 60 years before Maxwell made his theoretical predictions. The most convincing evidence had been discovered by Thomas Young in 1801. Young's two-slit experiment is illustrated in Figure 5.2.

The source of light on the left of Figure 5.2 emits light of a single wavelength. The light illuminates two parallel slits which lie perpendicular to the plane of the page, and the intensity of the transmitted light is observed on a photographic plate placed on the right-hand side of the diagram and perpendicular to the plane of the paper. Experimentally it turns out that there is a series of positions on the plate where the intensity of light is high, and a series of positions where the intensity is zero.

I have drawn wavefronts, which are the points in space where both the electric and magnetic vectors have their maximum values. The two fields vary in step with each other as they move through space. To understand what happens to the waves once they reach the slits, we regard each slit as a new source of radiation and so I have drawn sets of circular wavefronts spreading out from each of them.

Any point to the right of the slits receives light from both slits, and the combined effect is determined by a principle known as the Principle of Superposition. This means that the total electric field at any point will be a sum of the electric fields due to the light from either slit, as will the total magnetic field.

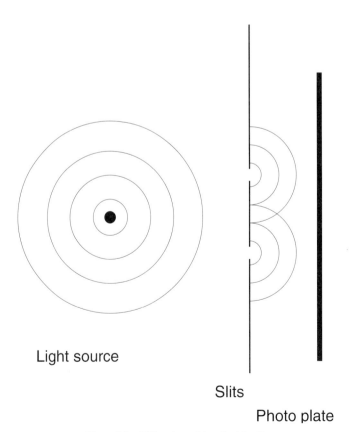

Light source

Slits

Photo plate

Figure 5.2   Diffraction with a double slit

At certain points the electric fields will cancel completely and at certain other points the fields will reinforce each other, and with this in mind you can understand the diffraction pattern that results on the photographic plate. You might be wondering what would happen if the intensity of light were to be reduced until only a single photon passed through the diffraction equipment. I have to tell you that the diffraction pattern is replaced by a single spot on the photographic plate. If we allow more and more photons through the diffraction equipment then the diffraction pattern is built up, and each photon strikes the plate in a random manner. We cannot determine the position that a photon will strike the plate, we can only assign a probability to the event. The points on the photographic plate corresponding to high intensity correspond to positions of high probability, the dark regions on the plate correspond to positions of low probability.

But how to calculate the probabilities? Well, a standard result from electromagnetic theory is that the intensity of light is proportional to the square of the electric (and magnetic) fields. So there ought to be a similar relationship for de Broglie waves. That is, if the amplitude of a particular de Broglie wave is A then the electron intensity at that point ought perhaps to be A^2. Because of this important relationship, de Broglie waves are sometimes referred to as probability waves. This connection between waves and probability was first suggested by Max Born.

5.5 WAVE PARTICLE DUALITY

Let us now return to the wave particle duality of electrons. According to quantum mechanics, it is not meaningful to ask whether an electron is a particle or a wave; for some types of measurement the wave model is appropriate, for other types of measurement the particle model is to be preferred. The simplest type of probability wave is the de Broglie wave that describes a free particle with a momentum of magnitude p. For the sake of argument, assume that the wave is moving along the x axis. This type of wave is an infinitely long wave of wavelength

$$\lambda_{dB} = \frac{h}{p}$$

According to our discussion above, the probability of finding the electron in a region along the x axis is proportional to the square of the amplitude of its probability wave in that region, and so it is equally likely that the electron will be found at any position along the x axis. So although we know the momentum exactly, we have no idea of the position of the electron.

It is also possible to make use of the principle of superposition and add together de Broglie waves of different wavelengths to give what is called a *wave packet*, a short burst of waves. Wave packets are strongly localized in space, but they do not have a definite momentum. By adding more and more waves with a spread of wavelengths close to the de Broglie wavelength it is possible to reduce the width of the wave packet, and make it as narrow as we wish in order to localize the electron.

So on the one hand, we can describe a free electron as an infinite de Broglie wave such that the momentum is known very precisely, but where we know nothing whatever about the position. On the other hand, we can describe the same free electron as a very localized wave packet, where the position is known precisely but the momentum is poorly defined.

Heisenberg's uncertainty principle relates (in this case) the uncertainties in the measurements of the position along the x axis and the component of the momentum along the x axis, p_x;

$$\Delta x \Delta p_x \geq \frac{h}{2\pi} \tag{5.11}$$

This is one of the most profoundly important ideas in quantum physics. It is often stated informally as Heisenberg's uncertainty principle

There is a fundamental limit to the precision with which the position x and momentum component p_x of a particle can be measured simultaneously.

Heisenberg's principle tells us that x and p_x cannot be measured simultaneously to an arbitrary degree of precision. Some pairs of properties can be measured simultaneously to arbitrary degrees of precision. For example, Heisenberg's principle also tells us that

$$\Delta x \Delta p_y = 0$$

and so the position along the x axis together with the y component of linear momentum can indeed be measured simultaneously to arbitrary precision.

Heisenberg's principle therefore tells us just which pairs of properties can be measured simultaneously to arbitrary precision and just which cannot. That is why I say that it is profound importance in physical science.

5.6 LINEAR OPERATORS

Before I proceed to set up the theory of quantum mechanics, I want to tell you a little about linear operators. These are mathematical entities that play a key rôle in our theory.

Loosely speaking, an operator $\hat{A}$ is any mathematical entity that operates on a function (written for example $f(x)$) and turns it into another function. To keep the discussion simple, I am going to deal mostly with functions of a single real variable that I will write as x.

It is conventional to write operators with the hat sign as in $\hat{A}$, to distinguish them from other mathematical entities. The simplest example of a linear operator is to take $\hat{A}$ itself to be a function of x, and so we might have

$$\hat{A}f(x) = xf(x) \tag{5.12}$$

Thus operating with $\hat{A}$ on $f(x)$ produces the new function $xf(x)$. This is not an equation that has to be solved for x, it is a rule showing what happens when $\hat{A}$ goes to work on a function $f(x)$. For example, $\hat{A} \sin(x) = x \sin(x)$.

Now consider the operator $\hat{B}$ which operates on a function $f(x)$ to give $df(x)/dx$. We write this as

$$\hat{B}f(x) = \frac{df(x)}{dx} \tag{5.13}$$

Notice that I used the word linear in an earlier paragraph; a linear operator $\hat{A}$ is one that satisfies the following two properties

$$\hat{A}[f(x) + g(x)] = \hat{A}f(x) + \hat{A}g(x)$$

$$\hat{A}[cf(x)] = c\hat{A}f(x)$$

(5.14)

where c is any scalar and $f(x)$ and $g(x)$ any two functions. Both operators $\hat{A}$ and $\hat{B}$ discussed above are linear operators.

Consider now the product of the operators $\hat{A}$ and $\hat{B}$, written $\hat{A}\hat{B}$. We define this as follows.

$$(\hat{A}\hat{B})f(x) = \hat{A}[\hat{B}f(x)]$$

(5.15)

This is mathematical shorthand for the following set of instructions. 'Take any function $f(x)$ and operate on it with $\hat{B}$. Then operate on the result of that with $\hat{A}$.' To use the specific examples of $\hat{A}$ and $\hat{B}$ above we have

$$(\hat{A}\hat{B})f(x) = \hat{A}[\hat{B}f(x)]$$

$$= \hat{A}\,\frac{df(x)}{dx}$$

$$= x\,\frac{df(x)}{dx}$$

whilst

$$(\hat{B}\hat{A})f(x) = \hat{B}[\hat{A}f(x)]$$

$$= \hat{B}xf(x)$$

$$= f(x) + x\,\frac{df(x)}{dx}$$

The effect of the operator product $\hat{A}\hat{B}$ is not the same as the effect of the operator product $\hat{B}\hat{A}$. We say that these operators do not commute. It matters which order we apply them. Some operators commute and some do not commute. Looking at the above operator equations, we can write the following operator identity

$$\hat{B}\hat{A} = \hat{A}\hat{B} + 1$$

(5.16)

It is an identity because it is true for every function $f(x)$ that is differentiable;

$$(\hat{B}\hat{A})f(x) = (\hat{A}\hat{B} + 1)f(x)$$

To each linear operator $\hat{A}$ there belongs a special set of functions $u_i(x)$ and numbers a_i such that

$$\hat{A}u_i(x) = a_i u_i(x)$$

(5.17)

That is to say, we operate with the operator $\hat{A}$ on some certain functions $u_i(x)$, and the effect is to give the functions back again unaltered apart from multiplication by a constant.

We say that the $u_i(x)$ are the *eigenvectors* of the operator $\hat{A}$ and that the constants a_i are the eigenvalues. The eigenfunctions of an operator are therefore special functions in that they remain unaltered under the operation of the operator, except that they get multiplied by the eigenvalue.

Depending on the operator, there can be a finite set of eigenvalues, a countably infinite set or the eigenvalues can be a continuous variable.

To finish this section, let me quote a standard result from the theory of linear operators.

Theorem

If the two linear operators $\hat{A}$ and $\hat{B}$ commute, then it is possible to find a set of eigenfunctions that are simultaneously eigenfunctions of both $\hat{A}$ and $\hat{B}$

The proof of this theorem is given in all the standard textbooks of quantum chemistry.

5.7 OBSERVATIONS

In classical physics it is tacitly assumed that the effect of making an observation on a system does not affect the system. The simplest type of observation we might imagine is where we look at an object, and in this case a photon has to be involved. The more accurately we want to pinpoint the object the shorter the wavelength of the radiation needed and hence the higher the momentum of the photon involved. So the effect of making a measurement on an atomic system can create a considerable disturbance.

According to Dirac:

There is a limit to the fineness of our powers of observation and the smallness of the accompanying disturbance—a limit that is inherent in the nature of things and can never be surpassed by improved technique

If the system is large enough for these disturbances to be negligible then classical physics applies. If on the other hand the disturbances are significant then the system is 'small' in an absolute sense.

Measurements have two properties.

(i) To each measurement there belongs a set of values, the possible results of the measurements. These values might run over a continuous range (as in the measurement of the energy of a free particle), or they might belong to a discrete set (as in the allowed energies of a H atom).

(ii) Suppose we want to make measurements of two quantities, denoted A and B. If I denote observation of A followed by B as BA, and observation of B followed by A as BA, then each observation may well interfere with the results of the other in a statistical manner, and so

$$AB \neq BA$$

Quantum mechanics is essentially a hard subject. It can be approached in very many ways, either by appeal to experiment or not. It is probably best if I launch straight into the *postulates of quantum mechanics*, a logical set of assumptions on which the whole theory is based, rather than spend any more time on the experimental evidence for the theory. Depending on which textbook you read, there are several different sets of postulates. I am going to give you a very simple subset, which ties in nicely with the experimental evidence given so far.

5.7.1 Postulate 1

To every observable A we can assign a linear operator $\hat{A}$.

5.7.2 Postulate 2

The possible results of a measurement of A are the eigenvalues of the associated linear operator $\hat{A}$.

5.7.3 Postulate 3

A measurement of property A on a system in a state that is an eigenfunction of $\hat{A}$ definitely leads to a result corresponding to the eigenvalue.

5.7.4 Postulate 4

The average value of repeated measurements of A on a set of systems, each one in an arbitrary state $\psi(x)$ (usually referred to as the *wavefunction*) is

$$\langle A \rangle = \frac{\int \psi^*(x)\hat{A}\psi(x)\,\mathrm{d}x}{\int \psi^*(x)\psi(x)\,\mathrm{d}x} \tag{5.18}$$

The final postulate needs some interpretation. If the system is in an eigenstate then we will certainly obtain the eigenvalue. If the system consists of a number of particles each in a different eigenstate then repeated measurements will give a statistical distribution of the different eigenvalues.

In the special case that $\psi(x) = u_i(x)$ then we have

$$\langle A \rangle = \frac{\int u_i^*(x)\hat{A}u_i(x)\,\mathrm{d}x}{\int u_i^*(x)u_i(x)\,\mathrm{d}x}$$

$$= \frac{\int u_i^*(x)a_i u_i(x)\,\mathrm{d}x}{\int u_i^*(x)u_i(x)\,\mathrm{d}x}$$

$$= a_i$$

which is consistent with Postulate 3.

Notice also that I have allowed for the possibility that eigenstates might be complex quantities. Just to remind you, if $z = x + jy$ is a complex number then the complex conjugate z^* is given by $x - jy$. The product $z^*z = (x - jy)(x + jy) = x^2 + y^2$ and so the size or modulus of the complex number z is

$$|z| = \sqrt{z^*z}$$

Notice also that linear operators follow very closely the properties of measurements; the set of unique numbers called the eigenvalues, the fact that the order of evaluation of operator products is significant, and the fact that sometimes pairs of operators have simultaneous eigenfunctions.

It should also be clear that there is one condition that quantum mechanics must satisfy if it satisfies no others; in the limit, for everyday systems such as colliding billiard balls and accelerating trains, the results of quantum mechanics must be identical to what we would expect from classical mechanics. This is called the *Correspondence principle*, and its interpretation is not obvious. At first sight, we might think to take a quantum mechanical result and substitute a large mass for an atomic one in order to get the classical result. The real meaning of the Correspondence principle is more subtle; what we have to do is derive the quantum mechanical result and then let the quantum number(s) get large. I will give you an example shortly.

5.7.5   Postulate 5: The Commutation Relations

We know from Heisenberg's uncertainty principle that

$$\Delta x \Delta p_x \geq \frac{h}{2\pi}$$

which means that the x component of position and the x component of linear momentum cannot be measured simultaneously to arbitrary precision. There are similar expressions for $\Delta y \Delta p_y$ and $\Delta z \Delta p_z$. However,

$$\Delta x \Delta p_y = 0$$

and so the x component of position and the y component of the linear momentum can indeed be measured simultaneously to arbitrary precision. In the language of linear operators, we must have

$$\hat{x}\hat{p}_x - \hat{p}_x\hat{x} \neq 0$$

The left-hand side is referred to as the *commutator* of the two operators, and our postulate tells us that

$$\hat{x}\hat{p}_x - \hat{p}_x\hat{x} = -j\frac{h}{2\pi} \tag{5.19}$$

with similar expressions for the other two Cartesian components. Also

$$\hat{x}\hat{p}_y - \hat{p}_y\hat{x} = 0 \tag{5.20}$$

with a similar expression for each other pair of operators.

5.7.6 Postulate 6: The Schrödinger Representation

We usually find it convenient to give some concrete form to the linear operators representing the position and momentum operators. In the Schrödinger representation, we substitute as follows:

For a position operator such as $\hat{x}$ we write x
For a momentum operator such as $\hat{p}_x$ we write $-j(h/2\pi)(\partial/\partial x)$.

   Application of these two operators to an arbitrary but differentiable function shows that they do indeed satisfy the correct commutation relation.

5.7.7 Angular Momentum

I discussed angular momentum in Chapter 3, and defined the angular momentum vector **l** of a particle with position vector **r** and linear momentum **p** as

$$\mathbf{l} = \mathbf{r} \times \mathbf{p}$$

In classical mechanics, we can measure simultaneously the size of the vector and all of its three Cartesian components to an arbitrary degree of precision. Things are very different for quantum mechanical systems, as I will now show you.
   We define the quantum mechanical vector operator $\hat{\mathbf{l}}$ by its Cartesian components

$$\hat{l}_x = \hat{y}\hat{p}_z - \hat{z}\hat{p}_y$$

$$\hat{l}_y = \hat{z}\hat{p}_x - \hat{x}\hat{p}_z$$

$$\hat{l}_z = \hat{x}\hat{p}_y - \hat{y}\hat{p}_x$$

and the operator corresponding to the scalar l^2 as

$$\hat{l}^2 = \hat{l}_x^2 + \hat{l}_y^2 + \hat{l}_z^2$$

Given the commutation rules between the various components of the position and momentum operators given by Postulate 5, it can be easily shown by direct calculation that the individual components of the angular momentum operator do not commute. In fact,

$$\hat{l}_x\hat{l}_y - \hat{l}_y\hat{l}_x = j\frac{h}{2\pi}\hat{l}_z$$

$$\hat{l}_y\hat{l}_z - \hat{l}_z\hat{l}_y = j\frac{h}{2\pi}\hat{l}_x \tag{5.21}$$

$$\hat{l}_z\hat{l}_x - \hat{l}_x\hat{l}_z = j\frac{h}{2\pi}\hat{l}_y$$

and so all three cannot be measured simultaneously to arbitrary degrees of precision.
   In addition, we find

$$\hat{l}_x\hat{l}^2 - \hat{l}^2\hat{l}_x = 0$$

$$\hat{l}_y\hat{l}^2 - \hat{l}^2\hat{l}_y = 0 \tag{5.22}$$

$$\hat{l}_z\hat{l}^2 - \hat{l}^2\hat{l}_z = 0$$

so the operator corresponding to the square of the size of the vector and all of the three components can be simultaneously measured to arbitrary precision. We conclude that it is only possible to measure simultaneously $\hat{l}^2$ and any one of the components, which we arbitrarily take to be the z component. According to the Theorem discussed above, we should therefore be able to find a set of simultaneous eigenfunctions.

It is most profitable for future chapters if I indicate the derivation and results in Schrödinger language. The x component of the angular momentum operator is

$$\hat{l}_x = -j\frac{h}{2\pi}\left(y\frac{\partial}{\partial z} - z\frac{\partial}{\partial y}\right) \tag{5.23}$$

and the other two are found by cyclic permutation of the variables x, y and z in this expression. It turns out to be best to rewrite the operators in spherical polar coordinates, as discussed in the Appendix. This gives

$$\hat{l}_z = -j\frac{h}{2\pi}\frac{\partial}{\partial\phi} \tag{5.24}$$

$$\hat{l}^2 = -\frac{h^2}{4\pi^2}\left(\frac{\partial^2}{\partial\theta^2} + \cot\theta\frac{\partial}{\partial\theta} + \frac{1}{\sin^2\theta}\frac{\partial^2}{\partial\phi^2}\right) \tag{5.25}$$

Both operators depend on the polar angles, but not on r and so it should be clear that the eigenfunctions should be spherical functions. The full derivation is given in many quantum mechanics textbooks, and we can write them as

$$\hat{l}^2 Y_{l,m}(\theta,\phi) = l(l+1)\frac{h^2}{4\pi^2}Y_{l,m}(\theta,\phi)$$

$$\hat{l}_z Y_{l,m}(\theta,\phi) = m_l\frac{h}{2\pi}Y_{l,m}(\theta,\phi) \tag{5.26}$$

The eigenfunctions are referred to as *spherical harmonics*, and they can be written

$$Y_{l,m}(\theta,\phi) = P_{l,m}(\theta,\phi)\exp(jm_l\phi) \tag{5.27}$$

where $P_{l,m}$ is a so-called *associated Legendre function*.

The quantum numbers l and m_l are restricted to certain discrete values; $l = 0,\ 1,\ 2,\ldots$ and for each value of l, m_l can take values of $-l$, $-l+1,\ldots,+l$. So for $l = 2$, m_l can take each of the five values -2, -1, 0, 1, 2.

Note that the values of l are restricted to non-negative integers. I will have more to say about angular momentum in Chapters 8 and 15.

5.7.8 Postulate 7: Spin

It turns out that particles can have an internal angular momentum, which is in addition to the angular momentum they might have because of their non-zero linear momentum. This internal angular momentum is referred to as *spin angular momentum*, and it is a fundamental property of the particle just like mass and charge. A significant difference between the two types of angular momentum is that the spin quantum number can be half integral as well as integral.

For an electron, we talk about electron spin and we write the two quantum numbers s and m_s. The electron spin operator is conventionally written $\hat{s}$ and it is a vector operator with three components, just like $\hat{l}$. The components of $\hat{s}$ satisfy the same commutation relations as $\hat{l}$. Experimentally, we find that $s = \frac{1}{2}$ and we conventionally label the two eigenfunctions α and β. Thus, for example

$$\hat{s}^2 \alpha = \frac{1}{2}\left(\frac{1}{2} + 1\right) \frac{h^2}{4\pi^2} \alpha$$

(5.28)

$$\hat{s}_z \alpha = \frac{1}{2} \frac{h}{2\pi} \alpha$$

Many nuclei also have a spin, and again by convention their angular momentum vector operator is written $\hat{I}$.

You may think it odd that I have introduced spin as a postulate. If so, then bear with me until Chapter 15.

5.8 TIME DEPENDENCE

As I mentioned above, there are many ways of teaching quantum mechanics. The main equation for this text is called the time-dependent Schrödinger equation, and I will write it, without any attempt at 'verification', in a simplified form that inter-relates the state of a single particle in one dimension (taken to be x) at time t, and the wavefunction $\Psi(x,t)$

$$\left[-\frac{h^2}{8\pi^2 m} \frac{\partial^2}{\partial x^2} + U(x,t)\right] \Psi(x,t) = j\frac{h}{2\pi} \frac{\partial \Psi(x,t)}{\partial t}$$

(5.29)

Notice that the equation contains j, the square root of -1, and a potential energy $U(x,t)$ that is taken to depend on time and distance along the x axis.

In three dimensions the equation is more complicated because both the potential U and the wavefunction Ψ might depend on the three dimensions x, y and z as well as time.

$$\left[-\frac{h^2}{8\pi^2 m}\left(\frac{\partial^2}{\partial x^2} + \frac{\partial^2}{\partial y^2} + \frac{\partial^2}{\partial z^2}\right) + U(x,y,x,t)\right] \Psi(x,y,z,t) = j\frac{h}{2\pi} \frac{\partial \Psi(x,y,z,t)}{\partial t}$$

(5.30)

This equation relates to a single particle. For many particles, the equation gets more complicated still, as will be shown in later Chapters.

A partial justification for the equation is sometimes given as follows.

Consider a classical particle of mass m moving along the x axis under the influence of a potential $U(x)$ that only depends on x and not on time.

The classical energy expression

$$\varepsilon = \frac{p_x^2}{2m} + U(x)$$

turns into an operator equation

$$\hat{\varepsilon} = \frac{\hat{p}_x^2}{2m} + U(\hat{x})$$

and then we substitute the Schrödinger representation of the operators into this equation in order to get the familiar energy. This energy operator is called the Hamiltonian operator.

$$\hat{H} = -\frac{h^2}{8\pi^2 m}\frac{\partial^2}{\partial x^2} + U(x) \tag{5.31}$$

and if we are interested in the energy eigenvalues then we must solve the eigenvalue problem

$$\hat{H}\psi(x) = \varepsilon\psi(x) \tag{5.32}$$

which is known as the time-independent Schrödinger equation. It only applies as written to a single particle of mass m located along the x axis.

This derivation does not explain the time dependence in the time-dependent equation, nor does it give us any clue as to how to interpret the wavefunction.

How then do we interpret the wavefunctions $\Psi(x, t)$, especially as I have told you that they can be complex quantities?

For a complex quantity $\Psi(x, t)$, the size or modulus of the quantity is given by the square root of $\Psi^*(x, t)\Psi(x, t)$ and according to Max Born's interpretation of quantum mechanics, the physical interpretation attaching to the wavefunctions is that $\Psi^*(x, t)\Psi(x, t)\,dx$ gives the probability of finding the particle between x and $x + dx$ (in one dimension) at time t: $\Psi^*(x, t)\Psi(x, t)$ is called the probability density.

For all the problems we will tackle in this book, the potential $U(x, t)$ will depend only on x and not on t. So can we somehow separate off the time dependence from the time-dependent Schrödinger equation? I am going to use the 'separation of variables' technique discussed above to solve the problem.

Let us start from the time-dependent Schrödinger equation for our single particle of mass m constrained to a region of the x axis by a potential $U(x, t)$

$$\left[-\frac{h^2}{8\pi^2 m}\frac{\partial^2}{\partial x^2} + U(x, t)\right]\Psi(x, t) = j\frac{h}{2\pi}\frac{\partial\Psi(x, t)}{\partial t}$$

Under what conditions can we write the total wavefunction $\Psi(x, t)$ as a product of independent functions that describe x and t? What we do is to substitute

$$\Psi(x, t) = \psi(x)g(t)$$

into the time-dependent Schrödinger equation above and see what conclusions we can draw. Thus

$$\left[-\frac{h^2}{8\pi^2 m}\frac{\partial^2}{\partial x^2} + U(x, t)\right]\Psi(x, t) = j\frac{h}{2\pi}\frac{\partial\Psi(x, t)}{\partial t}$$

becomes

$$\left[-\frac{h^2}{8\pi^2 m}\frac{\partial^2}{\partial x^2} + U(x, t)\right]\psi(x)g(t) = j\frac{h}{2\pi}\frac{\partial\psi(x)g(t)}{\partial t}$$

Notice that the differentials are partial ones and so they only operate on functions that contain x or t. We can simplify the above equation to

$$g(t)\left[-\frac{h^2}{8\pi^2 m}\frac{\partial^2}{\partial x^2} + U(x,t)\right]\psi(x) = \psi(x)j\frac{h}{2\pi}\frac{\partial g(t)}{\partial t}$$

or on dividing left and right by $\Psi(x,t)$

$$\frac{1}{\psi(x)}\left[-\frac{h^2}{8\pi^2 m}\frac{\partial^2}{\partial x^2} + U(x,t)\right]\psi(x) = \frac{1}{g(t)}j\frac{h}{2\pi}\frac{\partial g(t)}{\partial t}$$

Suppose now that the potential $U(x,t)$ depends only on x and not on t.

$$\frac{1}{\psi(x)}\left[-\frac{h^2}{8\pi^2 m}\frac{\partial^2}{\partial x^2} + U(x)\right]\psi(x) = \frac{1}{g(t)}j\frac{h}{2\pi}\frac{\partial g(t)}{\partial t} \tag{5.33}$$

Under these circumstances the left-hand side of this equation depends only on x and the right-hand side depends only on t. These two variables are completely independent of each other, and so the left- and right-hand sides must each equal a common constant. In differential equation theory this constant is usually referred to as the *separation constant*, but in our case I am going to use the fact that its dimensions are energy and call it the energy ε.

Thus I now have two (ordinary) differential equations

$$\left[-\frac{h^2}{8\pi^2 m}\frac{d^2}{dx^2} + U(x)\right]\psi(x) = \varepsilon\psi(x) \tag{5.34}$$

and

$$j\frac{h}{2\pi}\frac{dg(t)}{dt} = \varepsilon g(t) \tag{5.35}$$

The second equation can be solved easily to give

$$g(t) = A\,\exp\left(-\frac{2\pi j\varepsilon t}{h}\right) \tag{5.36}$$

where A is a constant of integration.

The full solution to the time-dependent Schrödinger equation is therefore

$$\Psi(x,t) = \psi(x)\,\exp\left(-\frac{2\pi j\varepsilon t}{h}\right) \tag{5.37}$$

where I have followed normal convention and absorbed the constant A into the spatial part $\psi(x)$.

5.9 FREE PARTICLES

For a free particle in one dimension, the potential $U(x,t)$ is a constant that may be taken to be zero.

The spatial equation is therefore

$$\left(-\frac{h^2}{8\pi^2 m}\frac{\mathrm{d}^2}{\mathrm{d}x^2}\right)\psi(x) = \varepsilon\psi(x) \tag{5.38}$$

which has solution that I will write as

$$\psi(x) = A\,\exp(jkx) \tag{5.39}$$

with

$$k = \frac{\sqrt{8\pi^2 m\varepsilon}}{h}$$

with any value of the energy $\varepsilon > 0$.

The time-dependent solution is therefore

$$\Psi(x,t) = A\,\exp(jkx)\,\exp\left(-\frac{2\pi j\varepsilon t}{h}\right) \tag{5.40}$$

which can be rewritten

$$\Psi(x,t) = A\,\exp[j(kx - \omega t)] \tag{5.41}$$

where $\omega = 2\pi\varepsilon/h$.

This is a classic wave equation as mentioned earlier in this chapter, corresponding to a de Broglie wave.

5.10 PARTICLES IN POTENTIAL WELLS

Now I want to show you how to go about solving the time-independent Schrödinger equation for particles that are acted on by very simple potentials. Authors often speak about 'particles in potential wells', and I am going to use this attractive terminology. Perhaps the easiest place to begin is with the particle in a so-called 'one-dimensional well' (Figure 5.3); a particle of mass m is confined to a region of the x axis by a potential that is everywhere infinite apart from $0 \leq x \leq L$, and inside this 'one-dimensional Schrödinger equation well' the potential is taken to be a constant U_0.

We say that the length of the 'well' is L, and we talk about a 'particle in a one-dimensional well'. You may find it odd to visualize such a one-dimensional well, so just for the minute think about the series of conjugated molecules.

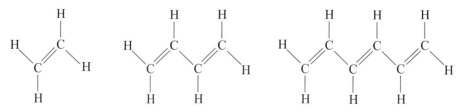

You probably know that we think of the electrons in such molecules as being of two 'types', the σ electrons and the π electrons. The σ electrons are thought of as to do with the C−H and the C−C single bonds, and they are localized in space in the chemical regions between the

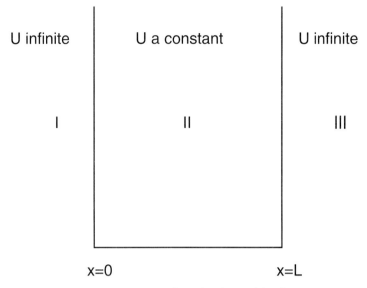

Figure 5.3 A one-dimensional potential well

atoms. The π electrons are to do with the double bonds, and they are delocalized over large regions of conjugated molecules. The σ electrons and the nuclei provide a potential U for the π electrons and as the chain gets bigger and bigger, you can perhaps imagine that the potential might get more and more constant. Do not worry for the minute about the physical reality of this model, but I will try to reassure you by saying that it can be applied successfully to several chemically interesting systems.

For a single particle, the time-independent Schrödinger equation in three dimensions

$$\frac{\partial^2 \psi}{\partial x^2} + \frac{\partial^2 \psi}{\partial y^2} + \frac{\partial^2 \psi}{\partial z^2} + \frac{8\pi^2 m}{h^2}(\varepsilon - U)\psi = 0$$

looks a little more friendly in one dimension

$$\frac{\mathrm{d}^2 \psi}{\mathrm{d}x^2} + \frac{8\pi^2 m}{h^2}(\varepsilon - U)\psi = 0$$

and the aim is to solve this equation subject to any boundary conditions. I am sorry to keep harping back to the boundary conditions, but you will soon get the idea that they can give quantization.

There is a standard way to solve problems of this kind, where the potential energy U has different but constant values in different regions of space. What we do is to consider the regions separately, and then try to match up the wavefunction ψ (and its first derivative if necessary) at the boundaries. The point is that the wavefunction (and its first derivative) must be continuous across boundaries.

Region I, $-\infty \leq x \leq 0$

Here the potential U is infinite and so the wavefunction is zero (in order to keep the Schrödinger equation finite).

Region III, $L \leq x \leq \infty$

Once again, the potential is infinite and so the wavefunction is zero.

Region II, $0 \leq x \leq L$

The potential is a constant (U_0) here, and the wavefunction is not necessarily zero. Schrödinger's equation becomes

$$\frac{\mathrm{d}^2\psi}{\mathrm{d}x^2} + \frac{8\pi^2 m}{h^2}(\varepsilon - U_0)\psi = 0$$

which can be rewritten

$$\frac{\mathrm{d}^2\psi}{\mathrm{d}x^2} + \beta^2\psi = 0$$

which is a standard second-order differential equation with general solution

$$\psi = A\,\sin(\beta x) + B\,\cos(\beta x)$$

where I have used the sybmol β to denote

$$\beta = \sqrt{\frac{8\pi^2 m(\varepsilon - U_0)}{h^2}}$$

The differential equation is a second-order one, and so on solution we get two constants of integration which I have called A and B. The next step is to consider the behavior of ψ at the potential boundaries ($x = 0$ and $x = L$). We know already that the wavefunction is zero at the boundaries, because it is zero for all values of x from $-\infty$ to 0, and from L to ∞.

Substituting $\psi = 0$ at $x = 0$ into the general solution shows that $B = 0$. Substituting $\psi = 0$ at $x = L$ into the general solution gives

$$0 = A\,\sin(\beta L)$$

where

$$\beta = \sqrt{\frac{8\pi^2 m(\varepsilon - U_0)}{h^2}}$$

This means that $\beta L = n\pi$, where n is an integer, which shows that the energy is quantized; it can only have certain fixed values, and cannot have an arbitrary amount.

Rearranging the above equation gives the expression

$$\varepsilon = U_0 + \frac{n^2 h^2}{8mL^2} \tag{5.42}$$

The energy is indeed quantized, and it generally happens that the imposition of boundary conditions leads to energy quantization. U_0 is normally set to zero.

We cannot use $n = 0$ because this gives a wavefunction that is zero everywhere, and according to the Born interpretation of quantum mechanics, $\psi^2\,\mathrm{d}x$ gives the probability of finding the particle between x and $x + \mathrm{d}x$. The probabilities have to sum to 1, which is clearly impossible if ψ is zero everywhere.

We therefore seem to be left with $n = \pm 1, \pm 2, \pm 3$.

So we are still left with two loose ends; are *all* values of n allowed on physical grounds, and how should we find the constant A?

We can find the constant A quite simply, by remembering the Born interpretation of quantum mechanics. The probability that we will find the particle between x and $x + dx$ is given by $\psi^*(x)\psi(x)\,dx$. Since the wavefunctions are real rather than complex, and probabilities must sum to 1, we have

$$\int \psi^2 \, dx = 1$$

This gives $A^2 = 2/L$ and so $A = \pm\sqrt{2/L}$. Which sign should we take? The answer is simple; the choice of sign is arbitrary. For suppose that ψ satisfies our one-dimensional Schrödinger equation

$$\frac{d^2\psi}{dx^2} + \frac{8\pi^2 m}{h^2}(\varepsilon - U_0)\psi = 0$$

then so does $-\psi$;

$$\frac{d^2(-\psi)}{dx^2} + \frac{8\pi^2 m}{h^2}(\varepsilon - U_0)(-\psi) = 0$$

If you want to think about this in terms of the Born interpretation, then the only physically measurable quantity is ψ^2 and so both $\pm\psi$ are acceptable solutions. In this case, you cannot take both because they are not independent. You can only take one set, and people conventionally take the positive ns.

So, the solutions of the time-independent Schrödinger equation for the particle in a one-dimensional well are

$$\psi_n = \sqrt{\frac{2}{L}} \sin\left(\frac{n\pi x}{L}\right)$$

$$\varepsilon_n = U_0 + \frac{n^2 h^2}{8mL^2}$$

Figure 5.4 shows an energy level diagram for a one-dimensional box. You might like to see some of the solutions for ψ_n. Figure 5.5 shows plots of $\sqrt{L}\,\psi_n$ *versus* x/L for $n = 1$ and 3.

According to the Born interpretation, which emphasizes the fact that $\psi^2\,dx$ is the probability that we will find the particle between x and $x + dx$, we should perhaps be studying plots of ψ^2 rather than ψ itself (Figure 5.6).

5.10.1 The Correspondence Principle

You might be wondering how these probabilities would relate to the case of a classical particle in such a potential well? Think about a pool ball that is set rolling on a frictionless pool table, at $90°$ to a table side (which defines the x axis). The ball will continue to bounce off opposing faces of the pool table for ever and with constant speed, and the chance of finding the ball at any instant in time between x and $x + dx$ will be the same.

How should our quantum mechanical model compare with the classical one? First of all, there is no time dependence in our quantum mechanical problem, we can only ask about probabilities.

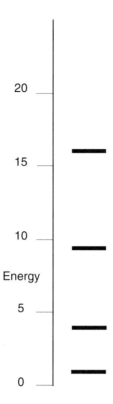

Figure 5.4 Energy level diagram for one-dimensional box

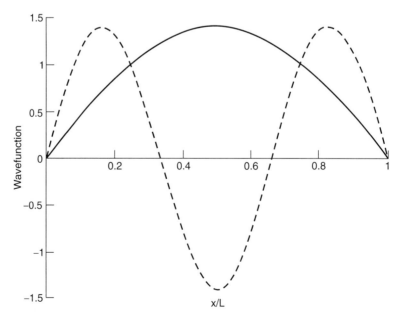

Figure 5.5 One-dimensional box wavefunctions for $n = 1$ and $n = 3$

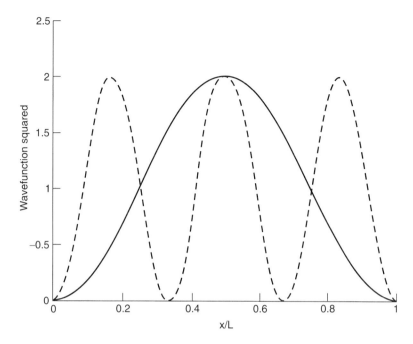

Figure 5.6 Squares of the above wavefunctions for $n = 1$ and $n = 3$

For a particle with $n = 1$ in a one-dimensional well, there is a maximum probability that we will find the particle at $L/2$, and zero probability that we will find the particle at $x = 0$ and $x = L$.

The *Correspondence principle* states that quantum and classical predictions have to agree with each other in the limit as the quantum number gets large. This is a statement that is true for all quantum systems, not just the particle in a well.

Nevertheless, in the case of a classical 'particle in a well', we would expect to find the particle at any point along the x axis with equal probability.

This is certainly not the case for the quantum states with low quantum numbers; for $n = 1$, there is a maximum chance of finding the particle at $L/2$, and for $n = 3$ there are certain points along the x axis where the probability is zero. We call these zero points *nodes*.

A plot of ψ_{30}^2 tells a different story (Figure 5.7). If we take a finite interval Δx along the x axis, then you can perhaps believe that the chance of finding the particle between x and $x + \Delta x$ becomes more and more equal. In the limit as the quantum number n increases further, the probability of finding the particle between x and $x + \mathrm{d}x$ becomes equal, in agreement with the Correspondence principle.

5.10.2   The Two-dimensional Well

Imagine now a two-dimensional version of the above problem (Figure 5.8). We consider a single particle of mass m, which is constrained to a region of the xy plane by a potential that is infinite everywhere except for a square region defined by $0 \leq x \leq L$, $0 \leq y \leq L$.

For a single particle, the time-independent Schrödinger equation in two dimensions becomes

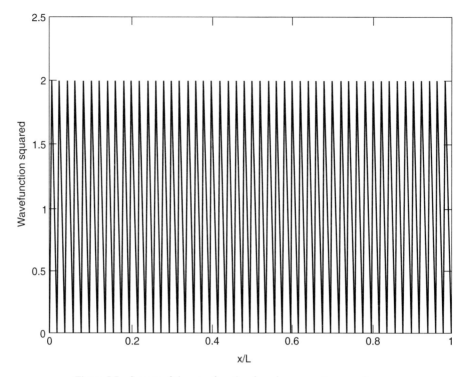

Figure 5.7   Square of the wavefunction for a large quantum number

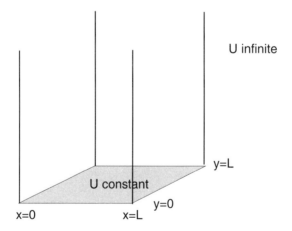

Figure 5.8   A two-dimensional box potential

$$\frac{\partial^2 \psi}{\partial x^2} + \frac{\partial^2 \psi}{\partial y^2} + \frac{8\pi^2 m}{h^2}(\varepsilon - U)\psi = 0 \tag{5.44}$$

Such differential equations are not usually easy to solve, but in this case we can be guided by the basic physics of the problem; since the potential is a constant, we might expect that the solution for the x direction is independent of the solution for the y direction.

Once again we reach for the separation of variables technique.

So, with a song and a prayer, let us see if we can use the separation of variables technique to find a solution to the differential equation

$$\frac{\partial^2 \psi}{\partial x^2} + \frac{\partial^2 \psi}{\partial y^2} + \frac{8\pi^2 m}{h^2}(\varepsilon - U)\psi = 0$$

that has the form $\psi(x, y) = X(x)Y(y)$ where $X(x)$ is a function of x only, and $Y(y)$ is a function of y only.

The first step is to substitute this possibility into the time-independent Schrödinger equation, to give

$$\frac{\partial^2 X(x)Y(y)}{\partial x^2} + \frac{\partial^2 X(x)Y(y)}{\partial y^2} + \frac{8\pi^2 m}{h^2}(\varepsilon - U)X(x)Y(y) = 0$$

This can be rearranged to give

$$\frac{Y(y)\mathrm{d}^2 X(x)}{\mathrm{d}x^2} + \frac{X(x)\mathrm{d}^2 Y(y)}{\mathrm{d}y^2} + \frac{8\pi^2 m}{h^2}(\varepsilon - U)X(x)Y(y) = 0$$

If we divide through by $\psi(x, y) = X(x)Y(y)$ we get

$$\left(\frac{1}{X(x)}\right)\frac{\mathrm{d}^2 X(x)}{\mathrm{d}x^2} + \left(\frac{1}{Y(y)}\right)\frac{\mathrm{d}^2 Y(y)}{\mathrm{d}y^2} + \frac{8\pi^2 m}{h^2}(\varepsilon - U) = 0$$

Now, x and y are independent variables; the first term above depends only on the variable x, whilst the second term depends only on the variable y. These two terms are in principle independently variable and when they are added together they have to come to a constant.

This means that each term has to equal a constant, and the two constants have to sum together to give $-(8\pi^2 m/h^2)(\varepsilon - U)$.

Let us call the two constants ε_x and ε_y, and so the two terms on the left-hand side become

$$\left[\frac{1}{X(x)}\right]\frac{\mathrm{d}^2 X(x)}{\mathrm{d}x^2} = \varepsilon_x$$

and

$$\left[\frac{1}{Y(y)}\right]\frac{\mathrm{d}^2 Y(y)}{\mathrm{d}y^2} = \varepsilon_y$$

Each of these differential equations is exactly what we found for the particle in a one-dimensional well, and so we can write the solutions for this problem as

$$\psi(x, y) = \frac{2}{L}\sin\left(\frac{n\pi}{L}x\right)\sin\left(\frac{k\pi}{L}y\right)$$

where n and k are quantum numbers which can each take values $1, 2, 3, \ldots$. The energy is given by the sum of ε_x and ε_y, and the final solution is

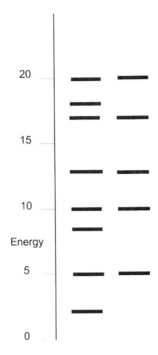

Figure 5.9   Energy level diagram for two-dimensional box

$$\psi_{n,k} = \frac{2}{L} \sin\left(\frac{n\pi}{L}x\right) \sin\left(\frac{k\pi}{L}y\right)$$

$$\varepsilon_{n,k} = \varepsilon_n + \varepsilon_k = (n^2 + k^2)\frac{h}{8mL^2}$$

(5.45)

An energy level diagram is given in Figure 5.9. It is quite different to the one-dimensional problem. The solution for $n = 1$, $k = 2$ has the same energy as that for $n = 2$, $k = 1$. We say that the wavefunctions $\psi_{1,2}$ and $\psi_{2,1}$ represent quantum states of the particle, and that they are *degenerate* (i.e. they have the same energy). A little thought will convince you that any linear combination $\psi_{1,2}$ of and $\psi_{2,1}$, for example $a\psi_{1,2} + b\psi_{2,1}$ where a and b are constants, is also a solution of the Schrödinger equation, with the same energy as either of the two quantum states.

There are two interesting features to the two-dimensional energy level diagram that are absent from the one-dimensional case. First of all, the one-dimensional energy levels get further apart as n rises. The two-dimensional quantum states tend to crowd together as the two quantum numbers n and k increase. Secondly, some of the two-dimensional energy levels are degenerate because the energy expression depends symmetrically on the quantum numbers n and k, but there are also certain combinations of $n^2 + k^2$ that give the same energy; for example, $2^2 + 9^2 = 85$ which is also equal to $6^2 + 7^2$.

We can visualize the solutions as either contour diagrams or as surface plots; Figure 5.10 shows such plots of the wavefunctions $L\psi_{3,1}$ *versus* x/L and y/L and the surface plot is shown in Figure 5.11.

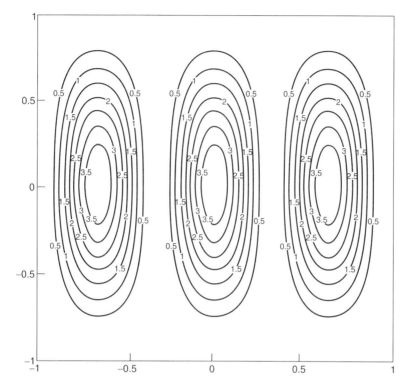

Figure 5.10 Contour plot for the square of the wavefunction with $n = 3$, $k = 1$

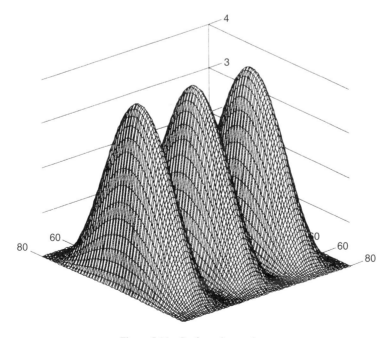

Figure 5.11 Surface plot as above

5.10.3   The Three-dimensional Well

We now proceed to a three-dimensional version of the same problem, that is a particle of mass m trapped by a potential U that is infinite everywhere outside a cubic three-dimensional region of space defined by $0 \leq x \leq L$, $0 \leq y \leq L$ and $0 \leq z \leq L$. Inside this region, we once again take the potential to equal a constant U_0.

The time-independent equation in three dimensions is

$$\frac{\partial^2 \psi}{\partial x^2} + \frac{\partial^2 \psi}{\partial y^2} + \frac{\partial^2 \psi}{\partial z^2} + \frac{8\pi^2 m}{h^2}(\varepsilon - U)\psi = 0 \tag{5.46}$$

and we seek to solve this by the separation of variables technique discussed above. The first step is to write $\psi(x, y, z) = X(x)Y(y)Z(z)$ where for example $X(x)$ only depends on the x coordinate.

I will leave you to demonstrate that substitution of this trial solution into the Schrödinger equation, followed by a little manipulation and thought leads to the following solution

$$\psi_{n,k,l} = \left(\frac{2}{L}\right)^{3/2} \sin\left(\frac{n\pi}{L}x\right) \sin\left(\frac{k\pi}{L}y\right) \sin\left(\frac{l\pi}{L}z\right)$$

$$\varepsilon_{n,k,l} = (n^2 + k^2 + l^2)\frac{h^2}{8mL^2} \tag{5.47}$$

where we now have three quantum numbers n, k and l, all of which can run from 1 to infinity.

The energy level diagram is shown in Figure 5.12.

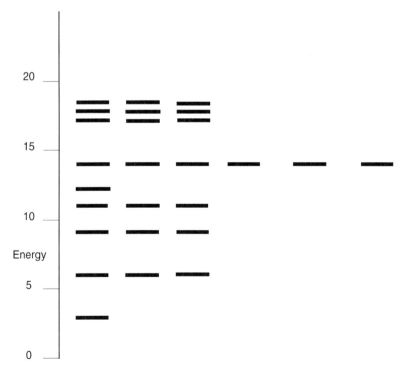

Figure 5.12   Energy level diagram for a three-dimensional box

Once again we notice the gradual crowding together of quantum states as the energy increases. This is due to two factors, the 'natural' degeneracies caused by the symmetry of the energy level expression, and the fact that there get to be very many combinations of integers n, k and l such that $n^2 + k^2 + l^2$ has the same value.

5.10.4 The Finite Well

The perfectly rigid rectangular potential well with infinite walls is a very simple but physically unrealistic model. No physical system has such a potential function, but it is a good place to begin the discussion.

Let us consider next a possible but crude model for a diatomic molecule (Figure 5.13) where we keep an infinitely steep potential for the close approach of the nuclei, but allow for a 'valence region' where the potential is a constant. In detail, U is infinite for $-\infty \le x \le 0$, a constant we will take it to be 0 for the region $0 \le x \le L$, and a constant D for $L \le x \le \infty$. So, our simple model should be good for a diatomic molecule with dissociation energy D.

What we do now is to divide up the x axis into three regions, solve the Schrödinger equation as far as possible for each of the three regions and then match up ψ and its first derivative $d\psi/dx$ at the boundaries of the three regions in order to find any constants of integration and so on.

In region I where the potential U is infinite, $\psi = 0$. More importantly, $\psi = 0$ also at the boundary between regions I and II. This is needed in order to make the wavefunction ψ continuous across the boundary.

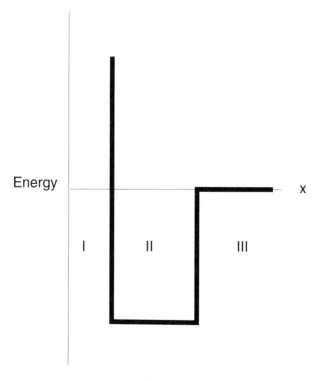

Figure 5.13 A finite well

In region II, where I have taken the potential to be a constant (zero), then we have

$$\frac{d^2\psi}{dx^2} + \frac{8\pi^2 m}{h^2}\, \varepsilon\psi = 0$$

which can be rewritten

$$\frac{d^2\psi}{dx^2} + \beta^2\psi = 0$$

which is a standard second-order differential equation with general solution

$$\psi = A\, \sin(\beta x) + B\, \cos(\beta x)$$

where I have used the symbol β to denote

$$\beta = \sqrt{\frac{8\pi^2 m\varepsilon}{h^2}}$$

In region III we have

$$\frac{d^2\psi}{dx^2} + \frac{8\pi^2 m}{h^2}(\varepsilon - D)\psi = 0$$

For the minute, let us interest ourselves in those solutions that have energies ε less than D. We call these solutions the *bound states*. As you will see shortly, there are solutions whose energies are greater than D and we call these the *unbound states*.

I will rewrite the differential equation above as

$$\frac{d^2\psi}{dx^2} - \frac{8\pi^2 m}{h^2}(D - \varepsilon)\psi = 0 \qquad (5.48)$$

because I am interested in the states for which D is greater than ε. I will write this as

$$\frac{d^2\psi}{dx^2} - \alpha^2\psi = 0$$

where $\alpha = \sqrt{8\pi^2 m(D - \varepsilon)}/h^2$ is a positive constant. The standard solution to this differential equation is usually written

$$\psi = E\, \exp(\alpha x) + F\, \exp(-\alpha x) \qquad (5.49)$$

where once again E and F are constants of integration.

I mentioned above that ψ had to be zero at $x = 0$, in order for it to be continuous across the $x = 0$ boundary. If $\psi = 0$ at $x = 0$, then $B = 0$ and so $\psi = A\, \sin(\beta x)$ for region II.

If we are only interested in bound states (where the energy ε is less than D), then E must be zero because otherwise the wavefunction would go to infinity as x got large. Unlike the 'one-dimensional potential well' problem, we cannot set the wavefunction ψ to 0 when $x = L$. What we do know however is that at $x = L$, the wavefunction in region II must equal the wavefunction in region III. Not only that, the first derivatives must be equal at that point.

This means that

$$A\, \sin(\beta L) = F\, \exp(-\alpha L)$$

and also

$$\beta A \cos(\beta L) = -\alpha F \exp(-\alpha L)$$

Dividing one equation by the other gives the equation for the allowed energies

$$\frac{1}{\beta} \tan(\beta L) = -\frac{1}{\alpha}$$

or, on simple rearrangement to give an explicit formula involving the energy ε

$$\tan\left(\sqrt{\frac{8\pi^2 m\varepsilon}{h}}\, L\right) = -\sqrt{\frac{\varepsilon}{D-\varepsilon}} \tag{5.50}$$

This places restrictions on the allowed values of the energy ε. The energy is quantized, and the precise values of ε depend on the details of the potential well (in particular, D and L).

5.11 UNBOUND STATES

The potential well above is referred to as a finite well, and finite wells have much more interesting properties than infinite wells. I mentioned earlier in the discussion that we would concern ourselves only with solutions where the energy ε was less than or equal to the 'depth' of the well D. I called such a solution a bound state; by this I mean a state that is somehow localized to the potential well in the sense that the probability of finding the particle between x and $x + \mathrm{d}x$ should get smaller, the further we move away from the potential well.

Depending on its width L and depth D, a potential well in one dimension can have one, many or an infinite number of bound state solutions.

Finite wells can also have unbound solutions, which are acceptable solutions of the Schrödinger equation in so far as ψ and its first derivative $\mathrm{d}\psi/\mathrm{d}x$ are finite and continuous for all values of x, but they do not have the property that they tend to 0 for large values of x.

A moments thought will show you that such unbound states cannot be normalized!

5.12 A MAJOR FAILURE OF THE CLASSICAL MODEL

If we irradiate a diatomic molecule such as HCl with infra-red radiation, then the molecule shows a strong absorption of radiation with wavenumber $2886\,\mathrm{cm}^{-1}$, a weaker absorption at $5668\,\mathrm{cm}^{-1}$ and a very weak absorption at $8347\,\mathrm{cm}^{-1}$. This is due to energy quantization within the molecule. Vibrational energy is quantized, as the spectroscopic experiment shows. This effect is completely lacking from the classical model. This has nothing whatever to do with my assumption of the perfect spring to represent the bond, as you will see shortly.

5.12.1 The Quantum Mechanical Treatment

I am now going to show you the quantum mechanical treatment of the vibrating diatomic, and once again I will assume that the potential energy is given by Hooke's law.

To keep the algebra simple, I am going to use the fact that the vibrations of such a diatomic (atomic masses m_1 and m_2) are exactly equivalent to the vibrations of a single particle with the reduced mass of the diatomic.

As I explained in an earlier chapter, there is a standard procedure for turning classical energy expressions into quantum mechanical wave equations.

1. Write down the classical energy expression in terms of the kinetic and the potential energy.
2. Construct the Hamiltonian operator $\hat{H}$.
3. Solve the wave equation for the system.

The classical energy of the vibrating atom (of mass μ) is

$$\varepsilon_{vib} = \frac{1}{2}\mu \left[\frac{d(R - R_e)}{dt} \right]^2 + \frac{1}{2}k_s(R - R_e)^2$$

Rather than keep writing the spring extension $R - R_e$, I am going to call it x and so

$$\varepsilon_{vib} = \frac{1}{2}\mu \left(\frac{dx}{dt} \right)^2 + \frac{1}{2}k_s x^2 \qquad (5.51)$$

In order to use the 'classical to quantum' recipe, I have to write the energy in terms of the linear momentum $p = \mu(dx/dt)$, giving

$$\varepsilon_{vib} = \frac{1}{2\mu}p^2 + \frac{1}{2}k_s x^2$$

The Hamiltonian operator for the system is therefore

$$\hat{H}_{vib} = -\frac{h^2}{8\pi^2\mu}\frac{d^2}{dx^2} + \frac{1}{2}k_s x^2 \qquad (5.52)$$

and the wave equation for the system is

$$\frac{d^2\psi}{dx^2} + \frac{8\pi^2\mu}{h^2}\left(\varepsilon_{vib} - \frac{k_s}{2}x^2 \right)\psi = 0 \qquad (5.53)$$

Solution of this differential equation is outside the scope of this simple text, it is covered in standard volumes such as *Quantum Chemistry* by Eyring, Walter and Kimball. To cut a long story short, imposition of the boundary condition $\psi = 0$ at $x = \infty$ gives rise to energy quantization described by a single quantum number that I will write as v, and it turns out that $v = 0, 1, 2, \ldots$.

The general formula for the allowed energies is

$$\varepsilon_{vib} = \frac{h}{2\pi}\sqrt{\frac{k_s}{\mu}}\left(v + \frac{1}{2} \right) \qquad (5.54)$$

Table 5.1   The first few hermite polynomials

v	$H_v(\xi)$
0	1
1	2ξ
2	$4\xi^2 - 2$
3	$8\xi^3 - 12\xi$
4	$16\xi^4 - 48\xi^2 + 12$
5	$32\xi^5 - 160\xi^3 + 120\xi$

You might recognize $1/2\pi\sqrt{(k_s/\mu)}$ as the classical vibration frequency, and so the quantum mechanical energies are $\frac{1}{2}, \frac{3}{2}, \ldots$ times the energy associated with the classical vibration frequency.

The normalized vibrational wavefunctions are given by the general expression

$$\psi_v(\xi) = \left(\frac{\sqrt{\beta/\pi}}{2^v v!}\right)^{1/2} H_v(\xi) \exp(-\xi^2/2) \tag{5.55}$$

where $\beta = 2\pi\sqrt{\mu k_s}/h$ and $\xi = \sqrt{\beta}x$. The polynomials H_v are called the hermite polynomials, and the first few are shown in Table 5.1.

Vibrational wavefunctions for the states $v = 0$ and $v = 1$ are shown in Figures 5.14 and 5.15. For the sake of illustration, I have taken numerical values appropriate to ^{1}H ^{35}Cl.

Note that the wavefunction has a maximum in the middle of the spring, in Figure 5.14. The wavefunction has a node at the midpoint in Figure 5.15.

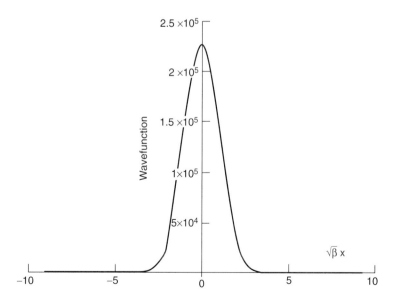

Figure 5.14   Vibrational wavefunction, $v = 0$

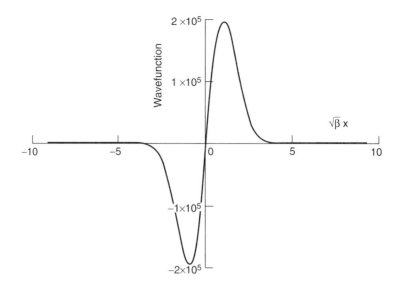

Figure 5.15   Vibrational wavefunction, $v = 1$

There are two points worth noticing about the quantum mechanical solutions.

1. The vibrational wavefunctions are alternately symmetrical and antisymmetrical.

2. The harmonic oscillator is not allowed to have zero energy. The smallest allowed value of vibrational energy is $(h/2\pi)\sqrt{k_s/\mu}(0 + \frac{1}{2})$ and this is called the *zero point energy*.

This finding is in accord with the Heisenberg uncertainty principle, for if the oscillating particle had zero energy it would also have zero momentum, and would therefore be located exactly at the position of minimum potential energy.

5.12.2   The Vibrational Spectrum of HCl

According to the harmonic model, the vibrational states should have energy $\frac{1}{2}\omega_e, \frac{3}{2}\omega_e, \ldots$, and the energy differences between the lowest state and higher ones would be exactly ω_e $2\omega_e$, $3\omega_e, \ldots$. An experimental infra-red spectrum shows lines at $2886\,\text{cm}^{-1}$, $5668\,\text{cm}^{-1}$, $8347\,\text{cm}^{-1}$, and so on. The reason why the wavenumbers are not exactly in the ratio $1:2:3,\ldots$ is because the harmonic potential is not exactly accurate. Figure 5.16 plots the potential energy for $^1\text{H}\,^{35}\text{Cl}$ (for which we can deduce a force constant $k_s = 480\,\text{N m}^{-1}$).

This harmonic potential energy is quite accurate for oscillations about the equilibrium minimum, but once the vibrational amplitude gets large then the perfect spring potential is no longer accurate. As we squeeze the nuclei in a diatomic molecule together, the electronic screening of the nuclei gets smaller and so the nuclei repel each other more strongly than would otherwise be the case. As we pull the nuclei apart, then the molecule will dissociate into atoms or ions as the case may be.

Over the years, very many people have attempted to give a more accurate representation to the potential. A famous (but old-fashioned) potential is that due to Morse

$$U(R) = D_e\{1 - \exp[-\alpha(R - R_e)]\}^2 \tag{5.57}$$

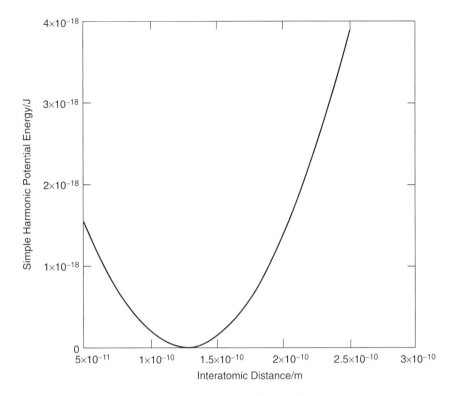

Figure 5.16   Harmonic potential energy for HCl

which has a more realistic shape in that it is steeper than the harmonic for small internuclear separations, and shows asymptotic behavior as the internuclear separation gets large (Figure 5.17).

The Morse potential is not particularly accurate, but it has the advantage over many other such potentials that the diatomic vibrational problem can be solved exactly to give an expression very similar to the harmonic one;

$$\varepsilon_{\text{vib}} = hc_0\omega_e(\nu + \tfrac{1}{2}) - hc_0 x_e(\nu + \tfrac{1}{2})^2 \tag{5.58}$$

where

$$\omega_e = \frac{\alpha}{\pi c}\sqrt{\frac{D_e}{2\mu}} \quad \text{and} \quad x_e = \frac{hc_0}{4D_e}\omega_e$$

The constant x_e is usually small (e.g. 1/100), and so the second term in the energy expression only becomes significant as the quantum number v gets large.

The Morse potential, although not particularly accurate, is widely quoted. More accurate potentials can be constructed by adding terms involving $(R - R_e)^3$, $(R - R_e)^4$, and so on, to the harmonic potential. The problem then is that the time-independent Schrödinger equation cannot be solved exactly.

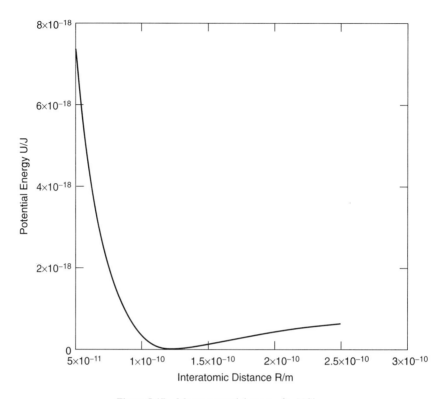

Figure 5.17 Morse potential energy for HCl

5.13 THE VARIATION THEOREM

I have just shown you how to solve the time-independent Schrödinger equation for several model systems. Most of the problems we encounter in real life do not have solutions that can be evaluated as closed expressions, but this problem is not unique to quantum mechanics. So for example the definite integral

$$\int_1^2 \frac{\mathrm{d}x}{\sqrt{(1+x)(1+x^2)}}$$

certainly exists but it cannot be evaluated in closed form. We have to resort to a numerical method such as Simpson's rule in order to evaluate it.

In this text, we are only interested in solving the time-independent Schrödinger equation, which I can write in Hamiltonian form as

$$\hat{H}\psi = \varepsilon\psi$$

Suppose that ψ_1 is the lowest energy eigenfunction with energy ε_1

$$\hat{H}\psi_1 = \varepsilon_1\psi_1$$

Let me now multiply this equation from the left by the complex conjugate of ψ_1 and integrate over all space

$$\int_{-\infty}^{\infty} \psi_1^* H \psi_1 \mathrm{d}x = \varepsilon_1 \int_{-\infty}^{\infty} \psi_1^* \psi_1 \, \mathrm{d}x$$

or on rearrangement

$$\varepsilon_1 = \frac{\displaystyle\int_{-\infty}^{\infty} \psi_1^* \hat{H} \psi_1 \, \mathrm{d}x}{\displaystyle\int_{-\infty}^{\infty} \psi_1^* \psi_1 \, \mathrm{d}x} \tag{5.59}$$

I mentioned expectation values earlier, and there is an obvious connection. At first sight, the expression above might give an alternative method for calculating the energy eigenvalue ε_1 once the eigenfunction ψ_1 was known but the expression involves an extra integration; if ψ_1 is genuinely an eigenfunction of the Hamiltonian then calculation of $\hat{H}\psi_1$ will give ψ_1 back again, multiplied by a constant. This constant is the energy.

Let us pretend that we do not know the exact lowest energy solution for the particle in a one-dimensional box, and so have to rely on an approximate method for solving the time-independent Schrödinger equation. We know from the boundary conditions that $\psi = 0$ when $x = 0$ and $\psi = 0$ when $x = L$, the length of the box. We also know from symmetry considerations that the wavefunction must be symmetrical or antisymmetrical about the mid point $x = L/2$.

The simplest function that springs to mind is the 'trial' wavefunction

$$\psi_t = x(L - x)$$

The energy expectation value is

$$\langle \hat{H} \rangle = \frac{\displaystyle\int_0^L x(L - x) \hat{H} x(L - x) \, \mathrm{d}x}{\displaystyle\int_0^L x^2 (L - x)^2 \, \mathrm{d}x}$$

which comes to $5h^2/(4\pi^2 mL^2)$. The main point to note is that the estimate of the energy is too high by a factor of $10/\pi^2 = 1.013$. In fact it can be proved that

$$\langle \hat{H} \rangle = \frac{\displaystyle\int_{-\infty}^{\infty} \psi_t^* \hat{H} \psi_t \, \mathrm{d}x}{\displaystyle\int_{-\infty}^{\infty} \psi_t \psi_t \, \mathrm{d}x} \geq \varepsilon_1 \tag{5.60}$$

for any Hamiltonian operator whose lowest energy eigenvalue is ε_1, provided that the trial wavefunction ψ_t satisfies the boundary conditions. This important result is known as the *Variation theorem*, or sometimes the *Variation principle*.

In order to increase the accuracy of the trial wavefunction, we would normally introduce a parameter for example

$$\psi_t = [x(L - x)]^k$$

where k is a real number. We know from the variation theorem that every value we choose for k will give an energy greater than or equal to the correct energy. Calculation of the expectation value

$$\langle \hat{H} \rangle = \frac{\displaystyle\int_0^L [x(L-x)]^k \, \hat{H}[x(L-x)]^k \, dx}{\displaystyle\int_0^L [x^2(L-x)^2]^k \, dx}$$

gives

$$\langle \hat{H} \rangle = \frac{h^2}{8\pi^2 mL^2} \frac{k(4k+1)}{2k-1} \tag{5.61}$$

and so the best possible value of k is when this expectation value has a minimum.

5.13.1 The Linear Variation Method

A difficulty with the variation method as formulated above is that every problem has a unique method of solution, leading to complicated integrals that are almost certainly unique to that problem. We then have to differentiate in order to find the lowest energy.

In the linear variational method, we express our trial function as a linear combination of simple possibilities, and use the variation principle to determine the extent of the mixing of these possibilities. For example, two possibilities for the particle in the one-dimensional well are

$$\psi_1 = x(L-x)^2$$

$$\psi_2 = x^2(L-x)$$

Neither by themselves are suitable because they emphasize different halves of the box, and so we take

$$\psi_t = c_1 \psi_1 + c_2 \psi_2 \tag{5.62}$$

where c_1 and c_2 are numerical coefficients that have to be determined.

Here is how we do it (and the proof is given in every quantum chemistry text). We calculate the matrices

$$\mathbf{H} = \begin{pmatrix} H_{11} & H_{12} \\ H_{21} & H_{22} \end{pmatrix}$$

$$\mathbf{S} = \begin{pmatrix} S_{11} & S_{12} \\ S_{21} & S_{22} \end{pmatrix}$$

with elements

$$H_{ij} = \int_0^L \psi_i^* \hat{H} \psi_j \, dx$$

and

$$S_{ij} = \int_0^L \psi_i^* \psi_j \, dx$$

We then solve the generalized matrix eigenvalue problem

$$\mathbf{Hc} = \varepsilon \mathbf{Sc} \tag{5.63}$$

Calculation gives

$$\mathbf{H} = \frac{h^2 L^5}{8\pi^2 M} \begin{pmatrix} \dfrac{2}{15} & \dfrac{1}{30} \\ \dfrac{1}{30} & \dfrac{2}{15} \end{pmatrix}$$

$$\mathbf{S} = L^7 \begin{pmatrix} \dfrac{1}{105} & \dfrac{1}{140} \\ \dfrac{1}{140} & \dfrac{1}{105} \end{pmatrix}$$

Solution using a numerical procedure or otherwise gives energies and eigenvectors

$$\varepsilon_1 = \frac{10}{\pi^2} \frac{h^2}{8mL^2} \quad \mathbf{c}_1 = \begin{pmatrix} \dfrac{1}{\sqrt{2}} \\ \dfrac{1}{\sqrt{2}} \end{pmatrix}$$

and

$$\varepsilon_2 = \frac{42}{\pi^2} \frac{h^2}{8mL^2} \quad \mathbf{c}_1 = \begin{pmatrix} \dfrac{1}{\sqrt{2}} \\ -\dfrac{1}{\sqrt{2}} \end{pmatrix}$$

The ratios $10/\pi^2 = 1.013$ and $42/\pi^2 = 4.255$ are to be compared with the exact solutions of 1 and 4 for the $n = 1$ and $n = 2$ states.

In the general case we might choose to use a linear combination of n suitable trial functions $\psi_1, \psi_2, \ldots, \psi_n$. In this case we have to solve the generalised matrix eigenvalue problem

$$\mathbf{Hc} = \varepsilon \mathbf{Sc} \tag{5.64}$$

where the matrices $\mathbf{H}$ and $\mathbf{S}$ are now n by n. According to MacDonald's theorem, the energies produced from this process approach from above the energies of each state. In other words, the more terms we add to the sum, the closer the approximate energies are to the true energies for each state in question, but the approximate energies are always greater than or equal to the true ones.

6 Electric Multipoles, Polarizabilities and Intermolecular Forces

In Chapter 4, I reminded you of the existence of electric charge. Electric charge comes in two kinds, which we call positive and negative, and I told you that electrons are responsible for negative charge and protons for positive charge. In the macroscopic world we rarely come across bodies that carry an excess of one kind of charge or another. In the molecular world, most molecules are electrically neutral but we do meet species with an imbalance of charge when we study ions in solution, ionic crystals, and so on.

In this chapter I want to give you a flavor for the way we deal with the forces between distinct molecular species. These are referred to as intermolecular forces; the forces between atoms in molecules, or between electrons in atoms and so on are usually referred to as intramolecular forces. I mentioned the Lennard–Jones and the Mie potentials in Chapter 4, but I did not explain their origin.

We already know that charged particles interact with each other according to Coulomb's law, and at first sight you might expect that two uncharged molecules would not interact with each other since each is exactly electrically neutral.

The fact remains that neutral atoms and molecules *are* seen to interact with each other. Theories of intermolecular forces often start from the classical description of charge distributions and the way that such charge distributions interact with one another, and I am going to progress down that path for the remainder of this chapter.

6.1 ELECTROSTATIC CHARGE DISTRIBUTIONS

Consider the set of n electric point charges $Q_1, Q_2, \ldots, Q_n$ at position vectors $r_1, \mathbf{r}_2, \ldots, \mathbf{r_n}$ relative to an arbitrary origin. This is said to define a distribution of charges. The next few sections discuss their properties.

6.1.1 Electric Multipoles

In order to describe a charge distribution, it proves useful to define certain so-called moments of this electric charge distribution as follows.

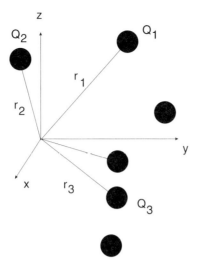

Figure 6.1   A distribution of point charges

The electric dipole moment $\mathbf{p}_e$ of the charge distribution shown in Figure 6.1 defined as

$$\mathbf{p}_e = \sum_{i=1}^{n} Q_i \mathbf{r}_i \qquad (6.1)$$

The electric dipole moment is therefore a vector quantity, and its components are

$$(\mathbf{p}_e)_x = \sum_{i=1}^{n} Q_i x_i$$

$$(\mathbf{p}_e)_y = \sum_{i=1}^{n} Q_i y_i$$

$$(\mathbf{p}_e)_z = \sum_{i=1}^{n} Q_i z_i$$

with respect to the coordinate origin as shown.

It is necessary to exercise caution when defining electric dipole moments, for the following reason. Suppose we move the coordinate axis to a new point in space such that each point charge Q_i has a position vector $\mathbf{r}'_i$ in the new axis system, where $\mathbf{r}_i = \mathbf{r}'_i + \Delta$ and Δ is a constant vector. From the definition of the electric dipole moment

$$\mathbf{p}_e = \sum_{i=1}^{n} Q_i \mathbf{r}_i$$

with respect to the first coordinate origin, and

$$\mathbf{p}'_e = \sum_{i=1}^{n} Q_i \mathbf{r}'_i$$

with respect to the new coordinate origin. If we substitute the expression $\mathbf{r}_i = \mathbf{r}_i' + \Delta$ into the first summation we see that

$$\mathbf{p}_e = \sum_{i=1}^{n} Q_i(\mathbf{r}_i' + \Delta)$$

$$= \mathbf{p}_e' + \Delta \sum_{i=1}^{n} Q_i \tag{6.2}$$

The two expressions are identical only if the sum of the charges is zero. The definition of the electric dipole moment gives a unique result only for a collection of charges that sums to zero. The electric dipole moment of a neutral molecule is independent of the coordinate origin, but the electric dipole moment of a molecular ion depends on the position arbitrarily chosen for the coordinate origin. The coordinate origin has to be quoted when discussing the dipole moment of a charged species.

In the case of two equal and opposite charges $+Q$ and $-Q$ distance d apart, the electric dipole moment has magnitude Qd and points along the direction of a vector drawn from the negative charge to the positive charge.

Returning to the distribution of point charges shown above, the six independent quantities

$$\sum_{i=1}^{n} Q_i x_i x_i, \ \sum_{i=1}^{n} Q_i x_i y_i, \ \sum_{i=1}^{n} Q_i x_i z_i, \ldots, \sum_{i=1}^{n} Q_i z_i z_i$$

are said to define the electric second moments, which we can collect into a real symmetric 3×3 matrix

$$\begin{pmatrix} \sum_{i=1}^{n} Q_i x_i x_i & \sum_{i=1}^{n} Q_i x_i y_i & \sum_{i=1}^{n} Q_i x_i z_i \\ \sum_{i=1}^{n} Q_i y_i x_i & \sum_{i=1}^{n} Q_i y_i y_i & \sum_{i=1}^{n} Q_i y_i z_i \\ \sum_{i=1}^{n} Q_i z_i x_i & \sum_{i=1}^{n} Q_i z_i y_i & \sum_{i=1}^{n} Q_i z_i z_i \end{pmatrix} \tag{6.3}$$

The matrix is real symmetric because of the equalities of the sums such as

$$\sum_{i=1}^{n} Q_i x_i y_i = \sum_{i=1}^{n} Q_i y_i x_i$$

The set of quantities $\sum_{i=1}^{n} Q_i x_i x_i x_i$ through $\sum_{i=1}^{n} Q_i z_i z_i z_i$ defines the electric third moment of the charge distribution, and so on.

There are many different definitions related to the second (and higher) moments in the literature. There is no uniformity of usage, and it is necessary to be clear about the definition and choice of origin when dealing with these quantities.

Most authors prefer to work with a quantity called the *electric quadrupole moment* rather than the second moments, but even then there are several different conventions. A common choice is to use Θ defined by

$$\Theta = \frac{1}{2} \begin{pmatrix} \sum_{i=1}^{n} Q_i(3x_ix_i - r_i^2) & 3\sum_{i=1}^{n} Q_ix_iy_i & 3\sum_{i=1}^{n} Q_ix_iz_i \\ 3\sum_{i=1}^{n} Q_iy_ix_i & \sum_{i=1}^{n} Q(3y_iy_i - r_i^2) & 3\sum_{i=1}^{n} Q_iy_iz_i \\ 3\sum_{i=1}^{n} Q_iz_ix_i & 3\sum_{i=1}^{n} Q_iz_iy_i & \sum_{i=1}^{n} Q_i(3z_iz_i - r_i^2) \end{pmatrix} \tag{6.4}$$

Notice that the diagonal elements of this matrix sum to zero and we say that the matrix has zero 'trace' (the trace being the sum of the diagonal elements).

Some authors ignore the factor of $\frac{1}{2}$.

I can illustrate these definitions by considering the five point charges shown in Figure 6.2. Their Cartesian coordinates and charges are given in Table 6.1.

Consider four simple cases. In each case I have adjusted the charge on the central point charge to make the overall charge distribution zero.

(i) $$Q_1 = Q_2 = Q_3 = Q_4 = +Q, \quad Q_5 = -4Q$$

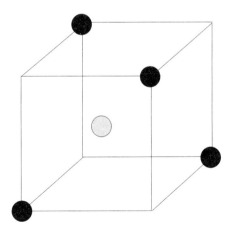

Figure 6.2   Five point charges symmetrically arranged

Table 6.1   Cartesian coordinates

x	y	z	Charge
a	$-a$	$-a$	Q_1
$-a$	a	$-a$	Q_2
a	a	a	Q_3
$-a$	$-a$	a	Q_4
0	0	0	Q_5

In this case the electric dipole moment is zero. The electric second moment matrix is

$$Qa^2 \begin{pmatrix} 4 & 0 & 0 \\ 0 & 4 & 0 \\ 0 & 0 & 4 \end{pmatrix}$$

and the quadrupole matrix is the null matrix.

$$Qa^2 \begin{pmatrix} 0 & 0 & 0 \\ 0 & 0 & 0 \\ 0 & 0 & 0 \end{pmatrix}$$

These matrices are typical for neutral systems having spherical symmetry. The quadrupole moment as I have defined it gives a measure of the deviation from spherical symmetry.

(ii) $Q_1 = Q_2 = Q_3 = +Q, \quad Q_4 = +2Q, \quad Q_5 = -5Q$

The dipole moment vector is $Qa(-1, -1, 1)$ and the electric second moment matrix is

$$Qa^2 \begin{pmatrix} 5 & 1 & -1 \\ 1 & 5 & -1 \\ -1 & -1 & 5 \end{pmatrix}$$

The electric quadrupole matrix is

$$Qa^2 \begin{pmatrix} 0 & 0.5 & -1.5 \\ 1.5 & 0 & -1.5 \\ -1.5 & -1.5 & 0 \end{pmatrix}$$

I can always ensure that such a matrix has diagonal form by rotating the Cartesian axes to form the so-called principal axes. To do this, I have to find the eigenvalues and eigenvectors of the matrix. Matrix diagonalization gives a matrix

$$Qa^2 \begin{pmatrix} -1.5 & 0 & 0 \\ 0 & -1.5 & 0 \\ 0 & 0 & 3 \end{pmatrix}$$

and a new set of coordinate axes related to the old x, y, z by the columns of the matrix

$$U = \begin{pmatrix} 0.4082 & -0.7071 & 0.5774 \\ 0.4082 & 0.7071 & 0.5774 \\ 0.8165 & 0.0000 & -0.5774 \end{pmatrix}$$

(iii) $Q_1 = Q_2 = +Q, \quad Q_3 = Q_4 = +2Q, \quad Q_5 = -6Q$

In this case the dipole moment vector is $Qa(0, 0, 2)$ and the electric second moment matrix is

$$Qa^2 \begin{pmatrix} 6 & 2 & 0 \\ 2 & 6 & 0 \\ 0 & 0 & 6 \end{pmatrix}$$

The quadrupole moment matrix is

$$Qa^2 \begin{pmatrix} 0 & 3 & 0 \\ 3 & 0 & 0 \\ 0 & 0 & 0 \end{pmatrix}$$

and this can be converted to diagonal form as

$$Qa^2 \begin{pmatrix} -3 & 0 & 0 \\ 0 & 0 & 0 \\ 0 & 0 & +3 \end{pmatrix}$$

with an eigenvector matrix of

$$U = \begin{pmatrix} -0.7071 & 0 & 0.7071 \\ 0.7071 & 0 & 0.7071 \\ 0 & 1 & 0 \end{pmatrix}$$

(iv) $Q_1 = +Q, \quad Q_2 = +2Q, \quad Q_3 = +3Q, \quad Q_4 = +4Q, \quad Q_5 = -10Q$

In this case there is less symmetry and the corresponding quantities are: dipole moment $= Qa(-2, 0, 4)$.

Second moment matrix is

$$Qa^2 \begin{pmatrix} 10 & 4 & 0 \\ 4 & 10 & -2 \\ 0 & -2 & 10 \end{pmatrix}$$

The quadrupole moment matrix is

$$Qa^2 \begin{pmatrix} 0 & 6 & 0 \\ 6 & 0 & -3 \\ 0 & -3 & 0 \end{pmatrix}$$

its eigenvalues are $-6.708Qa^2$, 0 and $+6.708Qa^2$ with an eigenvector matrix of

$$U = \begin{pmatrix} -0.6325 & 0.4472 & 0.6325 \\ 0.7071 & 0 & 0.7071 \\ 0.3162 & 0.8944 & -0.3162 \end{pmatrix}$$

The quadrupole moment gives a measure of deviations from spherical symmetry. It depends critically on the axis system chosen, and on the coordinate origin. The properties of the electric quadrupole matrix are best investigated with the matrix in its principal axis system.

At its simplest, a linear electric quadrupole (Figure 6.3) is created by three point charges $+Q \cdots -2Q \cdots +Q$ arranged such that the separation between the charges is a constant. We can think of such an electric quadrupole as the superposition of two simple electric dipoles; we take the first dipole, shift it along the axis, reverse the charges and add it to the first dipole.

It is possible to extend the concept of dipole and quadrupole indefinitely. A single charge is known as an electrical monopole. A dipole is obtained by displacing a monopole through a

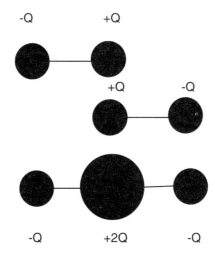

Figure 6.3 A linear quadrupole made from two dipoles

small distance and replacing the original monopole by another of the same magnitude but opposite sign. Likewise a linear quadrupole is obtained by displacing a dipole by a small distance and then replacing the original dipole by one of equal magnitude but opposite signs.

The electrostatic potential at field point **R** for a point charge at the origin varies as $1/R$. The electrostatic potential for a linear dipole varies as $1/R^2$. The electrostatic potential for a linear quadrupole varies as $1/R^3$, and so on. Higher moments such as the octupole and the hexadecapole (which are certain combinations of the third and fourth moments) are occasionally encountered in the literature.

6.2 CONTINUOUS CHARGE DISTRIBUTIONS

The discussion above was concerned only with a distribution of point charges. Sometimes it is necessary to consider distributions of charge that are continuous within certain regions of space.

This is illustrated in Figure 6.4. In order to deal with such charge distributions, we divide them up into differential volume elements as shown, and treat the charge contained within that differential element as a point charge.

For a charge density of ρ, the charge contained in the differential element $\mathrm{d}x\,\mathrm{d}y\,\mathrm{d}z$ shown is $\rho\,\mathrm{d}x\,\mathrm{d}y\,\mathrm{d}z$. Cartesian coordinates are an appropriate choice for a charge distribution having the symmetry of a cuboid. In the case of a charge distribution having spherical symmetry we would normally use spherical polar coordinates, for which the volume element is $r^2\sin\theta\,\mathrm{d}\theta\,\mathrm{d}\phi\,\mathrm{d}r$, and so on. We normally generalize, and write a volume element as $\mathrm{d}\tau$. The total charge enclosed by the region of the charge distribution is then written conventionally

$$Q = \int \rho\,\mathrm{d}\tau \tag{6.5}$$

which implies an integration (usually in three dimensions) over the coordinates of the problem in question.

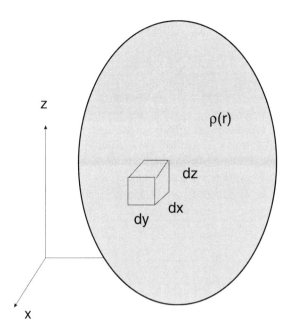

Figure 6.4   A continuous distribution of charge

In order to calculate the electric dipole moments of a charge distribution ρ, we have to evaluate the contributions from each differential charge element $\rho\,\mathrm{d}\tau$, giving

$$(\mathbf{p}_{\mathrm{e}})_x = \int \rho x \,\mathrm{d}\tau$$

$$(\mathbf{p}_{\mathrm{e}})_y = \int \rho y \,\mathrm{d}\tau$$

$$(\mathbf{p}_{\mathrm{e}})_z = \int \rho z \,\mathrm{d}\tau$$

with corresponding formulae for the higher moment contributions.

6.3   THE ELECTROSTATIC POTENTIAL

As discussed in Chapter 4, the electrostatic potential at field point $\mathbf{R}$ in space due to a point charge Q at the coordinate origin is

$$\phi(\mathbf{R}) = \frac{Q}{4\pi\epsilon_0 R}$$

If the point charge is instead situated at position $\mathbf{r}_i$ then the expression becomes

$$\phi(\mathbf{R}) = \frac{Q}{4\pi\epsilon_0 |\mathbf{R} - \mathbf{r}_i|}$$

and the potential at this position due to the array of point charges introduced at the start of the chapter is given exactly by

$$\phi(\mathbf{R}) = \sum_{i=1}^{n} \frac{Q_i}{4\pi\epsilon_0 |\mathbf{R} - \mathbf{r}_i|} \tag{6.6}$$

Although this expression is exact, it may sometimes contain more information than is strictly necessary for it to be useful. If the charges are all much closer together than to the field point $\mathbf{R}$ then we might expect that the potential would be determined mainly by some property of the origin and the distance from this origin to the field point $\mathbf{R}$. This idea forms the basis for the so-called *multipole expansion* of the potential.

A simple illustration of the multipole expansion is afforded by the linear dipole shown in Figure 6.5, which has Q_1 at position $-z_1$ along the z axis and Q_2 at position z_2. The electrostatic potential at field point P with position vector $\mathbf{R}$ is given exactly by

$$\phi(\mathbf{R}) = \frac{1}{4\pi\epsilon_0} \left(\frac{Q_1}{r_1} + \frac{Q_2}{r_2} \right) \tag{6.7}$$

which may also be written

$$\phi(\mathbf{R}) = \frac{1}{4\pi\epsilon_0} \left[\frac{Q_1}{\sqrt{(R^2 + z_1^2 + 2z_1 R \cos\theta)}} + \frac{Q_2}{\sqrt{(R^2 + z_2^2 - 2z_2 R \cos\theta)}} \right] \tag{6.8}$$

In cases where $R \gg z_1$ and $R \gg z_2$ the denominator can be expanded in terms of z_1/R and z_2/R in terms of the binomial theorem to give

$$\phi(\mathbf{R}) = \frac{1}{4\pi\epsilon_0} \left[\frac{(Q_1 + Q_2)}{R} + \frac{(Q_2 z_2 - Q_1 z_1) \cos\theta}{R^2} + \frac{(Q_1 z_1^2 + Q_2 z_2^2)}{2} \frac{(3 \cos^2 \theta - 1)}{R^3} + \cdots \right]$$

This is known as a multipole expansion of the potential.

It can also be written in terms of the total charge $Q = Q_1 + Q_2$, the electric dipole moment

$$p_e = Q_2 z_2 - Q_1 z_1$$

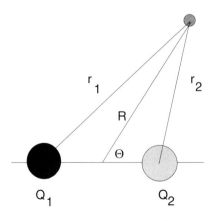

Figure 6.5 A simple dipole

and the second moment

$$\Theta = Q_1 z_1^2 + Q_2 z_2^2$$

as

$$\phi(\mathbf{R}) = \frac{1}{4\pi\epsilon_0} \left[\frac{Q}{R} + \frac{p_e \cos\theta}{R^2} + \frac{\Theta}{2} \frac{(3\cos^2\theta - 1)}{R^3} + \cdots \right] \tag{6.9}$$

The potential is therefore seen to be a sum of terms; each is the product of a moment of the charge and the other a function of the distance. The first term is the potential due to a point charge, the second term is the potential due to a dipole and so on. Successive terms fall off more rapidly with the separation r between the charge distributions, and sufficiently far from the origin the potential can be approximated as the potential due to the lowest non-zero multipole.

In the more general case of the charge distribution shown in Figure 6.1, the inverses of the distances can be expanded about field point P with position vector $\mathbf{R}$ to yield the Taylor series

$$\frac{1}{|\mathbf{R} - \mathbf{r}_i|} = \frac{1}{R} - \mathbf{r}_i \cdot \mathrm{grad}\left(\frac{1}{R}\right) + A + B + \cdots \tag{6.10}$$

The second term on the right is a sum of terms like $x_i \partial(1/R)/\partial X$, the third term A is a sum of contributions like $x_i x_j \partial^2(1/R)/\partial X\,\partial Y$, and so on. If I substitute this equation into the expression for $\phi(\mathbf{R})$ above, then the potential is seen to be a sum of terms each of which is a product of two factors. One factor is characteristic of the charge distribution, and one that is characteristic of the distance from the field point $\mathbf{R}$

$$4\pi\epsilon_0\phi(\mathbf{R}) = \left(\sum_{i=1}^{n} Q_i\right)\left(\frac{1}{R}\right) - \left(\sum_{i=1}^{n} Q_i \mathbf{r}_i\right) \cdot \mathrm{grad}\left(\frac{1}{R}\right) + A' + B' + \cdots \tag{6.11}$$

For this reason, classical theories of intermolecular forces always begin with the multipole expansion and consider charge...charge, charge...dipole, dipole...dipole...interactions individually.

When molecules are near enough to influence one another, forces of attraction and repulsion come into play. There are a number of types of intermolecular force, and most of them can be understood on the basis of classical electrostatics.

Consider now the mutual potential energies of two simple charge distributions A and B as shown in Figure 6.6.

The first charge distribution A (which consists of Q_1 and Q_2) is centred at the origin. The second charge distribution B (which consists of Q_1' and Q_2') has its center distance r away.

To calculate the mutual potential energy U of this array of four point charges, we start from the exact expression for the mutual potential energy of charge distribution B in an electrostatic potential due to charge distribution A

$$(4\pi\epsilon_0)U = Q_1'\phi(r_1) + Q_2'\phi(r_2)$$

Here r_1 and r_2 are the distances from the origin to Q_1' and Q_2'. The idea is to now express everything in terms of the variables r, θ_A, θ_B, Q, Q', the dipoles and the quadrupoles and

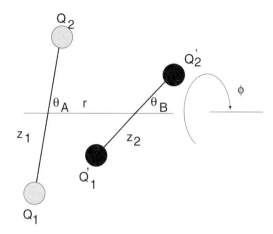

Figure 6.6   Two charge distributions A and B

so on. The answer turns out to be

$$(4\pi\epsilon_0)U = \frac{Q_A Q_B}{r} + \frac{1}{r^2}\left(Q_B p_A \cos\theta_A - Q_A p_B \cos\theta_B\right)$$

$$- \frac{p_A p_B}{r^3}\left(2\cos\theta_A \cos\theta_B - \sin\theta_A \sin\theta_B \cos\phi\right)$$

$$+ \frac{1}{2r^3}\left[Q_A \Theta_B(3\cos^2\theta_B - 1) + Q_B \Theta_A(3\cos^2\theta_A - 1)\right]$$

$$+ \cdots \tag{6.12}$$

where I have written the total charge of distribution A as $Q_A = Q_1 + Q_2$, the electric dipole p_A, and so on. If each charge distribution A and B corresponds to a neutral molecule then the leading term in this expression is seen to be the dipole–dipole interaction

$$(4\pi\epsilon_0)U_{\text{dip...dip}} = -\frac{p_A p_B}{r^3}\left(2\cos\theta_A \cos\theta_B - \sin\theta_A \sin\theta_B \cos\phi\right) \tag{6.13}$$

The sign and size of this dipole–dipole interaction depends very strongly on the orientations of the dipoles. Table 6.2 shows three simple examples (all of which have $\phi = 0$).

If U is averaged over all possible orientations of θ_A, θ_B and ϕ, and these orientations are all treated as equally important, then the average comes to zero. Such an averaging ignores the thermal motion and is not realistic. I will show you in later chapters that we have to include a so-called Boltzmann factor in order to allow for this thermal motion. The Boltzmann factor allows us to average out the thermal motions, and it is equal to $\exp(-U/kT)$. It therefore favors configurations with a negative energy relative to those with a positive energy. If I write the averaged potential as $\langle U \rangle_{\text{dip...dip}}$ then we find

$$\langle U \rangle_{\text{dip...dip}} = -\frac{2p_A^2 p_B^2}{3kT(4\pi\epsilon_0)^2}\frac{1}{r^6} \tag{6.14}$$

Table 6.2 Dipole – dipole interactions

θ_A	θ_B	Description of dipoles	Expression for dipole–dipole U
0	0	Parallel	$-\dfrac{2p_a p_b}{4\pi\epsilon_0 r^3}$
0	π	Antiparallel	$+\dfrac{2p_a p_b}{4\pi\epsilon_0 r^3}$
0	$\pi/2$	Perpendicular	0

Similar expressions may also be obtained for the dipole...quadrupole and quadrupole... quadrupole averages;

$$\langle U\rangle_{\text{dip...quad}} = -\frac{(p_A^2\Theta_B^2 + p_B^2\Theta_A^2)}{kT(4\pi\epsilon_0)^2}\frac{1}{r^8} \tag{6.15}$$

$$\langle U\rangle_{\text{quad...quad}} = -\frac{14\Theta_A^2\Theta_B^2}{5kT(4\pi\epsilon_0)^2}\frac{1}{r^{10}} \tag{6.16}$$

In all cases we see that the potential energies are negative and inversely proportional to the temperature. It also turns out that the distance dependence is the square of that for the corresponding fixed orientation case.

6.4 DIELECTRIC POLARIZATION AND POLARIZABILITIES

Charge distributions respond to the presence of other charge distributions, and we account for this in terms of the so-called response functions, the *polarizabilities*. It is therefore necessary to discuss such response functions in detail.

In 1837 Faraday showed that when the space between the plates of a parallel plate capacitor was filled with a substance such as glass or mica, its capacitance became greater. The multiplicative factor ϵ_r is called the *relative permittivity* of the medium. In the older literature, you will find it called the *dielectric constant*. Materials such as glass and mica differ from metals in that they have no free electrons which can move through the body of the material. We call materials such as glass and mica *dielectrics*, to distinguish them from metallic conductors.

Think of a two-dimensional slab of a simple dielectric which we can take to be composed of spherical particles comprising central nuclei surrounded by electron clouds (Figure 6.7). We now apply an electric field which points from the left to the right in Figure 6.7. Each (positively charged) nucleus will experience a force driving it to the right-hand side, and each electron cloud will experience a force driving it toward the left-hand side of the diagram.

Figure 6.8 gives a sense of this charge separation.

The macroscopic theory of dielectric polarization focuses attention on the separation of charge per small unit volume $d\tau$. Charge separation is equivalent to an electric dipole moment

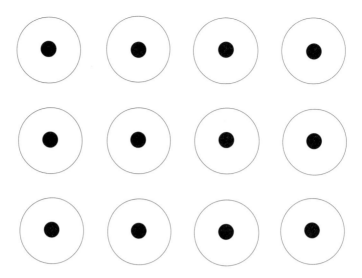

Figure 6.7 A dielectric material

and so we write

$$d\mathbf{p}_e = \mathbf{P}\,d\tau \tag{6.17}$$

where $d\mathbf{p}_e$ is the electric dipole induced in the volume $d\tau$. The proportionality constant (vector) $\mathbf{P}$ is called the dielectric polarization. It is an experimentally determined quantity.

At the microscopic level the applied field induces an electric dipole moment on each molecule making up the sample. The direction of this induced dipole moment need not be in the direction of the permanent electric dipole, and we usually write the dependence of the dipole moment $\mathbf{p}_e$ on applied electrostatic field $\mathbf{E}$ as

$$\mathbf{p}_e(\mathbf{E}) = \mathbf{p}_e(\mathbf{E} = \mathbf{0}) + \alpha \cdot \mathbf{E} + \cdots \tag{6.18}$$

The term on the left-hand side is the electric dipole moment in the presence of an applied field. The first term on the right-hand side is the permanent electric dipole moment. The second term is a product of the *dipole polarizability* α with the applied field, and higher terms include the dipole hyperpolarizabilities.

$\mathbf{p}_e$ and $\mathbf{E}$ are both vectors which are not necessarily parallel, and α is a tensor quantity that we can represent as a real symmetric 3×3 matrix

$$\alpha = \begin{pmatrix} \alpha_{xx} & \alpha_{xy} & \alpha_{xz} \\ \alpha_{yx} & \alpha_{yy} & \alpha_{yz} \\ \alpha_{zx} & \alpha_{zy} & \alpha_{zz} \end{pmatrix} \tag{6.19}$$

This matrix can always be written in the diagonal form by a suitable rotation of the Cartesian axes to give

$$\alpha = \begin{pmatrix} \alpha_{aa} & 0 & 0 \\ 0 & \alpha_{bb} & 0 \\ 0 & 0 & \alpha_{cc} \end{pmatrix}$$

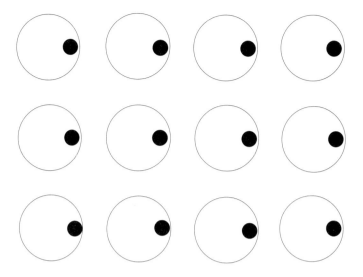

Figure 6.8   Polarization of a dielectric sample

where α_{aa}, α_{bb} and α_{cc} are the principal values of the polarizability. For molecules with symmetry, the principal axes of polarizability correspond to symmetry axes.

Thus for a linear molecule the polarizability matrix can be written

$$\alpha = \begin{pmatrix} \alpha_\perp & 0 & 0 \\ 0 & \alpha_\perp & 0 \\ 0 & 0 & \alpha_\| \end{pmatrix}$$

where $\alpha_\perp$ is called the perpendicular component of the polarizability and $\alpha_\|$ the parallel component. For an applied electrostatic field $E_\|$ parallel to a linear molecule, the induced dipole moment is $\alpha_\| E_\|$, and for an external field $E_\perp$ perpendicular to the molecular axis the induced dipole is $\alpha_\perp E_\perp$.

Very often, the quantity of interest is the *mean polarizability*, $\langle \alpha \rangle$ which is 1/3 the sum of the diagonal elements of the polarizability tensor.

A careful study of the interaction energy of two identical dipolar molecules each with permanent dipole p and mean polarizability $\langle \alpha \rangle$ gives the following expression for the so-called *induction energy*

$$\langle U \rangle_{ind} = -\frac{2p^2 \langle \alpha \rangle}{(4\pi\epsilon_0)^2} \frac{1}{r^6} \tag{6.20}$$

where $\langle \alpha \rangle$ is the mean molecular polarizability.

The energy is not temperature dependent, and it is only important when r is relatively large. The dipole–dipole interaction and the induction interaction are examples of long range inter-actions.

6.5 DISPERSION ENERGY

Finally we must consider the case of two interacting molecules, neither of which has a permanent dipole moment (or higher order moments). For such molecules, the most important source of attractive intermolecular energy is the dispersion energy, which was first identified by London in 1930. The dispersion energy cannot be described in classical terms, as its origins are quantum mechanical. The basis is this; electrons in atoms and molecules are never at rest, even in the lowest energy state. Although on average the electric dipole moment of an atom is zero, instantaneous dipole moments may occur. This instantaneous dipole can induce a dipole in a neighboring atom or molecule, and the net effect is attractive. An expression for the leading term in this dispersion energy for identical molecules is

$$U_{\text{disp}} = -\frac{3\langle\alpha\rangle^2\varepsilon_1}{4(4\pi\epsilon_0)^2}\frac{1}{r^6} \tag{6.21}$$

where ε_1 is the first ionization energy of the molecule.

The dispersion energy is seen to be attractive, and to again fall off as $1/r^6$.

6.5.1 Short Range Interactions

When two molecules approach so closely that their electron clouds overlap, the positively charged nuclei become less well shielded and so they repel each other. A generalized exponential form

$$U_{\text{rep}} = A\exp(-Br)$$

is frequently used to represent the short range mutual potential energy. The constants A and B have to be determined from experiment.

The total of attractive and repulsive terms would therefore be

$$U(r) = U_{\text{rep}} + U_{\text{dip}...\text{dip}} + U_{\text{ind}} + U_{\text{disp}}$$

which can be written

$$U(r) = A\exp(-Br) - C/r^6$$

This is known as the exp-6 model.

The best known potential for monatomic systems is the Lennard–Jones 12-6 potential

$$U_{\text{LJ}}(r) = \frac{A}{r^{12}} - \frac{B}{r^6}$$

The potential is often written as

$$U_{\text{LJ}}(r) = 4\varepsilon\left[\left(\frac{\sigma}{r}\right)^{12} - \left(\frac{\sigma}{r}\right)^6\right]$$

where the 'well depth' ε corresponds to the minimum value of U_{LJ} and the 'distance of closest approach' σ is the value of r for which U is zero.

Notice that at large r the energy falls off as $1/r^6$.

6.6 THE PAIR POTENTIAL

We normally refer to the mutual potential energy between a pair of bodies as the pair potential.

The construction of intermolecular potential energy functions is an important one, and a great deal of research has gone into this activity. It is a quantity that I will return to time and time again in this text.

7 Some Statistical Ideas

There are two ways to go about modeling. On the one hand, there is the macroscopic approach where we concern ourselves with the overall behavior of large amounts of matter. This approach involves a study of properties such as temperature, pressure and heat capacity, and leads us to equations of state such as the ideal gas equation $pV = nRT$ discussed in Chapter 1. In this context, the symbol n stands for the amount of substance.

On the other hand is the microscopic approach, where we study the properties of individual atoms and molecules. Scientific progress made in the twentieth century as led us to a position where we now understand many microscopic properties and can predict unknown ones with good accuracy, and we might guess that we should be able to extrapolate our microscopic theories to macroscopic systems.

There is a problem. A typical macroscopic system consists of 10^{25} interacting atoms or molecules, and our knowledge of the properties of microscopic systems *by themselves* is totally useless. For example, it is by no means evident from a study of the properties of a helium atom that helium gas can suddenly condense so as to form a liquid. And again, starting from the properties of simple atoms, one might not suspect that certain molecules can give rise to systems capable of biological growth. There is something missing, and we are going to have to investigate the missing ingredient.

The difference between macroscopic and microscopic systems requires that we should understand and reconcile different concepts. The problems that confront us are summarized in Table 7.1.

Table 7.1 illustrates the connection between microscopic and macroscopic quantities. Sometimes the connection is easy. The mass of 1 mol of water is 6.022×10^{23} times the mass of a single molecule.

But what about the internal energy? I have put a question mark in the right-hand column because we cannot always deduce the internal energy of a macroscopic sample from a knowledge of the energies of the constituent particles. The point is that these particles interact with each other, and so there is an intermolecular contribution that is absent for a single particle. The same comments also apply to the entropy, the Gibbs and the Helmholtz energies.

Then, there are properties where the connection is not at all obvious, and I have left blanks in the right- (or left-) hand columns accordingly. For example, what microscopic property corresponds to the temperature? How does an isolated molecule know the temperature?

Table 7.1 Microscopic Vs macroscopic properties

Microscopic properties	Macroscopic properties
Mass	Mass
Energy	? Internal energy
Momentum	
Spin	
	Density
	Temperature
	Pressure
	Entropy
	Gibbs energy
	Helmholtz energy
	Heat capacity

Temperature does not appear in Newton's laws, and it does not appear in the Schrödinger equation. I will show later that temperature is a concept that measures the way that energy is shared between the particles in a system. I hinted at this in the Introduction.

Things are not as bad as they might seem. What matters in a macroscopic sample is very often the average behavior of the atoms and molecules, and not their individual properties. We therefore concern ourselves with properties like the average value of the total energy and its fluctuation about the average rather than enquiring about the energy of every individual molecule. Boltzmann (1844–1906) realized that this was the way forward, and he helped lay the foundation for the branch of science known as *statistical mechanics*. A key concept of this field is that macroscopic phenomena such as the third law of thermodynamics can be explained by examining statistically the microscopic properties of a sample.

So, I had better say a little more about statistics.

7.1 STATISTICS

Statistics is a branch of maths that deals with the collection, organization and analysis of experimental data. The raw materials of statistics are sets of numbers that have been obtained by counting or measuring the results of experiments. Simple forms of statistics have been used since the beginning of civilization; the ancient Egyptians are said to have used primitive statistical methods to analyse the wealth and population of their kingdom before they began construction of the pyramids. The Domesday book was the result of a crude statistical census of England conducted in the year 1086.

7.1.1 Averages, Probabilities and Standard Deviations

Let me start the discussion with a very simple experiment; I take a six-faced die whose faces have the usual scores of $1, 2, \ldots, 6$, and throw the die 10 times (Figure 7.1). I will use the symbol N to denote the number of possible scores (6 in this case) and n to denote the number of experiments (10 in this case). The symbol S is used for the score.

Figure 7.1 A game of chance

My scores are recorded in Table 7.2.

Notice that all possible scores do not appear in this particular set of experimental data. I threw the die a further 10 times and came up with the set of scores shown in Table 7.3.

Usually it is easier to record and work with the *frequency* f_i of a given score.

For the 20 experiments above, the frequencies are given in Table 7.4.

These data may be given in the form of a histogram (Figure 7.2).

For a large number of experiments, the fractional frequencies are found to level off at 1/6 and the *probability* of each score is defined as the limiting value of the fractional frequency for a large number of experiments. Probabilities therefore sum to 1, and the symbol p is generally used for them. In this case, $p_i = 1/6$ for $i = 1, 2, \ldots, 6$.

After experimental data have been collected, analysis begins.

The first thing to do is to calculate the *mean* (or *average*) which is given as the sum of each set of numbers divided by the number of values. So the first set of data has a mean of 3.8 and the second set of data has a mean of 4.0. Advanced texts differentiate between the mean of a sample (the *sample mean*) and the mean of a very large number of readings (the *population mean*); it should be intuitively obvious that the population mean is $3\frac{1}{2}$ so the sample means give approximations to the population mean.

Table 7.2 Throwing a single die 10 times

Throw	1	2	3	4	5	6	7	8	9	10
Score S	5	4	4	2	2	5	1	5	5	5

Table 7.3 10 further tosses of a die

Throw	11	12	13	14	15	16	17	18	19	20
Score S	1	3	6	5	3	6	6	1	6	3

Table 7.4 Throwing a single die 20 times

Score S	1	2	3	4	5	6
Frequency	3	2	3	2	6	4
Fractional frequency	3/20	2/20	3/20	2/20	6/20	4/20

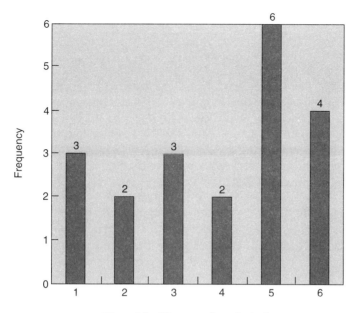

Figure 7.2   Histogram for a single die

If we put x_i for each of the values of the score S, then I am going to denote the mean value of the score x as $\langle x \rangle$, and so we have

$$\langle x \rangle = \frac{1}{n} \sum_{i=1}^{n} x_i \tag{7.1}$$

This can be rewritten in terms of the frequency data as

$$\langle x \rangle = \sum_{i=1}^{N} \frac{f_i}{n} x_i \tag{7.2}$$

or in terms of the probabilities (if we know them) as

$$\langle x \rangle = \sum_{i=1}^{N} p_i x_i \tag{7.3}$$

The second thing we might be interested in is the 'spread' of points about the average. Sometimes the values of x will turn out to be closely clustered about the average, sometimes they will be widely spread. To get a measure of the spread, we calculate the *standard deviation σ* for the n data points, defined as

$$\sigma = \sqrt{\sum_{i=1}^{N} \frac{f_i}{n} (x_i - \langle x \rangle)^2} \tag{7.4}$$

or in terms of probabilities

$$\sigma = \sqrt{\sum_{i=1}^{N} p_i(x_i - \langle x \rangle)^2}\qquad(7.5)$$

and so $\sigma = 1.47$ for the first set of throws and $\sigma = 1.95$ for the second set.

You will probably find that your pocket calculator has keys that will help you perform such simple statistical calculations. Be sure to press the key labelled σ**n** rather than σ**n** $-$ 1 if you want to reproduce my values for the standard deviation when using frequency data. The σ**n** key refers to the sample mean, the σ**n** $-$ 1 key refers to the population mean. For large enough samples, the two values are equivalent.

The standard deviation gives a measure of the spread of data about the mean. The higher the standard deviation, the larger the spread of the data about the mean.

You might have been expecting something rather different. The 20 throws of the die seem to have turned up a score of 5 rather more often than you might have anticipated, and so you might like to put money on the chance of a 5 coming up at the next throw? People have tried to make money from considerations like this since games of chance first made their appearance (when time began). The subject of *probability theory* is a branch of math that deals with determining the likelihood that an experiment will have a particular outcome. The probability of a particular outcome is represented by a number between 0 and 1, inclusive. An outcome with probability 0 means that the event will never occur, no matter how many times the experiment is tried. An outcome with probability 1 means that the event will always occur, to the exclusion of all others. So, for a six-faced die you were probably expecting that the scores would appear with equal regularity because the probabilities are all equal at 1/6 (we always arrange probabilities so that they sum to 1).

Throwing dice gets a bit tedious, and we need some way to model the experiment. Your pocket calculator probably has a button labelled RAN#; press it a number of times and you will see that it gives a series of numbers between 0 and 1.

Table 7.5 shows what I found when I pressed the RAN# key eight times on my CASIO:

This is a series of *pseudo-random* numbers; there is no discernible trend in the series, and the numbers are equally distributed over the range of values 0 to 1. If you switch off your pocket calculator, switch it back on again and repeat the experiment you will almost certainly find the same series of numbers. For this reason (amongst others), mathematicians call this a pseudo-random series. If you know about BASIC programming, or if you know about spreadsheets such as MS Excel, you will be able to identify pseudo-random number generators from your software manual. Do not worry about the name 'pseudo', these random numbers are fine for our purposes.

I am going to use pseudo-random numbers to model the process of tossing a die a number of times.

Table 7.6 gives my results for a further set of experiments, together with the means and standard deviations. I wrote a simple BASIC program to model the experiment. If you have

Table 7.5 Pressing the RAN# key 8 times

| 0.923 | 0.540 | 0.361 | 0.381 | 0.448 | 0.370 | 0.895 | 0.150 |

Table 7.6   Throwing a die many time

Throws	100	500	1000	10 000	1 000 000
$S = 1$	16	77	160	1651	166 721
2	17	93	184	1684	166 257
3	19	93	170	1689	167 324
4	20	81	161	1679	165 971
5	11	89	149	1669	166 981
6	17	68	176	1628	166 746
Mean	3.440	3.432	3.483	3.4915	3.500
σ	1.669	1.647	1.712	1.699	1.708

Table 7.7   Throwing a single die

Score x	1	2	3	4	5	6
Probability p	1/6	1/6	1/6	1/6	1/6	1/6
$x - \langle x \rangle$	-2.5	-1.5	-0.5	$+0.5$	$+1.5$	$+2.5$
$(x - \langle x \rangle)2$	6.25	2.25	0.25	0.25	2.25	6.25

enough patience, then you can repeat my experiments using a real die. You will not get the same numerical results that I got, but you will agree with all my conclusions on the basis of your data.

You can see that the larger the number of throws, the closer the number of occurrences of each score. As I explained above, we call the ratio of the score divided by the number of measurements the fractional frequency. As the number of experiments is increased, the ratio tends to the probability, and you can see that the frequency is fast approaching the theoretical probability of 1/6 for each score as the number of throws of the dice increases.

This is in accord with our everyday ideas about the laws of chance. So for a single die the results are shown in Table 7.7 which gives $\langle x \rangle = 3.5$ and $\sigma = 1.708$.

7.1.2   Two Dice

I am now going to repeat the experiment with two dice. Either die can come up with a score x of $1, 2, 3, \ldots, 6$. This is not the same as having a single die with 12 faces, as you will see shortly. If you want to model the experiment, you will have to use your random number generator for either die and then sum the score.

The scores S for two dice are shown in Table 7.8.

You can see the trend, the mean for the two dice looks like it ought to be twice the mean for a single dice, and the standard deviation looks like it is about $\sqrt{2}$ times the standard deviation for the single dice.

How do we go about understanding these results? What we have to do is the mathematical process of *enumeration*.

Table 7.8   Throwing two dice

Throws	100	500	1000	100 00	1 000 000
$S = 2$	4	14	26	285	27 793
3	5	29	45	569	55 143
4	6	38	87	846	83 099
5	11	48	111	1068	111 605
6	18	59	154	1410	138 889
7	14	94	179	1645	166 503
8	17	81	137	1333	138 786
9	11	52	107	1126	110 750
10	4	45	69	848	84 070
11	2	28	56	595	55 607
12	8	12	29	275	27 755
$\langle x \rangle$	7.010	7.074	6.980	7.010	7.003
σ	2.500	2.386	2.357	2.437	2.415

7.2   ENUMERATION

I am going to ask how many ways the two dice can each give the stated score, and work out the probabilities from that. I have summarized the individual dice scores (the two dice are labelled A and B) in Table 7.9

Table 7.9 is self-explanatory, and the theoretical probabilities are $1/36, 2/36, \ldots, 1/36$. The mean and standard deviation work out as $\langle x \rangle = 7$ and $\sigma = 2.404$.

I will tell you that if we toss p dice then the mean score is n times the mean for a single dice, and the standard deviation is $\sqrt{n}$ times the standard deviation for a single dice. To put it more formally; for n independent experiments, each with mean score $\langle x \rangle$ and standard deviation $\sigma(x)$, the total score S and the total standard deviation $\sigma(S)$ obey

$$\langle S \rangle = n\langle x \rangle$$
$$\sigma(S) = \sqrt{n}\sigma(x)$$

(7.6)

Table 7.9   Enumeration of the scores for two dice

Score S	Possible individual scores A + B $s_A + s_B$	Number of ways	Probability	
2	$1+1$	1	1/36	0.028
3	$1+2, 2+1$	2	2/36	0.056
4	$1+3, 3+1, 2+2$	3	3/36	0.083
5	$1+4, 4+1, 2+3, 3+2$	4	4/36	0.111
6	$1+5, 5+1, 2+4, 4+2, 3+3$	5	5/36	0.139
7	$1+6, 6+1, 2+5, 5+2, 3+4, 4+3$	6	6/36	0.167
8	$2+6, 6+2, 3+5, 5+3, 4+4$	5	5/36	0.139
9	$3+6, 6+3, 4+5, 5+4,$	4	4/36	0.111
10	$4+6, 6+4, 5+5$	3	3/36	0.083
11	$5+6, 6+5$	2	2/36	0.056
12	$6+6$	1	1/36	0.028

7.3 THE BINARY ALLOY

An alloy is a substance that has a distinct chemical composition, for example AB or $A_x B_{1-x}$, but does not necessarily have an ordered crystal structure.

Brass is such a substance, and it is made from equal parts of copper and zinc. I am going to refer to such a substance as a *binary alloy*. A binary alloy of chemical composition AB implies a chemical composition with as many A atoms as B atoms in a macroscopic sample of the material but no definite chemical compound AB.

The atoms of A and the atoms of B can occupy any of the sites in the material's crystal lattice, and there is no particular reason why any given site should be occupied by an A atom or a B atom. All that matters is that in a macroscopic sample the overall number of atoms of A is the same as the number of atoms of B.

The probability that any one site in the lattice will be occupied by atom A is $\frac{1}{2}$, and the probability that any one site in the lattice will be occupied by atom B is also $\frac{1}{2}$

The atoms are assumed to be distinguishable; certainly we can distinguish between the atoms of type A and type B, and for the minute I am going to assume that it is possible somehow to distinguish between the atoms of type A, and between the atoms of type B.

What I am going to do is to take a sample of sites from the alloy, and enumerate the number of atoms of type A and the number of atoms of type B. For example, if I take four sites then I might find the arrangement, as in Table 7.10 or I might find it as in Table 7.11.

I can model this behavior using a random number generator. I have taken the four sites above and done the calculation 40 times. The results for the 40 experiments are given in Table 7.12.

The integers in Table 7.12 are the number of A atoms and the number of B atoms, so for example the final line AAAA has 4 A atoms and 0 B atom in that particular sample.

In this particular set of experiments, there was a single occurrence of BBBB, where atom A was not present, 10 occurrences of types ABBB, BABB, BBAB and BBBA, where a single A atom was present, 16 occurrences where two atoms of A and two atoms of B were present, 7 occurrences of AAAB, AABA, ABAA and BAAA and so on.

Instead of presenting a table with 40 rows, it is much more convenient to present the data as a histogram (Figure 7.3).

Suppose we now run a larger number of experiments, and count the number of occurrences of atom A. The results are given in Table 7.13.

In Table 7.13 I have recorded the number of occurrences of atom A in the sample. So for example, after 40 experiments I found 1 occurrence with no A atoms present, 10 occurrences with 1 A atom present, 16 occurrences with 2 A atoms present and so on. A histogram is shown below in Figure 7.4 for the 500 experiments.

Table 7.10 Four sites

Site	1	2	3	4
Atom	A	A	A	B

Table 7.11 Four sites

Site	1	2	3	4
Atom	B	A	B	A

Table 7.12   Full enumeration for the four sites

BINARY ALLOY PROBLEM		
BBBA	1	3
ABAB	2	2
BBAA	2	2
BBAB	1	3
BABA	2	2
BBAB	1	3
AABB	2	2
AABB	2	2
BBBA	1	3
BBAB	1	3
ABBA	2	2
BAAB	2	2
AAAA	4	0
ABBA	2	2
AAAB	3	1
AABA	3	1
BAAB	2	2
AABA	3	1
BAAB	2	2
BBAA	2	2
BBBB	0	4
ABBA	2	2
BBBA	1	3
AAAA	4	0
ABAA	3	1
AAAA	4	0
ABBA	2	2
BBBA	1	3
ABAB	2	2
BBAA	2	2
AAAA	4	0
ABBB	1	3
BAAB	2	2
AAAB	3	1
BBAB	1	3
AAAA	4	0
AABA	3	1
BAAA	3	1
BBAB	1	3
AAAA	4	0

We might wonder about the theoretical probabilities? Surely we would expect that the most likely result of an experiment would be to find 2 A atoms and 2 B atoms in a sample?

7.3.1   Enumeration

To gain an insight into these questions I am going to ask how many ways the atoms can be distributed between the possible sites in the alloy. Each distinct arrangement is called a *configuration*. For a given site in the lattice there are two possible choices (the atom can be A or B)

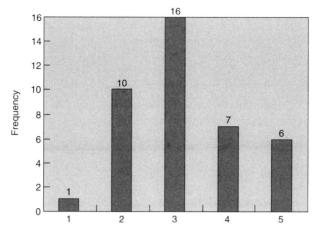

Figure 7.3 Binary alloy problem

Table 7.13 Larger number of experiments

No. expts	0	1	2	3	4
40	1	10	16	7	6
100	1	34	31	24	10
500	26	117	208	121	28

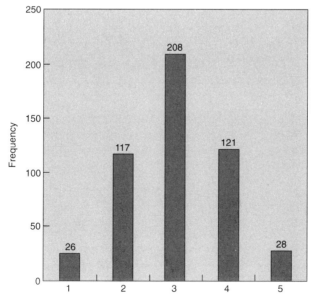

Figure 7.4 Binary alloy problem

and since the overall chemical composition is AB, there must be an equal probability of finding atom A or atom B at a given site (assuming that there is nothing about the atoms that would make them favor one particular site against another).

Table 7.14 lists all possible configurations for $N = 4$. I have labelled the lattice sites 1, 2, 3 and 4

There are 16 possible configurations, all of which are equally probable and of which several are equivalent to each other. The first configuration in Table 7.11 corresponds to all four lattice sites being occupied with an A atom, the next four configurations correspond to three A atoms and one B atom, and so on. For many purposes, I only want to know how many A-type atoms there are in the sample, and so I calculate probabilities of 1/16, 4/16, 6/16, 4/16 and 1/16. Note that only one of the configurations corresponds to finding all the atoms of type A. The probability is $(1/2)^4$.

The probabilities are therefore 1/16 (0.0625), 4/16 (0.2500), 6/16 (0.3750), 4/16 and 1/16 and these can be compared with my simulation above, where I found (for 500 experiments) 26/500 (0.0520), 117/500 (0.2340), 0.416, 0.2420 and 0.056.

We need to find a mathematical expression for the numbers of configurations, and this is given by the *binomial theorem*. For example, I can expand $(x + y)^4$ by the binomial theorem as follows

$$(x + y)^4 = x^4 + 4x^3y + 6x^2y^2 + 4xy^3 + y^4$$

The coefficients 1, 4, 6, 4 and 1 are examples of the *binomial coefficients*, and they are given generally by the coefficients in the expansion of $(x + y)^n$ as shown below

$$(x + y)^n = x^n + nx^{n-1}y + \frac{n(n - 1)}{2} x^{n-2}y^2 + \cdots + y^n \tag{7.7}$$

or in mathematical summation ($\sum$) notation

$$(x + y)^n = \sum_{t=0}^{n} \frac{n!}{(n - t)!t!} x^{n-t}y^t \tag{7.8}$$

Table 7.14 Enumeration

1	2	3	4	n_A	n_B	Number of configurations
A	A	A	A	4	0	1
A	A	A	B	3	1	
A	A	B	A	3	1	4
A	B	A	A	3	1	
B	A	A	A	3	1	
A	A	B	B	2	2	
A	B	A	B	2	2	6
A	B	B	A	2	2	
B	A	A	B	2	2	
B	A	B	A	2	2	
B	B	A	A	2	2	
A	B	B	B	1	3	
B	A	B	B	1	3	4
B	B	A	B	1	3	
B	B	B	A	1	3	
B	B	B	B	0	4	1

I have used the symbol $n!$ to mean factorial $n (n! = 1, 2, 3, 4, \ldots, n)$. Your pocket calculator might have a key to calculate factorials (look for $x!$), and if it does it will almost certainly have a key labelled nCr, which calculates the binomial coefficients. In the field of statistical physics, this expansion is often written

$$(x + y)^n = \sum_{t=0}^{n} g(n, t) x^{n-1} y^t \qquad (7.9)$$

where $g(n, t)$ is called the *multiplicity function*. In the field of statistical thermodynamics, g is referred to as the *statistical weight* of a configuration, and it is given the symbol Ω.

7.4 SHARPNESS OF THE MULTIPLICITY FUNCTION

It is a common experience in our macroscopic world that systems held at constant temperature have well-defined properties. This stability follows because the multiplicity function turns out to have an exceedingly sharp peak for large N; you might be able to see that this is the case by examining the binomial coefficients in $(1 + x)^n$ as n increases from $n = 4$ to $n = 10$ (Figures 7.5 and 7.6).

I will just sketch the mathematical derivation for large n, and leave you to fill in the blanks (should you so wish).

Our problem is to examine the behavior of the multiplicity function $g(n, t)$ for large n, typically 10^{23}. Since the multiplicity factor $g(n, t)$ involves factorials

$$g(n, t) = \frac{n!}{(n - t)! t!} \qquad (7.10)$$

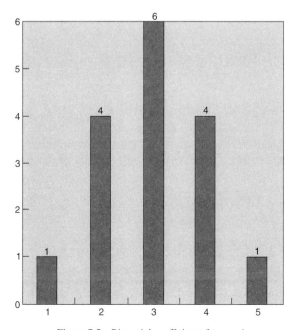

Figure 7.5 Binomial coefficients for $n = 4$

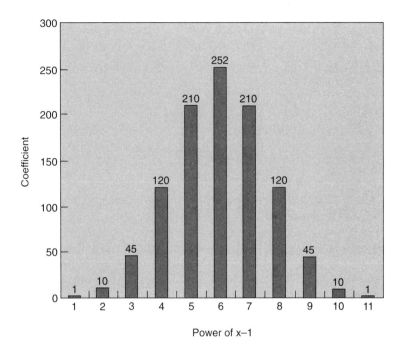

Figure 7.6   Binomial coefficients for $n = 10$

you will see that there is an immediate problem. Your pocket calculator will not be able to handle factorials of numbers greater than 70. So we need a method for estimating the factorials of large numbers and it turns out that Stirling's formula is valid for large n

$$n! \approx (2\pi n)^{1/2} n^n \exp\left[-n + \frac{1}{12n} + O\left(\frac{1}{n^2}\right) + \cdots\right] \qquad (7.11)$$

where I have used the symbol $O(1/n^2)$ to mean terms of order $1/n^2$. On taking logarithms of either side

$$\ln(n!) = \frac{1}{2}\ln(2\pi) + \left(n + \frac{1}{2}\right)\ln(n) - n + \frac{1}{12n} + O\left(\frac{1}{n^2}\right) + \cdots \qquad (7.12)$$

In practice the $1/(12n)$ term is usually omitted, together with higher terms.

My final result (which I have still to derive) looks a little neater if I write the multiplicity function

$$g(n, t) = \frac{n!}{(n - t)!\, t!} \qquad (7.13)$$

in terms of the average value of A and B atoms, $\frac{1}{2}n$, and

$$g(n, s) = \frac{n!}{\left(\frac{1}{2}n + s\right)!\left(\frac{1}{2}n - s\right)!} \qquad (7.14)$$

The variable s can be positive or negative, and it ranges from $-\frac{1}{2}n$ to $\frac{1}{2}n$. So for the binary alloy, the number of A atoms would be $\frac{1}{2}n + s$ and the number of B atoms would be $\frac{1}{2}n - s$.

My only reason for doing this is that the final result looks neater than it otherwise would. First then I take logarithms of each side of the expression above for $g(n, s)$

$$\ln[g(n, s)] = \ln(n!) - \ln\left(\frac{1}{2}n + s\right)! - \ln\left(\frac{1}{2}n - s\right)!$$

and then use the Stirling formula for each term, for example

$$\ln(n!) = \frac{1}{2}\ln(2\pi) + \left(n + \frac{1}{2}\right)\ln(n) - n + \cdots$$

On rearrangement we obtain terms that can be written typically $\ln(1 + x)$ where x is small; expansion of this natural log as

$$\ln(1 + x) = x - \frac{x^3}{2} + \cdots$$

leads to the final result

$$g(n, s) = g(n, 0) \exp\left(-\frac{2s^2}{n}\right) \tag{7.15}$$

where $g(n, 0)$ is a normalizing constant needed to make the probabilities add to 1. If you work through the derivation I have given above, you will find that

$$g(n, 0) = 2^n\sqrt{\frac{2}{n\pi}} \tag{7.16}$$

A more thorough treatment gives the correct value of

$$g(n, 0) = \frac{n!}{(\frac{1}{2}n)!(\frac{1}{2}n)!} \tag{7.17}$$

although for $n = 50$ these two expressions give 1.264×10^{14} and 1.270×10^{14}.

When $s^2 = \frac{1}{2}n$, the value of g is reduced by a factor of e from its maximum value. That is to say when the *fractional width* s/n is given by

$$\frac{s}{n} = \sqrt{\frac{1}{2n}} \tag{7.18}$$

The fractional width is used as a measure of the spread of the distribution. When we have to deal with collections of typically 10^{22} atoms or molecules, s/n is of the order of 10^{-11} and the distribution is extremely sharply defined.

7.5   THE BOLTZMANN DISTRIBUTION (i)

Let me now add a further complication. Consider a gas at low pressure in a thermally insulated constant volume container, such that no energy can be transmitted either to the surroundings or from the surroundings. Think of a thermos flask and you have got the right idea.

Because the pressure is low, the gas atoms are essentially free particles and they only collide with each other or the container wall on rare occasions. You know by now about energy quantization, so let me tell you that each atom has a very simple energy level diagram (Figure 7.7). It can have energies of 1Δ, 2Δ, 3Δ, and so on, until infinity.

These particular atoms are atypical because none of their energy levels are degenerate. We therefore have n distinguishable atoms in a thermally insulated container, and each of these atoms can have energies Δ, 2Δ, and so on. What distribution are we likely to find when we do a spot check on the atoms? Will each and every atom have the same energy, or will some have more than others?

In order to investigate the problem, we simply go about counting configurations again, but with a difference. If the container is thermally insulated then the internal energy is constant. Let me give you a couple of simple examples before I point you at the mathematical solution to the problem.

Take four particles designated A, B, C and D and assume that they each have the simple energy level diagram given in Figure 7.7. Let us assume that the internal energy of this system of four particles is fixed at 10Δ. In order to make up a total internal energy of 10Δ, we only need consider the possibilities where each particle has a maximum of 7Δ (in which case the others have 1Δ each). Even for such a simple system, the calculation is far from simple, and what we will do is make use of the binomial coefficients.

Table 7.15 below shows some of the possible distributions of energy amongst the allowed energy levels of the four particles in order to maintain a total energy of 10Δ. I have not listed all possible combinations of A, B, C and D that can be chosen to make up the total, just typical ones.

So for example, there are six possible ways (designated N in Table 7.15) in which I can arrange the four particles to have individual energies of 3Δ, 3Δ, 2Δ, 2Δ, four possible ways in

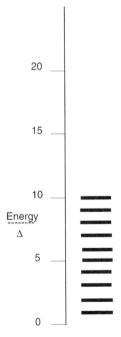

Figure 7.7   Simple energy level diagram

Table 7.15 Distinguishability (i)

A	B	C	D	N
3Δ	3	2	2	6
3	3	3	1	4
4	2	2	2	4
4	3	2	1	24
4	4	1	1	6
5	2	2	1	12
5	3	1	1	12
6	2	1	1	12
7	1	1	1	4

which I can arrange the four particles to have individual energies of 3Δ, 1Δ, and so on. These numbers depend on the fact that the particles are distinguishable, so there is an experimentally measurable difference between a configuration where particle A has 3Δ particle B has 1Δ, and a configuration where particle A has 1Δ, particle B has 3Δ. We might for example paint particle A white and particle B green, and so on, but we can certainly distinguish between the so-called identical particles.

You might like to mentally question my assumption about the distinguishability of particles at this stage.

I have missed the Δ from most of entries in Table 7.15 for clarity, and I have listed the total number of arrangements N for distributing the possible energies amongst the four particles.

Just to make the numbers clear, consider the situation where one particle (A in Table 7.15) has 4Δ, one particle (B in Table 7.15) has 3Δ, C has 2Δ and D has 1Δ. These add up to the constant internal energy of 10Δ, but of course the particles are distinguishable and so we should also consider configurations where (for example) particle A had 1Δ, particle B has 2Δ, particle C has 3Δ and particle 4 has 4Δ. I have listed all 24 possibilities in Table 7.16 and this number 24 is just the factorial of 4, 4! There are four ways of choosing the first particle, three ways of choosing the second particle, two ways of choosing the third and just one way of choosing the final particle.

All the other possibilities occur less times; take the case where two particles have 3Δ and the other two have 2Δ. The number of distinguishable configurations is now reduced from 24 as shown in Table 7.17.

The enumeration procedure is to take the maximum possible (4!) and divide it by the product of each of the factorials of the numbers of particles having equal energies.

7.5.1 The Boltzmann Distribution (ii)

Consider now a macroscopic system of fixed energy, comprising N identical and independent particles. We assume that the particles are distinguishable, and that their energy levels are non-degenerate, and we assume that N is very large.

Although we can fix the energy, we cannot fix how it is distributed amongst the available energy levels. The best we can do is to say that there are N_1 particles with energy ε_1, N_2 particles with energy ε_2, and so on. To make the following derivation easy, we will assume that

Table 7.16   Distinguishability (ii)

A	B	C	D
4	3	2	1
4	3	1	2
4	1	2	3
4	1	3	2
4	2	1	3
4	2	3	1
3	4	1	2
3	4	2	1
3	2	1	4
3	2	4	1
3	1	4	2
3	1	2	4
2	1	3	4
2	1	4	3
2	4	1	3
2	4	3	1
2	3	4	1
2	3	1	4
1	2	3	4
1	2	4	3
1	3	2	4
1	3	4	2
1	4	2	3
1	4	3	2

the maximum allowed energy level is ε_p. We talk about a *configuration* of the system, and there will be very many possible configurations. A general configuration where there are:

N_1 particles with energy ε_2

N_2 particles with energy ε_2

$\vdots$

N_p particles with energy ε_p

Table 7.17   Distinguishability (iii)

A	B	C	D
3	3	2	2
3	2	3	2
3	2	2	3
2	3	3	2
2	3	2	3
2	2	3	3

can be achieved in

$$W = \frac{N!}{N_1!N_2!\ldots N_p!} \tag{7.19}$$

different ways.

In this context, W is often called the weight, and it turns out that when N becomes large there is one configuration that dominates all the other possible ones. In order to find the appropriate values of N_i for this particular configuration we seek the values of the N_i that make W a maximum. This looks like a simple problem in differential calculus but there is however a sting in the tail of the derivation. The Ns cannot vary freely, they are constrained in the following two ways;

- if there are N particles then $N = N_1 + N_2 + \cdots N_p$
- if the system is isolated from the surroundings, and no energy can enter or leave, then the internal energy U is fixed and so $U = N_1\varepsilon_1 + N_2\varepsilon_2 + \cdots N_p\varepsilon_p$.

The way to solve such problems is to invoke *Lagrange's method of undetermined multipliers*. The derivation is given in detail in many more advanced texts; I will just give the key points without too much justification.

We want to find the values of $N_1, N_2, \ldots, N_p$ that make

$$W = \frac{N!}{N_1!N_2!\cdots N_p!}$$

a maximum, subject to the number of particles remaining constant, and the total energy remaining constant.

I am going to cater for the fact that the Ns are large by invoking Stirling's approximation for the factorials, and so we seek to maximize the natural log of W rather than W itself (because this procedure makes the algebra simpler). If this were just an ordinary problem in calculus to find the maximum of $\ln W$, we would differentiate, set the differential to zero and solve the resulting equation. We know from elementary rules of calculus that

$$\mathrm{d}\ln W = \sum_{i=1}^{p} \frac{\partial \ln W}{\partial N_i}\,\mathrm{d}N_i \tag{7.20}$$

but we have to cater for the constraints. We do this by differentiating the constraint equations to give

$$0 = \sum_{i=1}^{p} \mathrm{d}N_i$$

$$0 = \sum_{i=1}^{p} \varepsilon_i\,\mathrm{d}N_i$$

multiplying the constraint equations by arbitrary constants that I will call α and $-\beta$ and then adding the three equations together to give

$$\mathrm{d}(\ln W) = \sum_{i=1}^{p} \frac{\partial \ln W}{\partial N_i}\,\mathrm{d}N_i + \alpha \sum_{i=1}^{p} \mathrm{d}N_i - \beta \sum_{i=1}^{p} \varepsilon_i\,\mathrm{d}N_i$$

All the dN_i are now treated as independent and at a maximum (or minimum)

$$\frac{\partial \ln W}{\partial N_i} + \alpha - \beta \varepsilon_i = 0$$

The expression for $\ln W$ is

$$\ln W = \ln(N!) - \sum_{i=1}^{p} \ln(N_i!)$$

and if we use Stirling's formula but retain only the first two terms in each factorial

$$\ln W = N \ln N - \sum_{i=1}^{p} N_i \ln N_i$$

The final result turns out as

$$N_i = \exp(\alpha) \exp(-\beta \varepsilon_i) \tag{7.21}$$

I can easily evaluate the constant α by summing over all the allowed quantum states

$$\sum_{i=1}^{p} N_i = \exp(\alpha) \sum_{i=1}^{p} \exp(-\beta \varepsilon_i)$$

which gives

$$\exp(\alpha) = \frac{N}{\sum_{i=1}^{p} \exp(-\beta \varepsilon_i)} \tag{7.22}$$

But the value of β is more difficult, and what needs to be done is to use the distribution in order to calculate a measurable macroscopic quantity such as a heat capacity of an ideal gas. Comparison with experiment suggests that we can identify β with $1/k_B T$ where k_B is the Boltzmann constant. I am going to ask you to accept that the correct value of β is $1/k_B T$.

$$N_i = \frac{N \exp\left(-\frac{\varepsilon_i}{k_B T}\right)}{\sum_{i=1}^{p} \exp\left(-\frac{\varepsilon_i}{k_B T}\right)} \tag{7.23}$$

Remember that I set the maximum for the state quantum number as p in this case, hence the denominator summation runs from 1 to p.

Alternatively we can write the ratio of particles to be found with energies ε_2 compared to those with energy ε_1 as

$$\frac{N_2}{N_1} = \exp\left[-\frac{(\varepsilon_2 - \varepsilon_1)}{k_B T}\right] \tag{7.24}$$

7.6 THE PARTITION FUNCTION

The denominator in the Boltzmann expression turns out to be a useful quantity, and so it is given a special name and symbol.

$$Q = \sum_{i=1}^{p} \exp\left(-\frac{\varepsilon_i}{k_B T}\right) \tag{7.25}$$

Q is the partition function, and we will meet it again. Some authors use Z rather than Q, and lower case symbols q and z are often met in the literature when we want to emphasize that we are dealing with microscopic systems rather than macroscopic. Until we reach Chapter 14, I will just use Q.

7.6.1 Degeneracies

These formulae are correct even when the energy levels are degenerate (i.e. each energy level corresponds to a number of quantum states), but of course the sums run over the quantum states and not the energy levels. In the above formulae, N_i is the number of particles in one of the quantum states with energy ε_i and many authors choose to write the sums over the energy levels instead. For this purpose, they introduce a degeneracy factor g_i for each energy level and so

$$N_i = \frac{N g_i \exp\left(-\dfrac{\varepsilon_i}{k_B T}\right)}{\displaystyle\sum_{i=1}^{p} G_i \exp\left(-\dfrac{\varepsilon_i}{k_B T}\right)} \tag{7.26}$$

where it is understood that the sum runs over distinct energy levels and not the individual quantum states. I have used the symbol n in the summation to indicate the number of available energy levels.

7.6.2 Safety in Numbers

I have deliberately left a number of loose ends in the discussion. For example, you might be wondering how we can be sure that the Boltzmann distribution is correct, and that it will give reliable results? Where did the temperature come from, and what are the Ns in this distribution? Why does the internal energy of a sample stay so constant?

The first thing to say is that the Ns in expressions like

$$N \propto \exp\left(-\frac{\varepsilon}{kT}\right)$$

are *averages*. For the minute, I want to gloss over the difference between a time average and an average over a large number of identical systems. But nevertheless these numbers are averages and we should note that a macroscopic system contains far too many particles for us to ever know the true number having a given energy ε. So I should write these expressions as

$$\langle N \rangle \propto \exp\left(-\frac{\varepsilon}{kT}\right) \tag{7.27}$$

The average number of particles with energy ε fluctuates at random; the average population is given correctly by Boltzmann's law but the actual population fluctuates. It is sometimes greater than the average and sometimes smaller. The fluctuations about the mean are unpredictable. Do these fluctuations matter, when we want to predict macroscopic properties? To finish off this section, I want to show you that these fluctuations are so small that they are truly negligible.

First of all, let me establish that there is truly safety in numbers. Consider tossing a number of coins a certain number of times. I will call one side of the coin(s) the 'head' side (because it might well have a portrait of a monarch) and the other side the 'tail' side. In this experiment, I am going to throw the coin(s) in the air and record whether the coin lands 'heads' or 'tails' uppermost. I then repeat the experiment (say) 10 times. There is no particular reason for the coin to land either heads or tails uppermost, and so the theoretical probability of either event is $\frac{1}{2}$.

Table 7.18 gives typical results for a single coin to come up heads (H) or tails (T).

If I plot out the results (Figure 7.8) we see that the result is of course unpredictable. We cannot say what each result will be.

I now increase the number of coins to 10 and repeat the experiment again 10 times. That is to say, I take 10 coins and throw them into the air simultaneously. I record the result and then repeat the experiment a further nines times. The results of my experiments (which of course I simulated with a random number generator) are given in Table 7.19.

Table 7.18 Single coin results

Experiment	1	2	3	4	5	6	7	8	9	10
Outcome	T	T	T	H	H	T	H	T	T	T

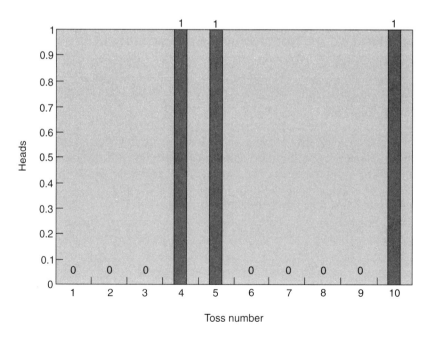

Figure 7.8 Single coin tossed 10 times

Table 7.19   Results from 10 coins

Experiment	1	2	3	4	5	6	7	8	9	10
Heads	3	5	5	2	6	7	7	5	2	8
Tails	7	5	5	8	4	3	3	5	8	2

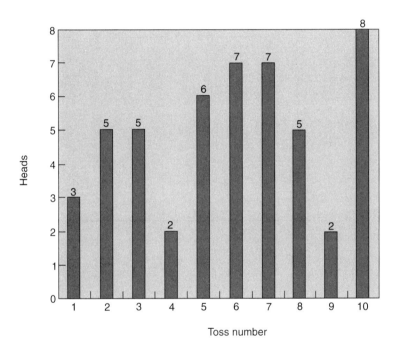

Figure 7.9   Ten coins tossed 10 times

Table 7.20   Results from 10 coins

Experiment	1	2	3	4	5	6	7	8	9	10
Heads	52	42	48	49	57	52	47	52	58	53
Tails	48	58	52	51	43	48	53	48	42	47

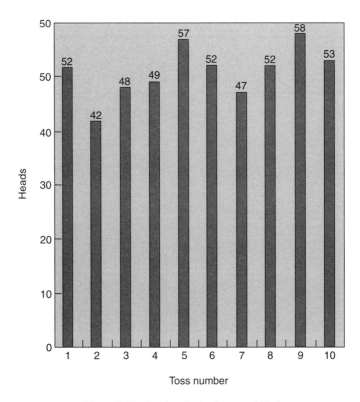

Figure 7.10   One hundred coins tossed 10 times

Figure 7.9 shows the results plotted.

We would expect an average of five heads per throw but we see a significant amount of scatter. I now repeat the 10 experiments with 100 coins. The results are given in Table 7.20 and Figure 7.10. You can see that there is much less scatter about the mean. The point is that there is safety in numbers; if we use 100 coins we can be pretty sure of getting a result between 40 and 60 heads.

To return to the energy problem, because of the very large number of particles that go to make up a macroscopic sample, we can be very certain that fluctuations about the mean are negligible.

8 Applications of the Boltzmann Distribution

In Chapter 7 I investigated the way in which energy is distributed amongst available quantum states for a collection of distinguishable particles in thermal equilibrium. The key formula is

$$N_i = \frac{N \exp\left(-\dfrac{\varepsilon_i}{k_B T}\right)}{\displaystyle\sum_{i=1}^{p} \exp\left(-\dfrac{\varepsilon_i}{k_B T}\right)} \tag{8.1}$$

which relates the number of distinguishable particles N_i in a quantum state of energy ε_i to the total number of particles N and the thermodynamic temperature T. The formula assumes that the quantum states $\varepsilon_1, \varepsilon_2, \ldots, \varepsilon_p$ are discrete rather than continuous, and that they are finite in number. There are correspondingly more complicated formulae otherwise.

As I explained in Chapter 7, it is often more convenient to write the above formula in terms of the energy levels, which I label $\varepsilon_1, \varepsilon_2, \ldots, \varepsilon_n$, and their degeneracy factors $g_1, g_2, \ldots, g_n$ rather than the quantum states. This gives a corresponding expression for the number of distinguishable particles N_i' in an energy level of energy ε_i.

$$N_i' = \frac{N g_i \exp\left(-\dfrac{\varepsilon_i}{k_B T}\right)}{\displaystyle\sum_{i=1}^{p} g_i \exp\left(-\dfrac{\varepsilon_i}{k_B T}\right)} \tag{8.2}$$

I have put N_i' rather than N_i to remind you that the two numbers so calculated will generally be different.

I also mentioned that the denominator had a special name and significance; it is the partition function Q given by

$$Q = \sum_{i=1}^{p} \exp\left(-\dfrac{\varepsilon_i}{k_B T}\right) \tag{8.3}$$

(in terms of the quantum states) or

$$Q = \sum_{i=1}^{n} g_i \exp\left(-\frac{\varepsilon_i}{k_B T}\right) \tag{8.4}$$

when written in terms of the energy levels.

I want to spend a little time in this chapter illustrating the use of this important formula; in particular, I want to show you how we can derive results that can be tested directly by experiment.

8.1 THE TWO-LEVEL QUANTUM SYSTEM

First of all, I want to consider a simple quantum system, which consists of just two non-degenerate energy levels.

In Chapter 4, we discussed the concept of a point charge and the fields due to such point charges. If we aggregate point charges together, we get electric multipoles. Electrical and magnetic phenomena have much in common, and indeed they can be rationalized by Maxwell's equations. One feature that differs in electrical and magnetic phenomena is that there do not appear to be any fundamental particles responsible for magnetism in the sense that electrons and protons are responsible for electric effects. The smallest magnetic 'unit' is a magnetic dipole. A simple example of a magnetic dipole is afforded by a small electrical current loop of area A as illustrated below. The magnetic dipole $\mathbf{p}_m$ associated with this small loop is

$$\mathbf{p}_m = IA\mathbf{n} \tag{8.5}$$

where $\mathbf{n}$ is a unit vector perpendicular to the current loop.

If the current loop (Figure 8.1) is due to a particle of charge Q and mass M moving with angular momentum $\mathbf{l}$, then we find

$$\mathbf{p}_m = \frac{Q}{2M}\mathbf{l} \tag{8.6}$$

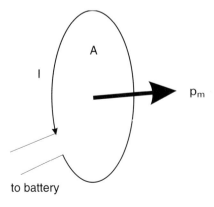

Figure 8.1   A current loop is a magnetic dipole

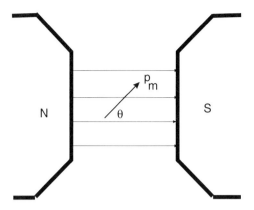

Figure 8.2   Magnetic dipole in external induction

In the presence of an external magnetic induction **B**, the magnetic dipole has an interaction energy

$$U = -\mathbf{p}_m \cdot \mathbf{B} \tag{8.7}$$

and each different alignment of the dipole with the induction therefore corresponds to a different potential energy (Figure 8.2).

Apart from any angular momentum that they might have on account of their motion in an atom, electrons possess an intrinsic angular momentum called electron spin, and hence they have an intrinsic magnetic dipole moment. The spin vector is written **s**, and we find that

$$\mathbf{p}_s = g \frac{(-e)}{2m_e} \mathbf{s} \tag{8.8}$$

where I have written $\mathbf{p}_s$ rather than $\mathbf{p}_m$ to indicate a spin result. The g-factor is needed to correct the classical expression given above for a quantum mechanical effect. It is called the Landé g-factor, and it is about 2.0023 for a free electron.

One of the many interesting results of quantum mechanical theory is that the angle θ between the electron spin dipole and the applied magnetic induction can only have one of two possible values, and the energies associated these alignments are given by

$$\varepsilon_{ms} = m_s \frac{h}{2\pi} g_e \frac{e}{2m_e} B \tag{8.9}$$

where m_s is the spin quantum number, which takes values of $\pm 1/2$. The lower energy state is therefore that with $m_s = -1/2$, the higher energy state that corresponding to $m_s = \pm 1/2$.

The energy differences between these two alignments are very small, even for the magnetic induction generated by a powerful electromagnet, and microwave radiation has to be used to investigate transitions between them. These considerations form the basis for the experimental technique known as electron spin resonance spectroscopy.

Consider then a sample of N non-interacting electrons (for example, a hydrogen atom has a single electron so I could use N hydrogen atoms) which I place in a constant magnetic induction. In order to keep the temperature constant, I will assume that the sample is surrounded by a heat bath. I will call the two energy levels ε_u and ε_l (upper and lower), and so $\Delta\varepsilon = \varepsilon_u - \varepsilon_l$ is fixed, since the temperature is constant.

If N_u and N_l are the number of electrons in the upper and lower energy levels, then

$$N = N_u + N_l$$

and the magnetic internal energy of the system is

$$U = N_u \varepsilon_u + N_l \varepsilon_l$$

We also know from the Boltzmann distribution law

$$N_i = \frac{N \exp\left(-\dfrac{\varepsilon_i}{k_B T}\right)}{Q} \tag{8.10}$$

In order to simplify the algebra, I will take the magnetic energy zero to be ε_l and the energy difference $\varepsilon_u - \varepsilon_l$ to be D. With this notation, explicit expressions for N_u and N_l are as follows

$$N_u = \frac{N \exp\left(-\dfrac{D}{k_B T}\right)}{1 + \exp\left(-\dfrac{D}{k_B T}\right)} \tag{8.11}$$

and

$$N_l = \frac{N}{1 + \exp\left(-\dfrac{D}{k_B T}\right)} \tag{8.12}$$

I have plotted N_u/N and N_l/N as a function of the reduced temperature $k_B T/D$ (which is dimensionless) in Figure 8.3. The upper curve corresponds to the lower energy state, and the lower curve corresponds to the higher energy state.

8.1.1   The Internal Energy

The internal energy U is just N times the average energy $\langle \varepsilon \rangle$ due to the magnetic interaction plus some undetermined constant U_0. Now

$$U = U_0 + N_l \varepsilon_l + N_u \varepsilon_u$$

If I ignore U_0 then substitution gives

$$\langle \varepsilon \rangle = \frac{D \exp\left(-\dfrac{D}{k_B T}\right)}{1 + \exp\left(-\dfrac{D}{k_B T}\right)} \tag{8.13}$$

A plot of $\langle \varepsilon \rangle/D$ *versus* the reduced temperature is identical to the plot for N_u/N *versus* the reduced temperature.

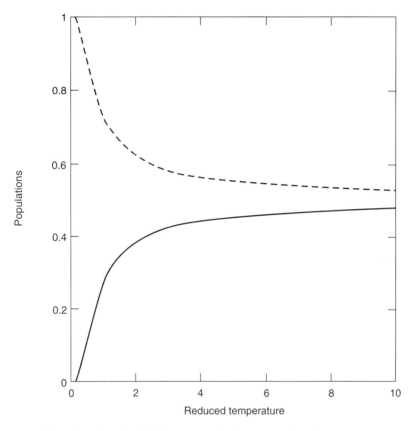

Figure 8.3 Fractional Boltzmann populations *versus* reduced temperature

8.1.2 The Heat Capacity

Scientists are rarely interested in the internal energy and its variation with temperature; a much more interesting quantity to study is the heat capacity at constant volume

$$C_V = \left(\frac{\partial U}{\partial V} \right)_V$$

So

$$C_V = DN \frac{\partial}{\partial T} \left[\frac{\exp\left(-\dfrac{D}{k_B T} \right)}{1 + \exp\left(-\dfrac{D}{k_B T} \right)} \right]_V$$

This gives

$$C_V = \frac{D^2 N}{k_B T^2} \left\{ \frac{\exp\left(-\dfrac{D}{k_B T} \right)}{\left[1 + \exp\left(-\dfrac{D}{k_B T} \right) \right]^2} \right\}$$

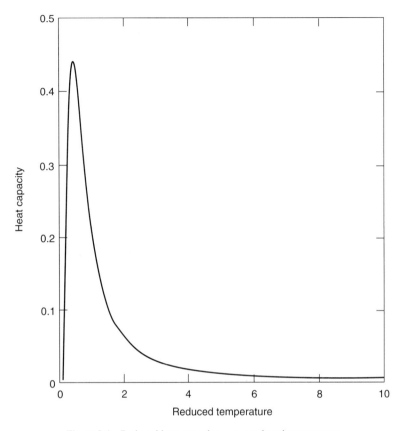

Figure 8.4   Reduced heat capacity *versus* reduced temperature

which I can write as

$$C_V = Nk_B \left(\frac{D}{k_B T}\right)^2 \left\{ \frac{\exp\left(-\dfrac{D}{k_B T}\right)}{\left[1 + \exp\left(-\dfrac{D}{K}\right)\right]^2} \right\} \tag{8.14}$$

A plot of the reduced heat capacity C_V/Nk_B *versus* the reduced temperature $k_B T/D$ is shown in Figure 8.4.

There is a very sharp hump in the heat capacity curve, and this is called a *Schottky anomaly*. Such anomalies have been observed in the heat capacity data of certain simple solids; they do not depend on the presence or otherwise of an external magnetic induction.

8.2   THE MAXWELL–BOLTZMANN DISTRIBUTION OF VELOCITIES AND SPEEDS

The thermal energies of particles manifest themselves as their kinetic energy of motion. Suppose we have a sample of particles at a fixed temperature, and we enquire about the distribution of velocities and speeds amongst these particles.

Maxwell first solved this problem in 1860, by making use of the Boltzmann distribution. For this reason, the solution is known as the *Maxwell–Boltzmann distribution of molecular speeds*. Maxwell solved the problem long before experimental data became available.

First of all, let us consider the simple one-dimensional case where each particle has a velocity

$$\mathbf{u}_i = u_i \mathbf{e}_x$$

and a kinetic energy

$$\varepsilon_i = \frac{1}{2} m u_i^2$$

I have assumed that the sample consists of particles each of mass m. Their velocities are distributed at random and so u_i can be positive (for a particle moving in the $+x$ direction) or negative (for a particle moving in the $-x$ direction).

First of all, I want to calculate the mean velocity. If the particles had (say) just two possible vector velocities $\mathbf{u}_1$ and $\mathbf{u}_2$ then the average vector velocity would be

$$(N_1 + N_2)\langle\mathbf{u}\rangle = N_1\mathbf{u}_1 + N_2\mathbf{u}_2$$

and we would find

$$\langle\mathbf{u}\rangle = \frac{\mathbf{u}_1 \exp\left(-\dfrac{mu_1^2}{2k_\mathrm{B}T}\right) + \mathbf{u}_2 \exp\left(-\dfrac{mu_2^2}{2k_\mathrm{B}T}\right)}{\exp\left(-\dfrac{mu_1^2}{2k_\mathrm{B}T}\right) + \exp\left(-\dfrac{mu_2^2}{2k_\mathrm{B}T}\right)} \tag{8.15}$$

by analogy with the energy formula given above. Since we are only considering for the minute a one-dimensional case, I can write

$$\langle u\rangle = \frac{u_1 \exp\left(-\dfrac{mu_1^2}{2k_\mathrm{B}T}\right) + u_2 \exp\left(-\dfrac{mu_2^2}{2k_\mathrm{B}T}\right)}{\exp\left(-\dfrac{mu_1^2}{2k_\mathrm{B}T}\right) + \exp\left(-\dfrac{mu_2^2}{2k_\mathrm{B}T}\right)}$$

The velocities are however continuous; they can take any value from minus infinity to plus infinity and the sums have to be replaced by integrals

$$\langle u\rangle = \frac{\displaystyle\int_{-\infty}^{\infty} u \exp\left(-\dfrac{mu^2}{2k_\mathrm{B}T}\right) \mathrm{d}u}{\displaystyle\int_{-\infty}^{\infty} \exp\left(-\dfrac{mu^2}{2k_\mathrm{B}T}\right) \mathrm{d}u} \tag{8.16}$$

The denominator is a standard integral, which can be found in any compilation of such standard integrals. We find then

$$\langle u\rangle = \left(\frac{m}{2\pi k_\mathrm{B}T}\right)^{1/2} \int_{-\infty}^{\infty} \exp\left(-\dfrac{mu^2}{2k_\mathrm{B}T}\right) \mathrm{d}u \tag{8.17}$$

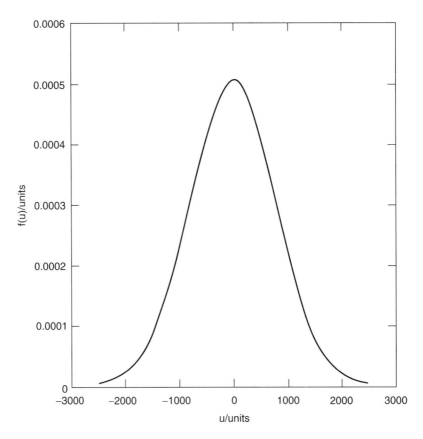

Figure 8.5   Helium velocity distribution function f(u) at 300 K

and this integral must obviously come to 0 by symmetry; there are as many particles moving in either direction with the same speed u.

That this integral is in fact zero can be checked once again from any standard compilation of integrals.

The integrand

$$f(u) = \left(\frac{m}{2\pi k_{B}T}\right)^{1/2} \exp\left(-\frac{mu^2}{2k_{B}T}\right) \tag{8.18}$$

is called the *velocity distribution function* f(u), and it is the distribution that was discovered by Maxwell and Boltzmann. The product f(u) du gives the fraction of the total number of particles having velocities in the range u to $u + \mathrm{d}u$. f(u) is plotted against u in Figure 8.5 for helium at 300 K.

It is just a Gaussian (or normal) distribution curve, and the average comes to zero. As mentioned above, the physical interpretation is that f(u) du gives the proportion of particles having velocities between u and $u + \mathrm{d}u$. The area under the curve must therefore come to 1 in order to represent a probability, as indeed it does.

In three dimensions the velocity vector can be written

$$\mathbf{u} = u_x\mathbf{e}_x + u_y\mathbf{e}_y + u_z\mathbf{e}_z$$

and since space is isotropic, each component should have a velocity distribution function identical to the one given above. In three dimensions, we are interested in the probability of the proportion of particles having simultaneously x components of their velocities between u_x and $u_x + du_x$, y components between u_y and $u_y + du_y$ and z components between u_z and $u_z + du_z$. The three velocity distributions must be independent of each other and so the velocity distribution function is

$$\mathrm{f}(u_x)\mathrm{f}(u_y)\mathrm{f}(u_z) = \left(\frac{m}{2\pi k_B T}\right)^{3/2} \exp\left(-\frac{mu_x^2}{2k_B T}\right) \exp\left(-\frac{mu_y^2}{2k_B T}\right) \exp\left(-\frac{mu_z^2}{2k_B T}\right)$$

or in other words

$$\mathrm{f}(u_x)\mathrm{f}(u_y)\mathrm{f}(u_z) = \left(\frac{m}{2\pi k_B T}\right)^{3/2} \exp\left(-\frac{mu^2}{2k_B T}\right)$$

where in this case

$$u^2 = u_x^2 + u_y^2 + u_z^2$$

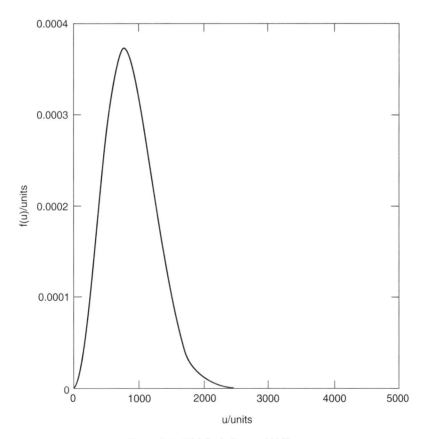

Figure 8.6 $\mathrm{F}(u)$ for helium at 300 K

and

$$f(u_x)f(u_y)f(u_z)\,\mathrm{d}u_x\,\mathrm{d}u_y\,\mathrm{d}u_z$$

gives the probability that the velocity vector **u** has components lying in the volume element $\mathrm{d}u_x\,\mathrm{d}u_y\,\mathrm{d}u_z$.

Most experimental measurements relate to the speed of the molecules rather than the velocities. The speed is the scalar magnitude of the velocity vector **u**. In particular, the quantity of experimental interest is the speed irrespective of the direction of motion. In order to find this, we have to sum all the probabilities given above over all volume elements subject to the restriction that the speed is constant.

Changing to polar coordinates and integrating over the polar angles we find a distribution given by

$$F(u) = 4\pi u^2 \left(\frac{m}{2\pi k_B T}\right)^{3/2} \exp\left(-\frac{mu^2}{2k_B T}\right) \tag{8.19}$$

and this is the famous Maxwell–Boltzmann distribution of molecular speeds.

Figure 8.6 gives the graph for helium at 300 K.

$F(u)$ depends on the mass of the particles and on the temperature. Figure 8.7 compares helium at 300 K and at 1000 K. Notice that the peak moves significantly to the right as the temperature increases.

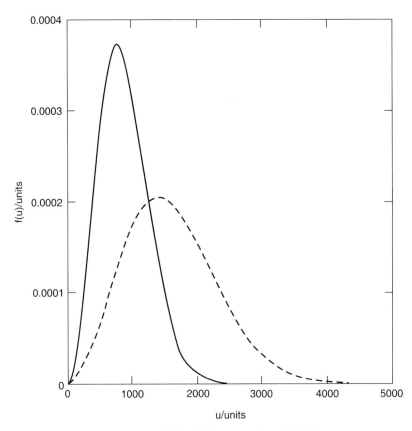

Figure 8.7 $F(u)$ for helium at 300 K and 1000 K

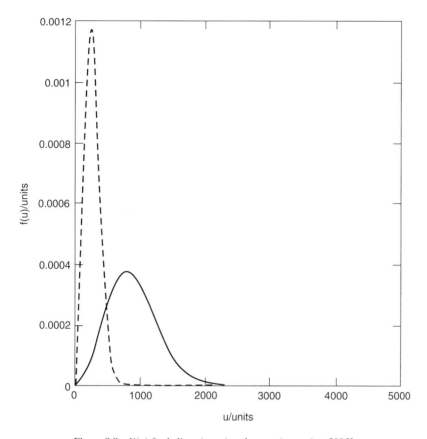

Figure 8.8 $F(u)$ for helium (——) and argon (- - - - -) at 300 K

Finally, Figure 8.8 compares the distribution curves for helium and argon at 300 K. These curves are in good agreement with experiment.

There are several ways of quoting the speeds of molecules. The *mean speed* is the mean of the speeds calculated using the Maxwell–Boltzmann distribution

$$\langle u \rangle = \int_0^\infty u F(u)\, du \tag{8.20}$$

and this works out to be

$$\langle u \rangle = \left(\frac{8 k_B T}{\pi m} \right)^{1/2} \tag{8.21}$$

The root mean square speed (usually written c, not to be confused with the speed of light) is the square root of the average values of u^2

$$c^2 = \int_0^\infty u^2 F(u)\, du$$

and this works out as

$$c = \left(\frac{3 k_B T}{m} \right)^{1/2} \tag{8.22}$$

The most probable speed (usually written $c*$) is the speed at which the Maxwell–Boltzmann distribution passes through a maximum. It turns out to be

$$c* = \left(\frac{2k_B T}{m}\right)^{1/2} \tag{8.23}$$

The three values differ by small numerical factors.

9 Modeling Simple Solids (ii)

For many years the subject of materials science consisted of nothing more than the accumulation of descriptive wisdom. It has been known since prehistoric times that metals differ from other materials in that they can be 'worked', but the underlying reasons for this behavior have only become understood in the last 100 years.

These days we want to find out *why* materials behave the way they do, and ideally we want to make use of this knowledge in designing new materials. The phrase 'materials engineering' has been coined to describe such endeavors.

9.1 YOUNG'S MODULUS

I described earlier the Hooke's law spring; we take a spring and subject its ends to a stretching (or *tensile*) force F_T. A plot of the force *versus* the extension x is sketched in Figure 9.1.

Not only is the extension directly proportional to the tensile force but the spring returns to its original length when the force is removed.

The results of such tests on a variety of materials show that there are three distinct ways in which a solid can respond to a tensile force; depending on the nature of the material and the magnitude of the force, the response is elastic, brittle or ductile.

(i) In *elastic* behavior the elongation is reversible and when F_T is removed the sample returns to its original shape. For many substances the extension is directly proportional to F_T as above, but this is not always the case. Rubber is an example of an elastic material whose extension does not depend linearly on the tensile force (Figure 9.2).

Most materials show elastic behavior for small applied forces.

(ii) In *brittle* behavior, the response of the sample is elastic up to a certain point called the yield point (denoted by an asterisk in Figure 9.3), where it suddenly breaks.

This behavior is shown by materials such as glass.

(iii) In *ductile* behavior, typical of metals, we find a typical F_T *versus* x curve (Figure 9.4).

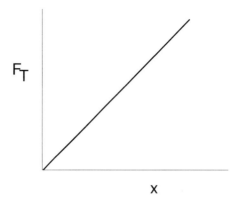

Figure 9.1   Hooke's law behavior

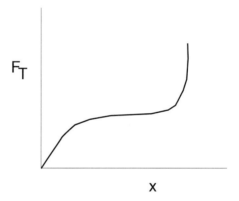

Figure 9.2   Behavior of rubber

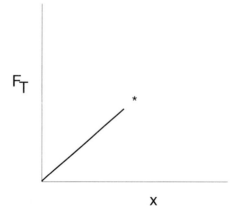

Figure 9.3   Brittle behavior

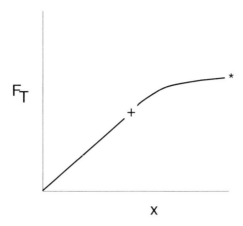

Figure 9.4 Ductile behavior

The extension is elastic and the extension is proportional to the tensile load until point '+'. After that point, further loads produce *plastic* deformations (where the sample does not return to its former length if the load is taken off) and finally the material breaks at the point denoted by an asterisk.

The first practical investigations of elastic phenomena were carried out by Robert Hooke, who found that for most materials the extension x was directly proportional to the tensile force F_T, and that the test samples recovered their original shape when the load was removed.

A fuller statement of Hooke's results is

$$\frac{F_T}{A} \propto \frac{x}{l_0} \tag{9.1}$$

which relates the tensile force per unit area to the fractional increase in length of the sample. A is the cross-sectional area of the sample, and l_0 the unstressed length.

The quantity $\epsilon_T = x/l_0$ is called the *tensile strain* whilst $\sigma_T = F_T/A$ is called the *tensile stress*. Hooke's results can be written

$$\sigma_T = Y\epsilon_T \tag{9.2}$$

where Y is called the Young's modulus of the material. Y is actually a tensor property, because the stress and the strain need not be in the same direction. For example, the Young's moduli of wood and graphite vary markedly with direction.

Materials can also be studied by the application of a couple or shearing torque, and this is quantified by the shear modulus G.

The specimen in Figure 9.5 below has been sheared by the application of a shearing torque, whose effect is to move the two parallel faces in opposite directions without noticeably changing the distance between them. The shear stress is defined as the force per unit area (shaded area)

$$\sigma_S = \frac{F_S}{A_S} \tag{9.3}$$

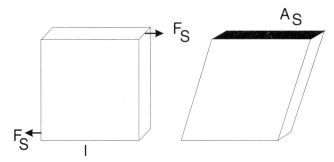

Figure 9.5   A shearing couple

whilst the shear strain is defined as the distance x moved by the heavily shaded edge divided by the length of the sample l,

$$\epsilon_S = \frac{x}{l}$$

Just as Hooke's law relates the elastic behavior of many materials that are subject to a tensile force, so there is a corresponding law relating the response of many materials to a shearing force

$$\sigma_S = G\epsilon_S \tag{9.4}$$

We often refer to Y and G as the elastic moduli of the material.

Table 9.1 shows values of the Young's modulus Y for a selection of substances. The obvious division of materials is into crystalline and amorphous materials. Rubber is a special case, and I have devoted Chapter 20 to modeling such polymeric materials.

An essential feature of crystalline materials such as metals or NaCl is that the molecules, atoms or ions are packed into a highly regular arrangement. When such a material is stretched or compressed, we can to a good approximation consider just the interatomic potential energy curves. As an empirical fact, the deeper the potential energy minimum the higher the melting point and the greater the Young's modulus. The shallower the curve, the lower the melting point and the smaller the Young's modulus.

To stretch a crystalline solid, every bond along the torsion direction has to be elongated. In an amorphous solid, the bonds are randomly oriented and to stretch the solid only a small proportion of the bonds need to be elongated. For this reason, glass behaves elastically for only very low strains and we normally think of glass as a brittle substance.

Table 9.1   Young's modulus for some simple materials

Substance	$Y\ (\mathrm{N\,m^{-2}})$
Diamond	83.0
Cu metal	13.0
Ag metal	7.8
Glass	7.1
NaCl crystal	4.4
Natural rubber	0.000 7

9.2   YOUNG'S MODULUS FOR A CRYSTALLINE SOLID

I am going to model the Young's modulus for a simple crystalline solid such as NaCl. Consider then a simple cubic ionic crystal comprising n columns of ions, each nearest pair of which is separated by distance a.

If the ionic radius is a, then roughly na^2 is the area A of the top of the crystal shown in Figure 9.6 and so $n = A/a^2$.

If the box is subject to a tensile force F_T along the vertical axis, then at equilibrium there will be an equal and opposite restoring force from within the crystal. The effect of this restoring force must be distributed across all the columns of the crystal (assuming that n is sufficiently large) and so

$$F_R = -\frac{F_T}{n}$$

or I can expand the restoring force as a Taylor series in $r - a$, the increase in ion spacing within the column,

$$F_R(r) = F_R(r = a) + \left(\frac{\mathrm{d}F_R}{\mathrm{d}r}\right)_{r=a}(r - a) + \frac{1}{2}\left(\frac{\mathrm{d}^2F_R}{\mathrm{d}r^2}\right)(r - a)^2 + \cdots$$

At equilibrium, when $r = a$, the force F_R is zero and so we find

$$\left(\frac{\mathrm{d}F_R}{\mathrm{d}r}\right)_{r=a}\frac{(r - a)}{a^2} = -\frac{F_T}{A} \tag{9.5}$$

which can be rearranged to give an expression for the Young's modulus in terms of the

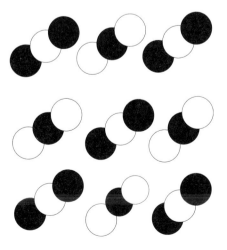

Figure 9.6   A set of nine columns each of three ions

restoring force

$$Y = -\frac{1}{a}\left(\frac{\mathrm{d}F_R}{\mathrm{d}r}\right)_{r=a} \tag{9.6}$$

All that remains is for me to link this with the pair potential of Chapter 5.

I showed you how to derive the potential energy of a single ion due to its interaction with the rest of the crystal

$$\varepsilon^{\mathrm{pot}} = -\frac{Q^2}{4\pi\epsilon_0}\frac{\alpha}{a} + z\frac{C}{a^{12}}$$

where a is the distance between nearest neighbor ions, z the coordination number (6 for NaCl), and α the Madelung constant.

First of all, let me eliminate one constant C from the expression, as follows. If I generalise the expression above to involve the distance r between nearest neighbor ions rather than a

$$\varepsilon^{\mathrm{pot}}(r) = -\frac{Q^2}{4\pi\epsilon_0}\frac{\alpha}{r} + z\frac{C}{r^{12}}$$

then the energy must be a minimum when $r = a$.

Differentiating $\varepsilon^{\mathrm{pot}}$ and setting the differential to zero at $r = a$ gives

$$C = \frac{Q^2 a^{11}\alpha}{4\pi\epsilon_0 12z}$$

and so we write

$$\varepsilon^{\mathrm{pot}}(r) = \frac{Q^2\alpha}{4\pi\epsilon_0}\left(-\frac{1}{r} + \frac{a^{11}}{12r^{12}}\right) \tag{9.7}$$

This potential arises because of the interaction of a single ion (of charge Q) with the rest of the crystal, it is certainly not the pair potential. If I make a rough and ready assumption that the only pair terms that contribute to $\varepsilon^{\mathrm{pot}}$ are the interactions between any one ion and its z nearest neighbors each at a distance a then

$$\varepsilon^{\mathrm{pair}}(r) = \varepsilon^{\mathrm{pot}}(r)/z$$

$$\varepsilon^{\mathrm{pair}}(r) = \frac{Q^2\alpha}{4\pi\epsilon_0 z}\left(\frac{a^{11}}{12r^{12}} - \frac{1}{r}\right) \tag{9.8}$$

Differentiating this expression twice with respect to r and substituting $r = a$ gives the final result

$$Y = \frac{11Q^2\alpha}{4\pi\epsilon_0 za^4}$$

or in other words, the Young's modulus is proportional to $1/a^4$, the interatomic spacing, and also to the square of the charge on each ion.

9.3   THE MELTING POINT OF A SIMPLE SOLID

A simple solid is one that comprises atoms, ions or spherical molecules that oscillate independently about their fixed equilibrium position. I am going to show how we can make predictions about the melting points of solids, using a very simple model of the vibrational motion.

I will use a Hooke's law model for the melting phenomenon, and so I will need to make an estimate for the force constant k_s. As described above, this can be obtained from a knowledge of the Young's modulus for the material. I can also find the mean distance between the particles in my simple solid from crystallographic studies.

A simple solid is depicted in Figure 9.7.

Each of the particles that goes to make up the solid is initially fixed at some point in the crystal lattice. At low temperatures, these particles vibrate about their equilibrium positions, but as we raise the temperature the oscillations become more violent until eventually the solid melts.

This simple observation gives a very clear description of the mechanism of melting, and so our model is a very simple one; we assume that the particles undergo simple harmonic motion in three dimensions about some equilibrium position. We need to find a temperature at which the solid melts.

Each atom vibrates about its equilibrium position under the influence of all the other atoms in the lattice. These other atoms generate the potential energy.

Just for the minute, consider the motion along one of the axes, say the x-axis.

The energy of a single atom is

$$\varepsilon_{tot} = \varepsilon_{kinetic} + \varepsilon_{vib}$$

$$\varepsilon_{tot} = \tfrac{1}{2}mv_x^2 + \tfrac{1}{2}k_s(x - x_e)^2$$

Figure 9.7   Vibration of a simple solid

where x_e is the equilibrium value of x. According to classical physics, the average translational energy of any atom or molecule, whether it is in the gaseous, liquid or solid phase is always $\frac{1}{2}k_B T$ per one-dimensional degree of freedom (do not confuse the Boltzmann constant k_B with the spring constant k_s).

But, because the particle is executing simple harmonic motion, the average value of the potential energy is equal to the average kinetic energy

$$\langle \varepsilon_{pot} \rangle = \langle \varepsilon_{trans} \rangle$$

and so the average value of the one-dimensional potential energy is also $\frac{1}{2}k_B T$.

By analogy with the one-dimensional problem, the potential energy of a three-dimensional simple harmonic oscillator is

$$U = \tfrac{1}{2}k_s[(x - x_e)^2 + (y - y_e)^2 + (z - z_e)^2]$$

and according to classical physics, the average value is $\frac{3}{2}k_B T$. This means that

$$\tfrac{1}{2}k_s(\langle(x - x_e)^2\rangle + \langle(y - y_e)^2\rangle + \langle(z - z_e)^2\rangle) = \tfrac{3}{2}k_B T \qquad (9.9)$$

Space is isotropic, and it is reasonable to suppose that the mean value $\langle(x - x_e)^2\rangle$ is exactly the same as $\langle(y - y_e)^2\rangle$ and of $\langle(z - z_e)^2\rangle$, and so

$$\tfrac{3}{2}k_s\langle(x - x_e)^2\rangle = \tfrac{3}{2}k_B T$$

or in other words

$$\langle(x - x_e)^2\rangle = \frac{k_B T}{k_s}$$

If we take the square root of both sides of this equation, we get an expression for the root mean square value of x, x_{rms}.

$$x_{rms} = \sqrt{\frac{k_B T}{k_s}} \qquad (9.10)$$

and this gives the temperature dependence of x_{rms} in a simple solid. This equation predicts that x_{rms} should increase with increasing temperature, in agreement with our simple physical intuition. At a certain temperature, the particles move about so much that the solid melts. All we have to do is to agree the value of x_{rms} which will define melting, and the figure is somewhat arbitrary. Most authors agree that melting occurs when x_{rms} achieves a percentage of the spacing d between the particles in the solid, typically $x_{rms} = 0.10\,d$.

This model is often referred to as the Lindemann theory of melting, and we can write the conclusion mathematically as

$$T_{melt} = k_s d^2 / (100\, k_B) \qquad (9.11)$$

Table 9.2. Predictions of Lindemann theory

Solid	d (pm)	k_s (N m^{-1})	T_{melt} (theory) (K)	T_{melt} (expt) (K)
Copper	230	30	1163	1356
Gold	260	20	991	1336
Lead	310	5	352	600
Silver	260	21	1041	1234
Tungsten	250	100	4582	3650

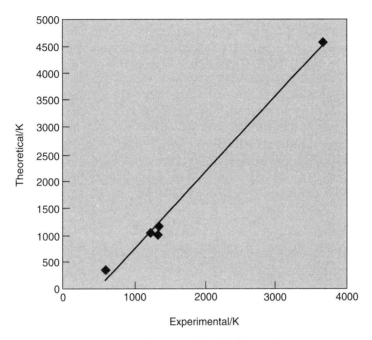

Figure 9.8 Linear regression, Lindemann theory

Does this agree with experiment? First of all, we have to estimate the force constant for our solids. We can do this indirectly by measuring the Young's modulus, as I explained earlier. Then, we have to choose a value of d. We can use crystal data for this, and a sample of data is shown in the Table 9.2. The agreement with experiment is not wildly exciting, but we seem to be on the right track (Figure 9.8).

9.3.1 A Consequence of the Zero Point Energy

The quantum mechanical treatment of a simple harmonic oscillator is certainly very different to the classical treatment; the quantum treatment gives energy quantization and in particular the zero point energy. What evidence is there for the zero point energy? The inert gases are very hard to solidify, even at very low temperatures and very high external pressures. The intermolecular forces are very small, and if we imagine a solid inert gas as a simple solid then the zero point energy will be large because of the small force constant. A consequence is that the atoms have a significant residual vibrational energy even at very low temperatures, and so they do not readily bind together to form a stable solid.

9.4   THE DULONG–PETIT RULE

It has been known for very many years that many simple solids have a molar heat capacity at constant volume of about $25\,\mathrm{J\,K^{-1}\,mol^{-1}}$ at room temperature. According to the classical equipartition of energy principle, an atom vibrating in three dimensions in a crystal should have internal energy $U = 1/2\,k_\mathrm{B}T$ ($1/2\,RT$ per mol) per degree of freedom. A particle of mass m vibrating in one dimension under the influence of a Hooke's Law potential has a total vibrational energy

$$\varepsilon_\mathrm{vib} = \frac{1}{2}\,m\left(\frac{\mathrm{d}x}{\mathrm{d}t}\right)^2 + \frac{1}{2}\,k_\mathrm{s}x^2$$

which depends on two 'squared' terms. In the simple solid, this is the total energy of each atom, and as we saw earlier the average translational and vibrational energies are equal. The molar internal energy of an atom vibrating in three dimensions in a crystal should therefore be six times $\frac{1}{2}k_\mathrm{B}T$ giving

$$U_\mathrm{m} = 3RT$$

The molar heat capacity at constant volume should therefore be

$$C_{V,\mathrm{m}} = 3R \approx 25\,\mathrm{J\,K^{-1}\,mol^{-1}}$$

and this simple result is known as the Dulong and Petit law. Accurate experimental tests of the Dulong and Petit law only became available during Einstein's time, when it became apparent that many substances deviated strongly from the Dulong and Petit law, especially at low temperatures. Every metal studied was found to have a molar hear capacity at constant volume much less than the value $3\,R$, and this heat capacity appeared to approach zero as the temperature was reduced to $0\,\mathrm{K}$.

We can use the principles learned in earlier sections to calculate this heat capacity. Starting from the Boltzmann distribution law

$$N_i = \frac{N \exp\left(-\dfrac{\varepsilon_i}{k_\mathrm{B}T}\right)}{Z}$$

and a knowledge of the vibrational energy levels, what we have to do is to calculate the total energy using the quantum mechanical formula

$$U = \sum_{\nu=0}^{\infty} \varepsilon_\nu N_\nu$$

and then partially differentiate in order to get the heat capacity at constant volume

$$C_V = \left(\frac{\partial U}{\partial T}\right)_V$$

Note that I have started the summation at 0, and written the summation index as v; this is because one-dimensional harmonic vibrational energies go as

$$\varepsilon_v = (v + \tfrac{1}{2})hc_0\omega_e$$

where the vibrational quantum number v runs from 0 to infinity.

Three-dimensional harmonic vibrational energies can likewise be written

$$\varepsilon_{vx,vy,vz} = (v_x + v_y + v_z + \tfrac{3}{2})hc_0\omega_e$$

where the three quantum numbers v_x, v_y and v_z individually run from 0 to infinity. The quantity ω_e is the fundamental vibrational wavenumber. The one-dimensional vibrational energy can also be written in terms of the fundamental vibrational frequency f_e

$$\varepsilon_v = (v + \tfrac{1}{2})hf_e$$

rather than the fundamental vibrational wavenumber.

The U summations look complicated, but they can be done analytically. Remembering the factor of 3 (for the three-dimensional problem), the total vibrational energy for a sample of N atoms turns out as

$$U = \frac{3}{2}\,Nhf_e + \frac{3Nhf_e \exp\left(-\dfrac{hf_e}{k_B T}\right)}{1 - \exp\left(-\dfrac{hf_e}{k_B T}\right)} \tag{9.12}$$

In order to find the heat capacity at constant volume, we have to differentiate with respect to temperature, keeping the volume constant. This gives

$$C_{V,m} = 3R\left(\frac{hf_e}{k_B T}\right)^2 \frac{\exp\left(-\dfrac{hf_e}{k_B T}\right)}{\left[1 - \exp\left(-\dfrac{hf_e}{k_B T}\right)\right]^2} \tag{9.13}$$

and this is the formula attributed to Einstein. I have plotted it in Figure 9.9 as a function of the reduced variable $x = k_B T/hf_e$. The quantity hf_e/k_B is very often called the *Debye temperature*.

The asymptotic behavior is clearly shown, and it can be deduced by expanding the

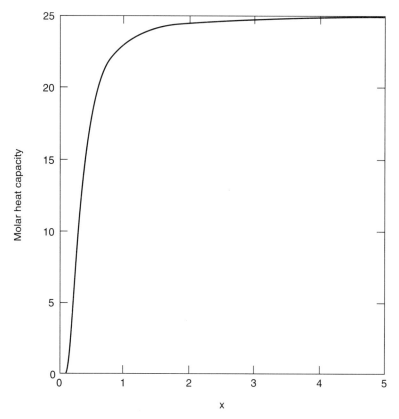

Figure 9.9 Einstein formula

exponentials in terms of $1/x$. This gives

$$
C_{V,\mathrm{m}} = 3Rx^2 \frac{\exp\left(-\dfrac{1}{x}\right)}{\left[1 - \exp\left(-\dfrac{1}{x}\right)\right]^2}
$$

$$
= 3Rx^2 \frac{1 - \dfrac{1}{x} + \cdots}{\left(1 - 1 + \dfrac{1}{x} + \cdots\right)^2}
$$

(9.14)

which becomes $3R$ in the limit of large x (high temperature).

This is consistent with the Dulong–Petit formula for atomic heat capacities.

The relationship between the fundamental harmonic vibrational wavenumber ω_e, the mass μ (or reduced mass, for a diatomic) and the Hooke's law force constant k_s is

$$
\omega_\mathrm{e} = \frac{1}{2\pi c_0} \sqrt{\frac{k_\mathrm{s}}{\mu}}
$$

(9.15)

and it is interesting to compare the simple solid with the vibrating diatomic. Table 9.3 summarizes our experimental data. There is a very big difference between the vibrational energies of a diatomic molecule and the vibrational energies of the atoms in a simple crystal. The force constants are an order of magnitude different as are the fundamental vibrational frequencies.

The important thing is to compare energy differences with 'thermal energy', k_BT. At 300 K, k_BT comes to 4.142×10^{-21} J.

Table 9.3 Harmonic force constants

	Mass (kg)	Harmonic force constant ($N\,m^{-1}$)	ω_e (cm^{-1})
$^1H^{35}Cl$	1.627×10^{-27}	480	2880
^{63}Cu	1.045×10^{-25}	30	90

Table 9.4 Vibrational energy levels for HCl and Cu metal

Energies (10^{-21} J)	$^1H^{35}Cl$	^{63}Cu
$v = 0$	28.645	0.893
$v = 1$	85.935	2.680
$v = 2$	143.23	4.467
$v = 3$	200.52	6.254
$v = 4$	257.81	8.041
$v = 5$	315.10	9.827

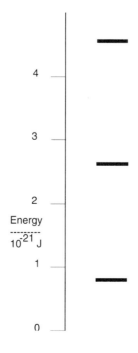

Figure 9.10 Vibrational energy levels for a simple solid

The energy levels of a one-dimensional simple harmonic oscillator with quantum number v are given by

$$\varepsilon_\nu = \frac{h}{2\pi} \, \omega_e \left(\nu + \frac{1}{2} \right) \tag{9.16}$$

The first few vibrational energy levels for HCl and Cu metal are given in table 9.4.

The important thing to note is that molecular vibrational energy differences are much larger than $k_B T$ but the corresponding quantities for atoms in simple crystals are smaller than $k_B T$ (Figure 9.10).

9.5   WHEN DO WE NEED TO USE QUANTUM MECHANICS?

I want to state two simple criteria that we can use to decide whether quantum mechanics or classical mechanics might be the more appropriate, when studying a given problem.

9.5.1   The de Broglie Criterion

A particle whose linear momentum is p has an associated de Broglie wavelength

$$\lambda_{dB} = \frac{h}{p}$$

In order to diffract such a particle, we need a grating whose spacing is of the order of the de Broglie wavelength. Thus, electrons accelerated by kilovolts have wavelengths that are typically picometers and so can be used in electron diffraction experiments.

A short calculation of the momentum of the melting solid problem discussed above shows that the de Broglie wavelength is much smaller than the natural spacing of the problem, which is in this case equal to the spacing between the atoms at rest.

9.5.2   The Boltzmann Criterion

This criterion deals with the spacing between energy levels, and the ratio of this spacing to $k_B T$. If the spacing between the energy levels is significantly less than $k_B T$, then classical mechanics should suffice. If the spacing between the energy levels is significantly greater than $k_B T$ then quantum mechanics will probably be needed.

10 Molecular Mechanics

Let me recap a few important points from previous chapters, where it was shown how to model the vibrations of a diatomic molecule such as HCl in terms of the chemically appealing idea of two balls connected by a spring. We looked at the classical description in Chapter 3 and the quantum mechanical description in Chapter 5. For many purposes, the classical description turns out to be satisfactory, but we need recourse to the quantum mechanical treatment in order to discuss experiments that depend explicitly on discrete energy levels (as in spectroscopy), or implicitly on things like the zero-point energy, which is a pure quantum mechanical effect.

Figure 10.1 is a graph of the Hooke's law (or harmonic) potential against bond length, copied from Chapter 5 (Figure 5.16).

If the potential is written

$$U(R) = U_0 + \frac{1}{2}k_s(R - R_e)^2$$

(where U_0 is a constant of integration) then we have

$$\frac{\mathrm{d}U}{\mathrm{d}R} = k_s(R - R_e)$$

and

$$\frac{\mathrm{d}^2U}{\mathrm{d}R^2} = k_s$$

Note that the equilibrium bond distance R_e corresponds to the value of R at the bottom of the potential curve, where the first derivative is zero. Note also that the force constant is given by the second derivative of the curve which happens to be a constant for all values of R in the special case of Hooke's law.

For more complicated potentials we need to delve a little more deeply into theory and a good place to start is with the Taylor expansion for a diatomic potential that depends on the

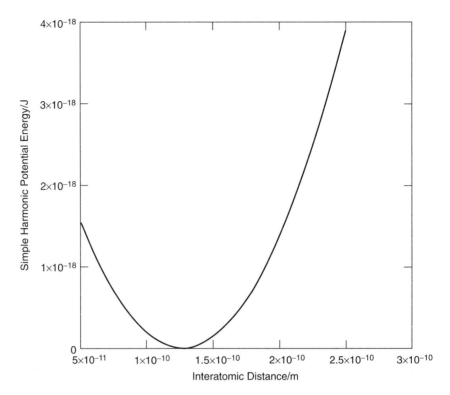

Figure 10.1 Simple harmonic potential for HCl

internuclear separation R about some minimum point R_e. We write

$$U(R) = U(R_e) + (R - R_e)\left(\frac{dU}{dR}\right)_{R=R_e} + \frac{1}{2}(R - R_e)^2\left(\frac{d^2U}{dR^2}\right)_{R=R_e} + \cdots \tag{10.1}$$

$U(R)$ is obviously equal to $U(R_e)$ when $R = R_e$, and this fixes the constant of integration. The equation is sometimes written as

$$U(R) - U(R_e) = (R - R_e)\left(\frac{dU}{dR}\right)_{R=R_e} + \frac{1}{2}(R - R_e)^2\left(\frac{d^2U}{dR^2}\right)_{R=R_e} + \cdots$$

or even

$$U(R) = (R - R_e)\left(\frac{dU}{dR}\right)_{R=R_e} + \frac{1}{2}(R - R_e)^2\left(\frac{d^2U}{dR^2}\right)_{R=R_e} + \cdots \tag{10.2}$$

where it is understood that $U(R)$ is measured relative to the potential energy minimum (that is to say, we take the zero as $U(R_e)$).

The quantity dU/dR is referred to as the gradient of U. The second derivative evaluated at the minimum where $R = R_e$ is called the (harmonic) force constant. Although there will be in general an infinite number of terms in the expansion, experience suggests that only the first few are important.

Take for example the Lennard–Jones 12-6 potential

$$U = 4\varepsilon\left[\left(\frac{\sigma}{R}\right)^{12} - \left(\frac{\sigma}{R}\right)^{6}\right] \tag{10.3}$$

where ε and σ are the two parameters that characterize the potential. U is zero when $R = \sigma$ and when $R = \infty$. The first derivative can be written

$$\frac{\mathrm{d}U}{\mathrm{d}R} = -\frac{4\varepsilon}{R}\left[12\left(\frac{\sigma}{R}\right)^{12} - 6\left(\frac{\sigma}{R}\right)^{6}\right]$$

which is zero when R is infinite and when $R = (2)^{1/6}\sigma$. When R is infinite, the potential is zero and when $R = (2)^{1/6}\sigma$ the potential is a minimum (which we denote R_{e}) with value $-\varepsilon$. The second differential is

$$\frac{\mathrm{d}^2 U}{\mathrm{d}R^2} = \frac{4\varepsilon}{R^2}\left[12 \times 13\left(\frac{\sigma}{R}\right)^{12} - 6 \times 7\left(\frac{\sigma}{R}\right)^{6}\right]$$

which becomes, at $R = (2)^{1/6}\sigma$

$$\left(\frac{\mathrm{d}^2 U}{\mathrm{d}R^2}\right)_{R=R_{\mathrm{e}}} = \frac{72\varepsilon}{\sigma^2 2^{1/3}} \tag{10.4}$$

and this is the harmonic force constant by definition.

Similarly, calculation of the first and second derivatives for the Morse potential

$$U(R) = D_{\mathrm{e}}\{1 - \exp[-\alpha(R - R_e)]\}^2 \tag{10.5}$$

shows that $\mathrm{d}U/\mathrm{d}R = 0$ at $R = R_{\mathrm{e}}$ and that the second derivative $\mathrm{d}^2 U/\mathrm{d}R^2$ has a value of $2D_{\mathrm{e}}\alpha^2$ when $R = R_{\mathrm{e}}$. This means that the force constant predicted by the Morse potential is $2D_{\mathrm{e}}\alpha^2$.

To calculate the minimum, i.e. the equilibrium bond length, and the force constant for an arbitrary potential, we therefore have to find the first and second derivatives.

For diatomic molecules, there is only one degree of vibrational freedom and only one normal mode of vibration. But for a non-linear polyatomic molecule with N atoms there are $3N - 6$ degrees of vibrational freedom, and so $3N - 6$ normal modes of vibration. For a linear molecule, there are $3N - 5$ degrees of vibrational freedom.

As I showed in Chapter 3, each normal mode of vibration has a corresponding normal coordinate, which is a linear combination of the Cartesian coordinates. All possible vibrations of a molecule can be described in terms of these normal coordinates, they are absolutely fundamental to studies of vibrational motion.

A detailed description of these normal modes can be obtained once the *force field* is known.

A spectroscopist would study a polyatomic molecule by measuring all possible vibrational bands using both infra-red and Raman techniques, and then deducing a force field. At this point, it should be stated that spectroscopists and molecular modelers have two different aims. A spectroscopist would measure all the vibrational frequencies of a material and then fit them to a potential. The spectroscopist would spare no pains to get the best fit between theory and experiment, and would be quite happy to report an accurate potential for a given molecule.

Molecular modelers have a different aim, which is to have a set of force constants and so-called 'reference' geometries (for example, the values of R_{e} for each bond) that can be transferred from molecule to molecule, and that are capable of predicting accurately the molecular equilibrium geometry of an 'unknown' molecule. They make use of the bond concept, and appeal to traditional chemists' ideas that a molecule is a sum of bonded atoms, and that molecular properties can often be written as sums of contributions from each bond.

10.1   IMPLEMENTATION OF MOLECULAR MECHANICS

Molecular mechanics (MM) aims to model the vibrations of a complex system in terms of classical mechanics. The primary objective of MM calculations is usually to predict an equilibrium geometry, and such calculations refer to the molecular system when it is completely at rest (i.e. at $0\,K$).

So, think of a small organic molecule such as adrenaline, whose two-dimensional drawing is shown in Figure 10.2.

Adrenaline ($C_9H_{13}O_3N$) is a non-steroid hormone. It is active only when given by injection; it raises the blood pressure and is used locally to stop haemorrhage.

This simple organic molecule has a planar benzene ring, and a variety of bonds.

The mutual potential energy U is built up as follows.

10.1.1   A Contribution from each Chemical Bond

This is often taken as a Harmonic term $\frac{1}{2}k_s(R - R_e)^2$ for each bond type, and we regard the force constants k_s and the equilibrium bond distances R_e as parameters that have to be fixed by modeling large numbers of related molecules, and comparing properties predicted with those measured by experiment. So for example, we can see examples of $C(sp^3)$–$C(sp^3)$ and aromatic C–C bonds in the sample molecule, and also a $C(sp^3)$–C(aromatic). There are also C–N, C–O and O–H bonds, even in this simple organic molecule.

10.1.2   A Contribution from each Bending Mode

The above molecule is taken to have bending contributions from (e.g.) C–C–H, C–C–C, and so on.

These are usually written as $\frac{1}{2}k_\theta(\theta - \theta_e)^2$ where θ_e is the equilibrium angle and k_θ the relevant force constant.

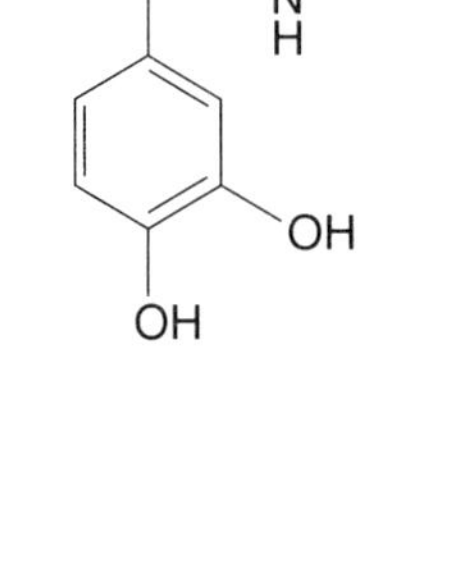

Figure 10.2   Two-dimensional drawing of adrenaline

10.1.3   A Contribution from each Dihedral Motion

If we have four atoms A, B, C and D arranged as follows:

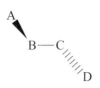

then the angle between the plane containing atoms ABC and the plane containing atoms BCD is called the dihedral angle of ABC and D. You can imagine a relative motion where atom A moves out of plane and atom D moves into the plane (and then the motion is reversed).

10.1.4   An Electrostatic Contribution

This is often added to the force field, depending on the whim of the author. In classical terms, one ought to take account of the differing electronegativities of the constituent atoms in one's target molecule above, and so we put

$$U_{AB} = \frac{Q_A Q_B}{4\pi\epsilon_0 r_{AB}}$$

where Q_A and Q_B are the formal charges associated with atoms A and B. For example, we would expect that the oxygen atoms would have negative Qs in the target molecule.

10.1.5   Non-bonded Interactions

It soon became clear in MM calculations that non-bonded interactions had to be included, in order to make the force field transferable from molecule to molecule. Without the non-bonded interactions, equilibrium geometry predictions were very poor.

Many commercial MM packages take a Lennard–Jones 12-6 potential

$$U = 4\varepsilon\left[\left(\frac{\sigma}{r}\right)^{12} - \left(\frac{\sigma}{r}\right)^{6}\right]$$

as discussed above. But there are many variations on a theme. The most important thing to note is that the Lennard–Jones parameters ε and σ relate to a pair of atoms. So, the parameters will be different for $C \cdots C$, for $C \cdots O$ and for $C \cdots H$. Whilst it is possible in principle to parameterize for all possible pairs of atom types, most authors prefer to parameterize for pairs of identical atoms, and then seek to express the Lennard–Jones potential in terms of these parameters. Let me illustrate this by writing the atom–atom potential U_{ij} between atoms i and j as

$$U_{ij} = \left(\frac{A_{ij}}{r_{ij}^{12}} - \frac{B_{ij}}{r_{ij}^{6}}\right)$$

where for example atom i might be a H and atom j a C(sp). The idea is to express the A and B parameters for the pair in terms of properties of atom types i and j separately. Two schemes from the literature are

$$A_{ij} = \left(\frac{r_i^*}{2} + \frac{r_j^*}{2}\right)^{12} \sqrt{\varepsilon_i \varepsilon_j}$$

$$B_{ij} = 2\left(\frac{r_i^*}{2} + \frac{r_j^*}{2}\right) \sqrt{\varepsilon_i \varepsilon_j}$$

(10.6)

and

$$A_{ij} = 4(\sigma_i \sigma_j)^6 \sqrt{\varepsilon_i \varepsilon_j}$$

$$B_{ij} = 4(\sigma_i \sigma_j)^3 \sqrt{\varepsilon_i \varepsilon_j}$$

(10.7)

where r_i^* is the minimum energy separation for two atoms of type i, σ_i the distance of closest approach and ε_i the well depth. These rules are known as the Berthelot mixing rules; they have been used in other applications by chemical engineers for very many years.

For a large molecule, there are many more non-bonded interactions than bonded interactions, and their calculation can take up much of the time in a MM calculation. Some commercial packages allow you to switch off the non-bonded interactions until the calculation is well on its way to an energy minimum, and others include a cut-off distance beyond which the non-bonded interactions are ignored.

10.1.6 Conjugated Systems

It was realized many years ago that the presence of a conjugated system could be used to simplify MM calculations, as conjugated systems are usually planar. Most commercial MM packages test for the existence of a conjugated system, and use this to simplify the geometry prediction. Those parts of the system that are conjugated are automatically assumed to be planar, and models more appropriate to conjugated systems are used to predict the bond lengths and angles in those parts of the molecular system.

10.2 AN MM CALCULATION

To see how it is done, let us follow our target molecule adrenaline through a simple MM geometry optimization.

The first step is to produce a starting geometry. There are several ways to get started, the most usual being to take a related crystal structure from the literature. Alternatively we can start with a two-dimensional chemical drawing such as the CHEMDRAW one in Figure 10.2 and let the modeling package generate a three-dimensional structure as best it can.

There is a variety of different file formats; the most transferable one is the protein database file format (computer files often have a .PDB extension), and Table 10.1 gives part of a .PDB file for adrenaline.

It is fairly simple to understand. The first record gives a title to the calculation. Records starting HETATM define each atom in turn by giving them a serial number and an atomic type label, together with the Cartesian coordinates. The records starting CONECT show how a chemist would join the atoms together to form bonds. It gives the so-called *connectivity data*;

Table 10.1   The .PDB file for adenaline

COMPND adren.pdb							
HETATM	1	C		-1.5878	-0.0240	1.7869
HETATM	3	C		-0.3580	-2.1540	1.7869
HETATM	4	C		0.8717	-1.4439	1.7869
HETATM	5	C		0.8716	-0.0239	1.7869
HETATM	6	C		-0.3581	0.6860	1.7800
HETATM	7	O		2.0452	-2.1213	1.7752
HETATM	8	O		-0.3580	-3.5089	1.7752
HETATM	9	C		-0.3578	2.1830	1.7788
HETATM	10	O		-1.6795	2.6506	1.7850
HETATM	11	C		0.3540	2.6895	0.5313
HETATM	12	N		-0.3300	2.2092	-0.6390
HETATM	13	C		0.3422	2.6874	-1.8169
HETATM	26	H		-2.5403	0.5260	1.8017
HETATM	25	H		-2.5403	-1.9940	1.7774
HETATM	24	H		1.8242	0.5261	1.7827
HETATM	14	H		2.5404	-1.8781	2.5754
HETATM	15	H		0.1082	-3.8023	0.9744
HETATM	23	H		0.1168	2.5533	2.7149
HETATM	16	H		-1.6371	3.5916	1.7840
HETATM	21	H		1.3984	2.3049	0.5195
HETATM	22	H		0.3400	3.8023	0.5235
HETATM	17	H		-0.3300	1.1893	-0.6381
HETATM	18	H		1.3892	2.3100	-1.8106
HETATM	19	H		0.3578	3.8000	-1.7942
HETATM	20	H		-0.1810	2.2892	-2.7150
CONECT	1	2	6	26		
CONECT	2	1	3	25		
CONECT	3	2	4	8		
CONECT	4	3	5	7		
CONECT	5	4	6	24		
CONECT	6	1	5	9		
CONECT	7	4	14			
CONECT	8	3	15			
CONECT	9	6	10	11	23	
CONECT	10	9	16			
CONECT	11	9	12	21	22	
CONECT	12	11	13	17		
CONECT	13	12	18	19	20	
CONECT	26	1				
CONECT	25	2				
CONECT	24	5				
CONECT	14	7				
CONECT	15	8				
CONECT	23	9				
CONECT	16	10				
CONECT	21	11				
CONECT	22	11				
CONECT	17	12				
CONECT	18	13				
CONECT	19	13				
CONECT	20	13				
END						

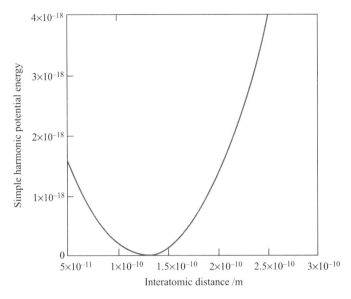

Figure 10.3 Screen grab for HyperChem run on adrenaline

if we focus on the first CONECT record, it says that atom number 1 (a carbon) is joined to those atoms with serial numbers 2, 6 and 26.

The next step is input to a MM program. I am going to illustrate the calculation by using the commercially available package HyperChem. Figure 10.3 shows a screen grab.

Each package has its own strengths and weaknesses, some find favor with users and some do not. It is a personal matter, and I have no axe to grind. Like many other packages, HyperChem runs under M/S Windows. It has an intuitive graphical user interface (GUI).

The display option in Figure 10.3 is often known as 'sticks'. Most packages have a variety of molecular display options, for example 'disks', (Figure 10.4), 'spheres' (Figure 10.5),

Figure 10.4 Disks representation

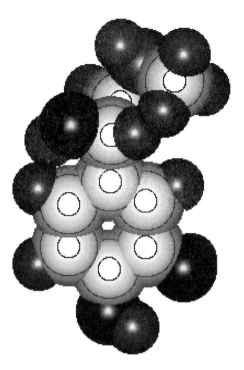

Figure 10.5 Spheres representation

three-dimensional, and so on. Sometimes it is desirable to turn off the hydrogen atoms from the display, particularly for a large molecule.

The next step is the geometry optimization. Depending on the force field chosen, lone pairs may be added as appropriate. These are often treated as pseudo atoms, with their own parameter set. Some force fields treat (for example) a CH_3 group as a 'composite' atom.

The (partial) screen grab in Figure 10.6 shows a typical output after geometry optimization.

I have probably raised more problems than I have answered. How does the program find the minimum, and why did I choose the particular starting structure that I did? Does it matter, surely I will get the same result no matter where I start from? Why didn't I start from the two-dimensional structure in Figure 10.7, where I have made a couple of simple modifications to the original structure?

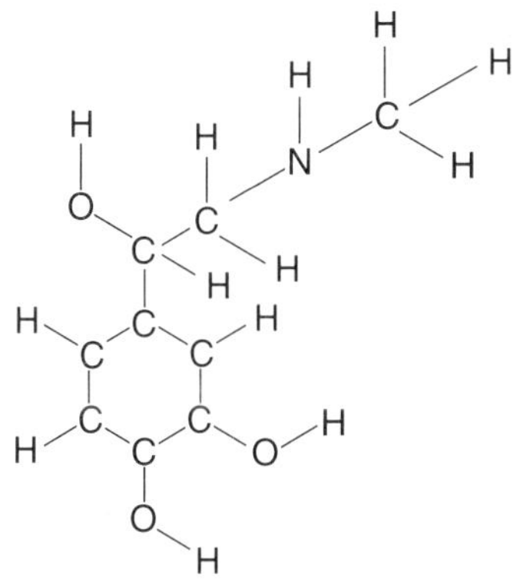

Figure 10.6 MM+ Geometry optimization

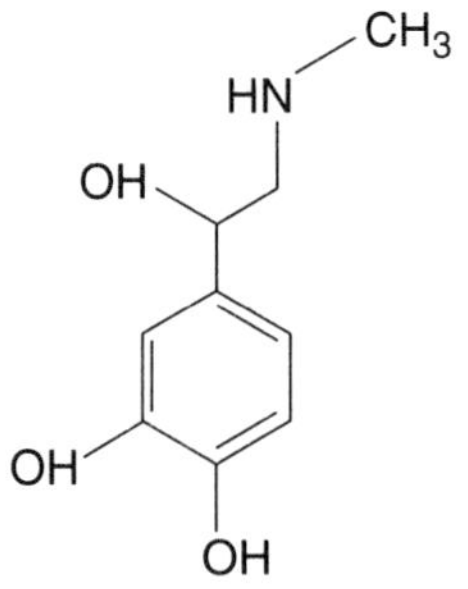

Figure 10.7 Modified starting
structure

The quick answer is that this starting structure also leads to a minimum, but with a different energy than the structure originally shown. In fact there are very many potential energy minima.

You might also have noticed that I have made no mention about temperature, and because the model is a classical one we do not have to worry about quantized energies. Equally, there is no zero point energy to worry about; in fact there are no vibrations at all to worry about.

10.3   FEATURES OF PE SURFACES

When we have to deal with a non-linear polyatomic molecule or material made up from N atoms, then the potential energy depends on the $3N$-6 normal coordinates q_1, q_2, ... Simple potential energy curves for diatomics such as the one given for HCl become very complicated, and they show some complicated features.

10.3.1   Multiple Minima

We do not have to look at anything so complicated as adrenaline in order to investigate multiple minima. Consider 'rigid' ethane C_2H_6, where for the sake of argument we imagine that the potential energy does not depend on the C–C bond length, the C–H bond length or anything else but the rotation about the C–C axis.

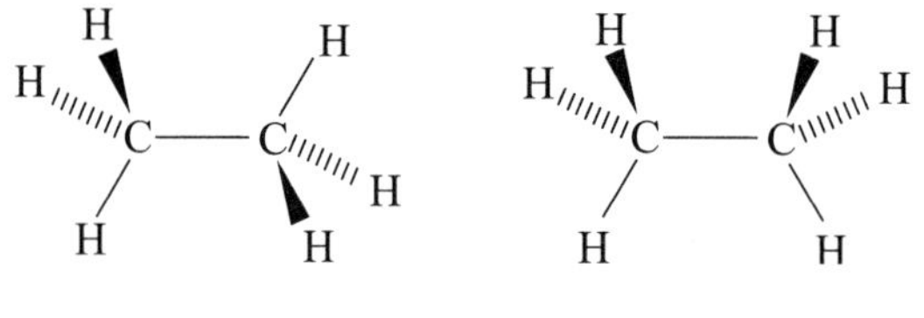

As we rotate the molecule about the C–C axis, there will be three equivalent positions where the C–H bonds eclipse each other, and three equivalent positions where the C–H bonds are as far apart from each other as possible (the staggered positions) as shown.

The three staggered configurations have a lower energy than the three eclipsed configurations, and a rough calculation of the potential energy dependence on the HCCH dihedral angle is shown in Figure 10.8.

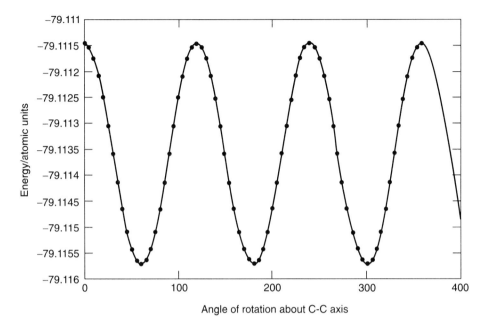

Figure 10.8   Three-fold rotation axis in ethane

10.3.2   Local and Global Minima

Suppose that we now substitute and make CH_2FCH_2Cl.

There is still a barrier to internal rotation, but the curve is a little different. The highest maximum corresponds to the position where the C—F and the C—Cl bonds oppose each other. As we rotate about the C—C axis from there, there will be two equivalent minima when the C—F and the C—Cl bonds are at (about) $120°$ to each other.

This is an example of a potential energy curve showing three minima, but with one minimum lower in energy than the other two. The point corresponding to $180°$ is called the global minimum, and the other two minima are called local minima (Figure 10.9).

10.3.3   Saddle Points

For many polyatomic molecules we can also find a feature called a saddle point, which at its simplest is a maximum in one dimension and a minimum in another. This is illustrated for the two-dimensional function $z = x^2 - y^2$ in Figure 10.10.

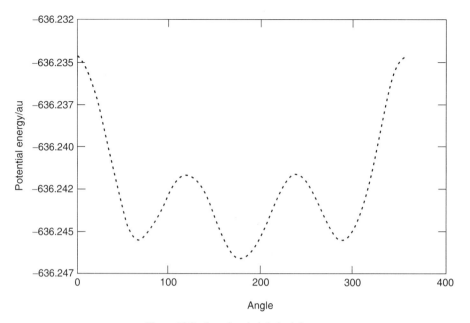

Figure 10.9 Local and global minima

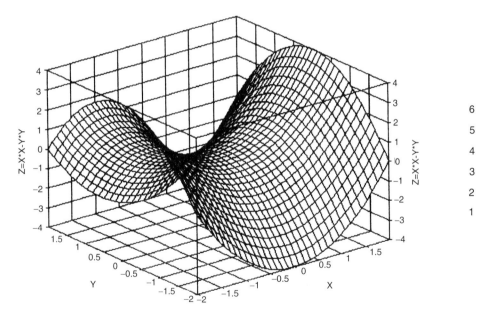

Figure 10.10 Saddle point

The saddle point is at the origin; the function $z = x^2 - y^2$ has a minimum with respect to x at the origin when y is kept constant, and a maximum at the origin when x is kept constant. Saddle points are important in chemistry because they correspond to transition states.

We refer to all such points where the gradient is zero as *stationary points*, and in general a potential energy surface for a polyatomic species will have many local minima, maxima and

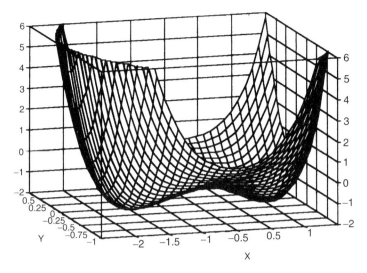

Figure 10.11   Saddle point and minima

saddle points. For example, the function $z = x^4 + 4x^2y^2 - 2x^2 + 2y^2 - 1$ shows two minima (at $x = 1$, $y = 0$ and $x = -1$, $y = 0$) and a saddle point at the origin (Figure 10.11).

The overall aim of a MM investigation is invariably to locate the global minimum, if possible. We therefore need to investigate two separate problems.

- How do we locate the stationary points, and characterize them?
- How do we find the global minimum?

10.4   LOCATING STATIONARY POINTS

The simplest way forward is to choose a suitable range of values for each of the bond lengths, bond angles and dihedral angles, together with suitable increments, and calculate the potential energy for all possible combinations of these. The combination of variables that corresponds to the lowest energy is then taken blindly to be the global minimum.

If we use the example of adrenaline discussed earlier, and do our potential energy search in terms of the bond lengths (25 in all), bond angles (24) and dihedral angles (23) and for the sake of argument take five values of each variable, then we will have to calculate the potential energy $5 \times 25 \times 24 \times 23 = 644\,000$ times. However, the energy calculation is particularly simple.

A refinement is to treat each variable in turn, and keep the remaining variables constant. So we would choose values for each of the $25 + 24 + 23 = 72$ variables, and then optimize each one in turn keeping the other 71 constant. Having optimised each of the 72 variables in turn we would then go back to the beginning of the calculation, but using the optimized values. Eventually the whole calculation usually converges on the point which we once again take to be the global minimum. The effort needed for such a calculation can be calculated as follows. For comparability we assume that five points are sufficient to locate the energy minimum for each variable with all the other variables kept constant.

For each of the 72 variables we need five energy calculations, and so a complete cycle needs $72 \times 5 = 360$ energy evaluations.

This is not a problem with MM calculations, where the energy evaluation is particularly simple. What we normally observe is a large number of local minima, all very close together in energy. These can be ranked in importance in comparison with $k_B T$.

10.5   GRADIENT METHODS

If you look back to Figure 10.9, which illustrates saddle point behavior, you can see a simple analogy with walking in a hilly countryside. If the aim of your walk is to descend a hill, then you look for the direction that will take you down as quickly as possible. The word 'direction' implies a vector quantity, so I need to discuss the concept of a gradient.

Look back to the potential energy graph for HCl (Figure 10.1). If I draw a gradient at any point on the graph and measure the slope then

- a positive gradient means that the function is increasing
- a negative gradient means that the function is decreasing

In order to calculate the gradient of $y = f(x)$ at some point x_0 then what we do is calculate dy/dx and substitute the value of x_0 into the expression. This gives the gradient at the point x_0.

In three dimensions, we proceed as follows.

Suppose we have a function $f(x, y, z)$ that depends on the three variables x, y and z. We take unit vectors $\mathbf{e}_x, \mathbf{e}_y$ and $\mathbf{e}_z$, calculate the partial derivatives $\partial f/\partial x, \partial f/\partial y$ and $\partial f/\partial z$. The gradient of f, grad f, is defined as

$$\text{grad } f = \frac{\partial f}{\partial x}\,\mathbf{e}_x + \frac{\partial f}{\partial y}\,\mathbf{e}_y + \frac{\partial f}{\partial z}\,\mathbf{e}_z \tag{10.8}$$

Going back to our MM calculation, we have an energy that depends on very many variables, the bond lengths, bond angles and dihedral angles. If we write U for the energy and x_1, $x_2, \ldots x_n$ for these variables then we can define the gradient of U with respect to the variables $x_1, x_2, \ldots x_n$ as

$$\text{grad } U = \frac{\partial U}{\partial x_1}\,\mathbf{e}_1 + \frac{\partial U}{\partial x_2}\,\mathbf{e}_2 + \cdots + \frac{\partial U}{\partial x_n}\,\mathbf{e}_n \tag{10.9}$$

where the $\mathbf{e}$s are unit vectors associated with the variables.

The gradient (grad U) is a vector quantity, and it points in the direction of an increase in U. The negative of this, $-\text{grad } U$ therefore points in the direction of a decrease in U. The idea is that if we follow $-\text{grad } U$ along a potential energy surface, then we will get to a local potential energy minimum.

Mathematicians generalize the idea of a gradient to cover the case where they have a function of many variables, and they refer to the vector

$$\mathbf{g} = \begin{pmatrix} \dfrac{\partial U}{\partial x_1} \\[2mm] \dfrac{\partial U}{\partial x_2} \\[1mm] \vdots \\[1mm] \dfrac{\partial U}{\partial x_n} \end{pmatrix} \tag{10.10}$$

as the gradient vector.

10.6 STEEPEST DESCENTS

The negative of the gradient, when evaluated at a point on the potential energy surface, points in the direction of steepest descent down the surface and away from the point in question. In the steepest descents technique, we calculate the gradient and then choose somehow a step size in order to progress to the bottom of the potential energy curve. Each steepest descent step defines simultaneous changes in all the variables. This is analogous to walking down a hill using the steepest path. The problem with the method of steepest descent is that the step size is critical. Too large a step size and you pass the minimum, too small a step size and the calculation takes forever. Near the minimum, the method of steepest descents tends to oscillate about the minimum. Experience shows that the first few steps in a steepest descent calculation usually give very rapid progress to the minimum, but that the optimization slows down when the molecular conformation reaches a flatter part of the potential energy surface.

Sometimes we need a knowledge of the second derivatives in addition to the first derivatives. The second derivatives are conveniently collected into an n by n matrix called the hessian $\mathbf{H}$.

$$\mathbf{H} = \begin{pmatrix} \dfrac{\partial^2 U}{\partial x_1^2} & \dfrac{\partial^2 U}{\partial x_1 \partial x_2} & \cdots & \dfrac{\partial^2 U}{\partial x_1 \partial x_n} \\[2ex] \dfrac{\partial^2 U}{\partial x_2 \partial x_1} & \dfrac{\partial^2 U}{\partial x_2^2} & \cdots & \dfrac{\partial^2 U}{\partial x_2 \partial x_n} \\[2ex] \vdots & \vdots & \cdots & \vdots \\[2ex] \dfrac{\partial^2 U}{\partial x_n \partial x_1} & \dfrac{\partial^2 U}{\partial x_n \partial x_2} & \cdots & \dfrac{\partial^2 U}{\partial x_n^2} \end{pmatrix} \tag{10.11}$$

10.7 SECOND DERIVATIVE METHODS

Consider a potential energy U that depends only on a single variable, $U = U(x)$, and we want to find a solution of the equation $U(x) = 0$.

Newton's method gives a route to the solution of this problem, as follows. We start with some guess at the zero, call it $x^{(1)}$, and then fit a tangent line to $U(x)$ at this point as illustrated in Figure 10.12.

A simple calculation shows that the point $x^{(2)}$ at which this tangent line cuts the x axis is

$$x^{(2)} = x^{(1)} - \left(\frac{\mathrm{d}U(x^{(1)})}{\mathrm{d}x}\right)^{-1} U(x^{(1)}) \tag{10.12}$$

where we have to calculate the derivative at the point $x^{(1)}$. This is illustrated figure 10.12. The point $x^{(2)}$ will generally be closer to zero than $x^{(1)}$, and the process is repeated until the desired accuracy is attained.

The Newton–Raphson method is formulated so as to give the roots of the equation, i.e. solution of $U(x) = 0$. Our minimization problem is different in that we want to find a minimum of $U(x)$, and at the minimum we know that $\mathrm{d}U/\mathrm{d}x = 0$.

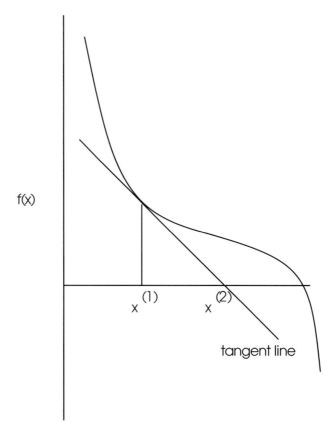

Figure 10.12 Newton–Raphson technique

The formula given above thus becomes

$$x^{(2)} = x^{(1)} - \left[\frac{\mathrm{d}^2 U(x^{(1)})}{\mathrm{d}x^2} \right]^{-1} \frac{\mathrm{d} U(x^{(1)})}{\mathrm{d}x} \qquad (10.13)$$

and we have to calculate the first and second differentials at the point x_1.

This can be generalised into a method for many variables. We start at a point $\mathbf{x}_1$ on the potential energy surface, calculate the gradient vector $\mathbf{g}$ and the hessian matrix $\mathbf{H}$ at this point and then step along to the new point $\mathbf{x}_2$ given by

$$\mathbf{x}_2 = \mathbf{x}_1 - \mathbf{H}^{-1}\mathbf{g} \qquad (10.14)$$

The difficulty with this method is that we need to evaluate the hessian $\mathbf{H}$ and the gradient $\mathbf{g}$ at each point in the calculation. A number of ingenious methods are currently available that update $\mathbf{H}^{-1}$ as the calculation progresses, without having to calculate it from scratch.

10.8 CHARACTERIZATION OF STATIONARY POINTS

For a diatomic molecule, the force constant is equal to the second derivative calculated at the minimum of the potential energy curve. In the case of a maximum on the potential energy curve, the second derivative would be negative and it is zero for a point of inflection. This gives us a clue for the characterization of the stationary points.

For a polyatomic (non-linear) molecule with N atoms there are $3N$-6 force constants; what we have to do is to calculate these force constants at each stationary point, and investigate whether they are positive or negative.

Depending on the way the calculation is performed, there may be $3N$ force constants of which six are zero; these six correspond to translation and rotation.

The force constants can be found as eigenvalues of the hessian matrix. If the force constants are all positive, then we have an energy minimum. If one of the force constants is negative, then we have a saddle point, and so on. As a molecular system moves from one minimum to another, the energy increases to a maximum at the transition state and then falls. The number of negative force constants is used to characterize the saddle point; an nth-order saddle point has n negative eigenvalues. We are usually most interested in first-order saddle points because they correspond to chemical transition states.

Negative eigenvalues give imaginary frequencies (in the complex number sense).

10.9 PROTEINS AND DOCKING

Proteins perform a wide variety of functions in organisms. There are tens of thousands of different types of proteins all with different biological functions. Table 10.2 illustrates the point.

Proteins are polypeptides, made by linking together many hundred or thousand amino acid monomers by so-called *peptide bonds*. A protein is formed when one or more polypeptides coil up in certain repeatable ways to form a certain three-dimensional structure with certain biological properties.

Twenty amino acid monomers are found in proteins. All amino acids have the same fundamental chemical structure but all are rather different in character, and that is why stringing them together in different combinations produces proteins with such different biological functions.

Table 10.2 Some proteins and their function

Protein	Function
Pepsin	Digestive enzyme in stomach
Hexokinase	Enzyme in cytosol of cells
Immunoglobin	Antibody in blood serum
Haemoglobin	Oxygen carrier in red cells
Insulin	Hormone that controls blood glucose levels
Cholera toxin	Toxin produced by bacteria
Actin	Protein of muscle
Collagen	Fibrous protein of bone and tendon
Keratin	Fibrous protein of hair and feather
Silk	Fibrous protein of insects

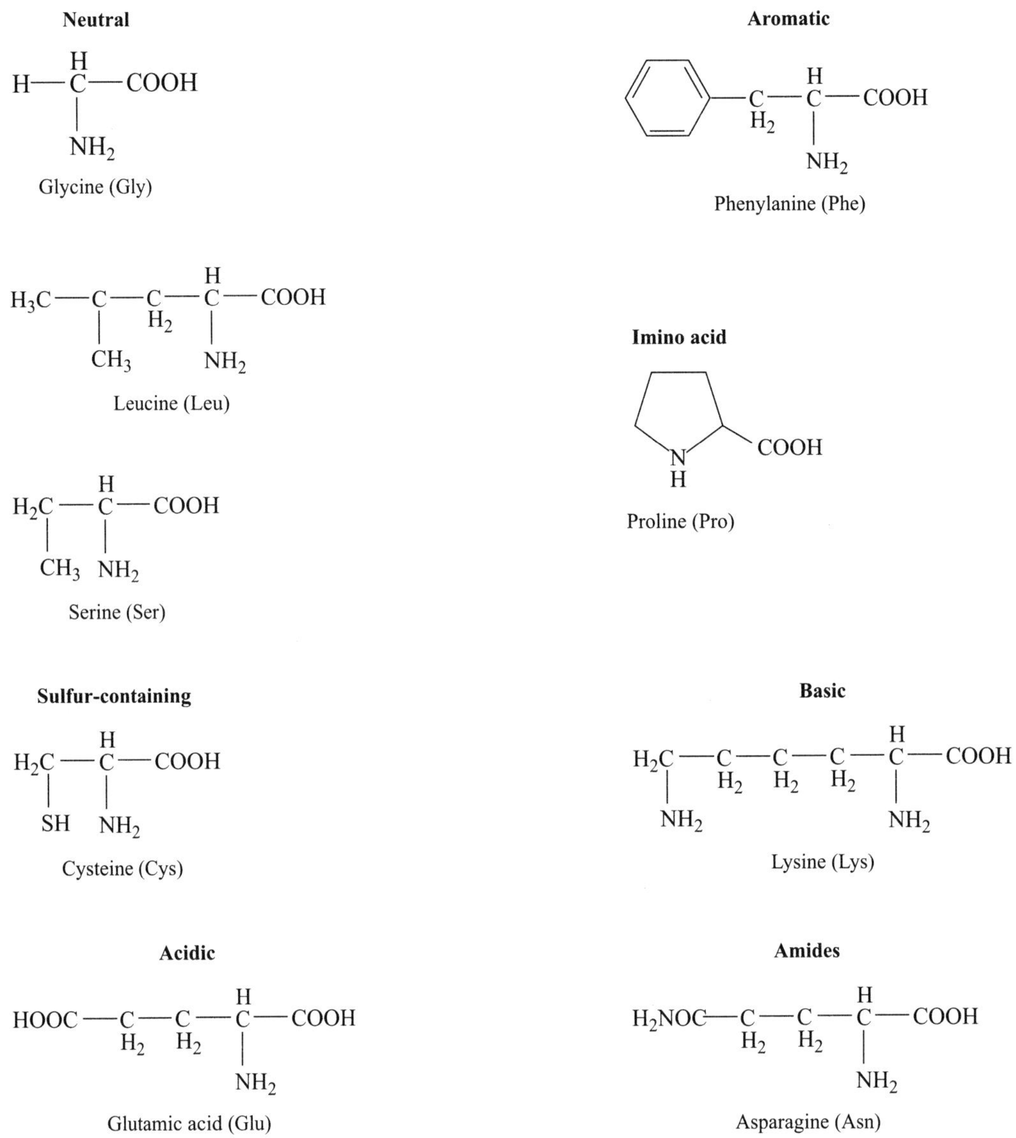

Neutral

$$H\!-\!\underset{\underset{\displaystyle NH_2}{|}}{\overset{\overset{\displaystyle H}{|}}{C}}\!-\!COOH$$

Glycine (Gly)

$$H_3C\!-\!\underset{\underset{\displaystyle CH_3}{|}}{C}\!-\!\underset{\displaystyle H_2}{C}\!-\!\underset{\underset{\displaystyle NH_2}{|}}{\overset{\overset{\displaystyle H}{|}}{C}}\!-\!COOH$$

Leucine (Leu)

$$H_2C\!-\!\underset{\underset{\displaystyle CH_3\ \ NH_2}{|\ \ \ \ |}}{\overset{\overset{\displaystyle H}{|}}{C}}\!-\!COOH$$

Serine (Ser)

Sulfur-containing

$$H_2C\!-\!\underset{\underset{\displaystyle SH\ \ NH_2}{|\ \ \ \ |}}{\overset{\overset{\displaystyle H}{|}}{C}}\!-\!COOH$$

Cysteine (Cys)

Acidic

$$HOOC\!-\!\underset{\displaystyle H_2}{C}\!-\!\underset{\displaystyle H_2}{C}\!-\!\underset{\underset{\displaystyle NH_2}{|}}{\overset{\overset{\displaystyle H}{|}}{C}}\!-\!COOH$$

Glutamic acid (Glu)

Aromatic

$$\text{(benzene ring)}\!-\!\underset{\displaystyle H_2}{C}\!-\!\underset{\underset{\displaystyle NH_2}{|}}{\overset{\overset{\displaystyle H}{|}}{C}}\!-\!COOH$$

Phenylanine (Phe)

Imino acid

Proline (Pro)

Basic

$$H_2C\!-\!\underset{\underset{\displaystyle NH_2}{|}}{C}\!-\!\underset{\displaystyle H_2}{C}\!-\!\underset{\displaystyle H_2}{C}\!-\!\underset{\underset{\displaystyle NH_2}{|}}{\overset{\overset{\displaystyle H}{|}}{C}}\!-\!COOH$$

Lysine (Lys)

Amides

$$H_2NOC\!-\!\underset{\displaystyle H_2}{C}\!-\!\underset{\displaystyle H_2}{C}\!-\!\underset{\underset{\displaystyle NH_2}{|}}{\overset{\overset{\displaystyle H}{|}}{C}}\!-\!COOH$$

Asparagine (Asn)

Figure 10.13   A sample of the 20 amino acids

Figure 10.13 shows a selection of these 20 amino acids, together with their three-letter abbreviations.

Polypeptides can be very large entities, and many of them have been studied by X-ray crystallographers and NMR spectroscopists. Just to take one example, consider the polypeptide enterostatin. Enterostatin is formed in the trypsin-catalyzed cleavage of the precursor procolipase which forms colipase and enterostatin in the duodenum. Enterostatin has been widely studied by biochemists, and it has some interesting biological effects. In particular, it inhibits the intake of diets high in fat but not of carbohydrates. Enterostatin may therefore act as a possible feedback regulator to prevent excessive intake of fat (in rats, at least).

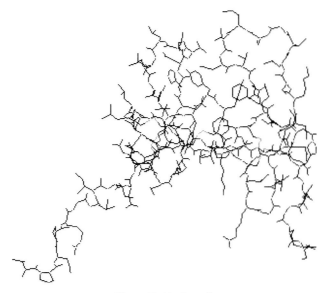

Figure 10.14   Procolipase

I was able to find a geometry for porcine pancreatic procolipase in the Protein Database, as determined by NMR spectroscopy. Such structures are normally reported without data for the hydrogen atoms, but many MM packages will automatically add hydrogen atoms on request. Figure 10.14 gives the procolipase structure, as shown by HyperChem.

It is obviously a very large molecule, difficult to visualize or to present on screen.

Many MM packages allow for the presentation of such large molecules in terms of 'ribbons'. Ribbons are used to display large proteins and attention focuses on the protein backbone (Figure 10.15).

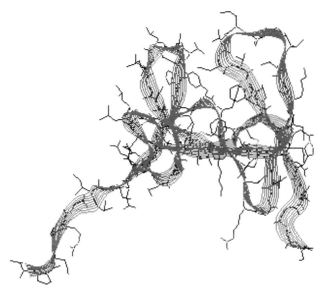

Figure 10.15   Ribbon representation of enzyme

10.9.1 The Effect of pH

Glycine is a very simple amino acid:

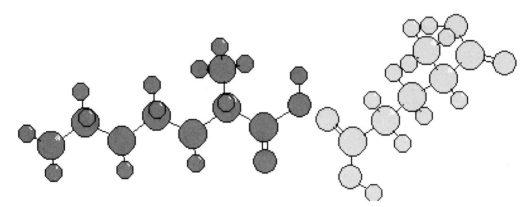

(it is shown in unionized form on the left.) Amino acids tend to be ionized at the pH values found in cells, and we call the ions *zwitterions*. I have shown the glycine zwitterion on the right. Many MM packages will automatically switch an amino acid into its zwitterion form.

Although the sequences of many thousands of proteins are now known, the prediction of the three-dimensional shape for a given protein is very difficult. We refer to this as the 'protein folding' problem.

10.10 PROTEIN DOCKING

Apart from protein folding (a conformational study of proteins), there is much interest in how proteins interact with each other and we refer to this as the 'protein docking' problem. To give an example, consider the lysine...glutamic acid (Lys...Glu) pair.

Figure 10.16 shows Lys and Glu, and we now keep the molecular geometries frozen but allow the molecules to interact with each other.

In *vector docking* we choose a point in either molecule and then examine how the interaction energy changes as the two molecules approach along the line joining these two points. In *unconstrained docking* we simply let the two fragments move around in space until an energy minimum is reached (Figure 10.17).

10.11 MOLECULAR PROPERTIES

Many MM packages will calculate certain simple molecular properties, apart from the local energy minimum. Lysine is shown in Figure 10.18.

Figure 10.16 Before docking. Lys (left) and Glu (right)

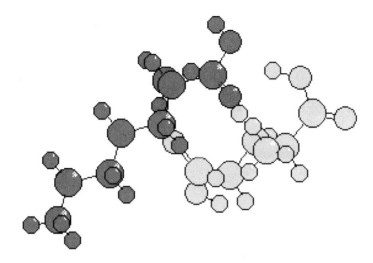

Figure 10.17   After unconstrained docking

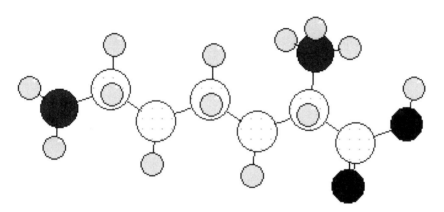

Figure 10.18   Lysine

A property of interest is the electrostatic potential (not to be confused with the mutual potential energy of a charge distribution) $\varphi(\mathbf{r})$ at points in space $\mathbf{r}$ surrounding the molecule. This is often used as a rough and ready indication of the reactive sites in a molecule, and some MM packages will calculate it (Figure 10.19).

Approximations to the volume and to the surface area can often be calculated. These are often used in Quantitative Structure and Activity Relationship (QSAR) analysis.

10.12   THE EFFECT OF SOLVENTS

Isolated molecules in the gas phase are the easiest to treat by MM, because we only need take account of the intramolecular interactions. Most chemical reactions are performed in condensed phases; in a medium with relative permittivity ε_r, the force between two point charges

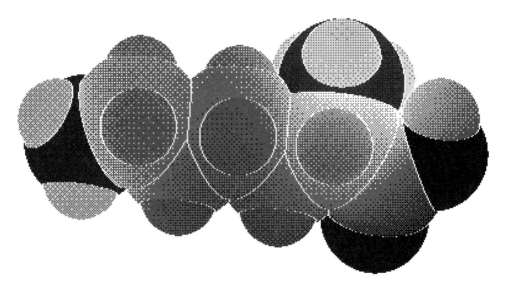

Figure 10.19 Electrostatic potential around lysine

Q_A and Q_B is reduced by a factor of ε_r and the mutual potential energy becomes

$$U_{AB} = \frac{Q_A Q_B}{4\pi\varepsilon_0\varepsilon_r r_{AB}} \tag{10.15}$$

The old-fashioned name for relative permittivity is the dielectric constant, usually written D or occasionally K. The relative permittivity is temperature dependent, and it is usually reported as a power series in T

$$\varepsilon_r = a + b\left(\frac{T}{K}\right) + c\left(\frac{T}{K}\right)^2 + \cdots \tag{10.16}$$

where $a, b, c, \ldots$ are numerical coefficients determined from experiment. A few values of the relative permittivity at room temperature are shown in Table 10.3.

What was done in the early days was to reduce the ionic interactions by an arbitrary quantity, vaguely related to the relative permittivity. Some early calculations used 2.5 for water.

A more obvious approach is to add a number of solvent molecules around the molecule of interest, in order to model the solvent. Many of the early MM force fields were developed at a time when it was not computationally feasible to add such a number of solvent molecules, but

Table 10.3 Relative permittivities

Substance	ε_r
Free space	1
Air	1.0006
Glass	6
Water	81

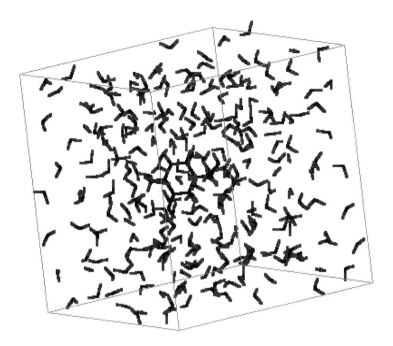

Figure 10.20   A solvent box

this is now routine. What we do is to create a box of solvent molecules that surround the molecule of interest, and then optimize the geometry.

The number of water molecules depends on the size of the box taken, but the number density is carefully arranged so that the density models the density of liquid water.

A solvent box for adrenaline is shown in Figure 10.20.

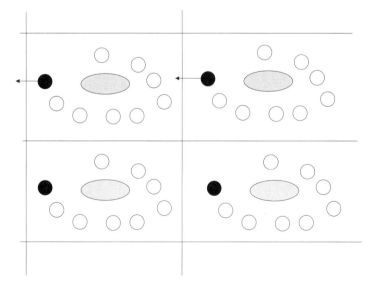

Figure 10.21   A solvent molecule and its image(s)

As the MM geometry optimization proceeds, water molecules can get placed outside the box by the optimization algorithm. Be clear that this is a MM calculation at 0 K, and that nothing is moving. Solvent molecules do not 'move' out of the box under their own kinetic energy; they get placed outside the box by the geometry minimization algorithm in order to find an arrangement that corresponds to a lower energy.

If solvent molecules do 'leave' the box, then the density changes. In order to keep the density constant, we imagine that our box is one of a very large number of boxes that all adjoin each other. I can illustrate the idea of a periodic box best in two dimensions (Figure 10.21).

Focus attention on the top left-hand box in Figure 10.21, which is a two-dimensional representation of the solute plus water molecules. Suppose that the geometry optimization procedure places the top left-hand black solvent molecule outside the box. We permit an equivalent black solvent molecule from the top right-hand box to enter, in order to keep the density constant. We refer to this as the 'periodic box'. The idea of a periodic box will appear repeatedly in later chapters.

11   Molecular Dynamics and Monte Carlo Techniques

In Chapter 10, I discussed molecular mechanics (MM). The aim of a MM calculation is usually to find the lowest energy conformation of a given molecular system; either a single molecule, or a system of interacting ones. MM calculations refer to 0 K, and because zero point energies are absent from classical mechanics, the predicted geometries refer to molecular systems at rest. The generic phrase 'energy minimization' is often used to describe such calculations.

For very many applications we need to know about the effect of motion and of temperature. One way to do this is to solve the classical equations of motion for the atoms and molecules, given the intermolecular and the intramolecular potentials, and this is the field of molecular dynamics (MD).

11.1   MOLECULAR DYNAMICS

In previous chapters I spoke about the force exerted by atom B on atom A, $\mathbf{F}_{\text{B on A}}$, or indeed the force exerted by molecule B on molecule A. Although I discussed MM in terms of potentials rather than forces, the two are related mathematically by the gradient

$$\mathbf{F} = -\operatorname{grad} U \qquad (11.1)$$

The force $\mathbf{F}$ is a vector field, whilst the potential U is a scalar field, and both are defined at all points in space. In particular, if we want to know the force on atom A at position vector $\mathbf{r}_A$ then we have to evaluate grad U at that point in space. This differentiation can be done numerically or analytically.

Let us return to the adrenaline problem of Chapter 10, and once again treat an isolated molecule in the gas phase (Figure 11.1).

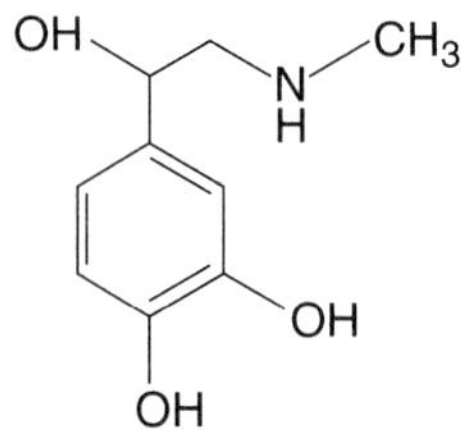

Figure 11.1   Adrenaline

Each atom in the molecule experiences an intramolecular force due to the remaining atoms, and this force is zero at an equilibrium geometry. The total force on a given atom A can be calculated as a sum of the forces due to the remaining atoms

$$\mathbf{F}_A = \sum_{B \neq A} \mathbf{F}_{B \text{ on } A} \tag{11.2}$$

and this is related by one of Newton's equations to the acceleration of atom A by

$$\mathbf{F}_A = m_A \frac{d^2 \mathbf{r}_A}{dt^2} \tag{11.3}$$

where m_A is the mass of atom A and $\mathbf{r}_A$ its position vector. Calculating the trajectories of N atoms therefore involves the solution of a set of $3N$ second-order differential equations. For any set of N particles it is possible to find three coordinates that correspond to translation of the center of mass of the system, and three coordinates that correspond to rotations about three axes that pass through the center of mass.

It is also possible to work with the generalized coordinates q_i and momenta p_i discussed in Chapter 5. If the classical Hamiltonian for the system is H, then the $6N$ first-order differential equations of motion are

$$\frac{dq_i}{dt} = \frac{\partial H}{\partial p_i}$$

$$\frac{dp_i}{dt} = -\frac{\partial H}{\partial q_i} \tag{11.4}$$

These are known as *Hamilton's equations of motion*. Provided that the potential U is independent of time, the Hamiltonian is equal to the energy of the system.

Ordinary differential equations such as the ones given above are very common in science and engineering. A standard method for their solution is afforded by the technique of finite differences. The general idea is that given the atom positions, velocities, and so on, at time t, we try and obtain the positions, velocities, and so on, at time $t + \delta t$ to a sufficient degree of accuracy. A review of the literature suggests that there are very many algorithms to accomplish this task, but many of these algorithms turn out to have features in common and so I will just tell you about a couple of popular ones.

First of all the *predictor–corrector method*. If the position ($\mathbf{r}$), velocity ($\mathbf{v}$), acceleration ($\mathbf{a}$) and time derivative of the acceleration ($\mathbf{b}$) are known at time t, then these quantities can be obtained at $t + \delta t$ by a Taylor expansion

$$\mathbf{r}^p(t + \delta t) = \mathbf{r}(t) + \delta t \mathbf{v}(t) + \frac{1}{2}(\delta t)^2 \mathbf{a}(t) + \frac{1}{6}(\delta t)^3 \mathbf{b}(t) + \cdots$$

$$\mathbf{v}^p(t + \delta t) = \mathbf{v}(t) + \delta t \mathbf{a}(t) + \frac{1}{2}(\delta t)^2 \mathbf{b}(t) + \cdots \tag{11.5}$$

$$\mathbf{a}^p(t + \delta t) = \mathbf{a}(t) + \delta t \mathbf{b}(t) + \cdots$$

$$\mathbf{b}^p(t + \delta t) = \mathbf{b}(t) + \cdots$$

To do this we ignore terms higher than those shown explicitly, and calculate the 'predicted' terms starting with $\mathbf{b}^p(t)$. However, this procedure will not give the correct trajectory because we have not included the force law. This is done at the 'corrector' step. We calculate from the new position $\mathbf{r}^p$ the force at time $t + \delta t$ and hence the correct acceleration $\mathbf{a}^c(t + \delta t)$. This can be compared with the predicted acceleration $\mathbf{a}^p(t + \delta t)$ to estimate the size of the error in the prediction step

$$\Delta \mathbf{a}(t + \delta t) = \mathbf{a}^c(t + \delta t) - \mathbf{a}^p(t + \delta t) \tag{11.6}$$

This error, and the results from the predictor step, are fed into the corrector step to give

$$\mathbf{r}^c(t + \delta t) = \mathbf{r}^p(t + \delta t) + c_0 \Delta \mathbf{a}(t + \delta t)$$

$$\mathbf{v}^c(t + \delta t) = \mathbf{v}^p(t + \delta t) + c_1 \Delta \mathbf{a}(t + \delta t)$$

$$\mathbf{a}^c(t + \delta t) = \mathbf{a}^p(t + \delta t) + c_2 \Delta \mathbf{a}(t + \delta t) \tag{11.7}$$

$$\mathbf{b}^c(t + \delta t) = \mathbf{b}^p(t + \delta t) + c_3 \Delta \mathbf{a}(t + \delta t)$$

These values are now better approximations to the true position, velocity, and so on. The constant coefficients c_0 through c_4 are available in the literature.

This summarizes the simple predictor–corrector algorithm.

Second, the *Verlet algorithm*. If we start from the Taylor expansion about $\mathbf{r}(t)$ then

$$\mathbf{r}(t + \delta t) = \mathbf{r}(t) + \delta t \mathbf{v}(t) + \frac{1}{2}(\delta t)^2 \mathbf{a}(t) + \cdots$$

$$\mathbf{r}(t - \delta t) = \mathbf{r}(t) - \delta t \mathbf{v}(t) + \frac{1}{2}(\delta t)^2 \mathbf{a}(t) + \cdots \tag{11.8}$$

the Verlet algorithm for advancing the position vector from its value at time t to time $t + \delta t$ is

$$\mathbf{r}(t + \delta t) = 2\mathbf{r}(t) - \mathbf{r}(t - \delta t) + (\delta t)^2 \mathbf{a}(t) \tag{11.9}$$

The acceleration is calculated from the force on the atom at time t. The velocity does not appear in this expression, but it may be obtained from the formula

$$\mathbf{v}(t) = \frac{\mathbf{r}(t + \delta t) - \mathbf{r}(t - \delta t)}{2\delta t} \tag{11.10}$$

The timescale of many molecular processes is a femtosecond (10^{-15} s), and this is a natural choice for the time interval δt.

11.1.2   Collection of Statistics

A MD calculation generates information about atomic positions, velocities, and so on. The conversion of this information into macroscopic variables such as the temperature and pressure is the business of statistical mechanics. Statistical mechanical averages of energetic or structural properties are obtained as averages over the time steps. So if we solve the equations of motion at times t_1, $t_2 = t_1 + \delta t$, t_3, and so on, and calculate $\mathbf{r}_A$ at each of these n time intervals (which I will denote by $\mathbf{r}_A(t_1)$, $\mathbf{r}_A(t_2)$, $\mathbf{r}_A(t_3)$, and so on) then the statistical mechanical average $\langle \mathbf{r}_A \rangle$ is just

$$\langle \mathbf{r}_A \rangle = \frac{1}{n} \sum_{i=1}^{n} \mathbf{r}_A(t_i) \tag{11.11}$$

For the purpose of exploring fluctuations about the mean, we can also calculate the standard deviation.

For a system whose potential energy function depends only on the magnitude of the separation between the constituent atoms, then both the total linear momentum and the total angular momentum have constant magnitude.

The total kinetic energy can be calculated as a sum of the kinetic energy of the individual atoms, whilst the evaluation of the potential energy involves a sum over the pairs of atoms present. The temperature can be calculated by averaging the kinetic energies of the N individual atoms, and setting this to $3Nk_B T$.

11.1.3   Heating

One often starts a MD calculation from a structure appropriate to 0 K. The first step in the calculation is to bring the system to temperature T and this is often called the heating step. By the end of the heating step, the system should have lost all its knowledge of the starting point. The initial velocities are often randomly assigned in such a way that the Maxwell–Boltzmann distribution is reproduced. The velocities are often scaled so that the atomic kinetic energy is exactly $3/2 \, k_B T$.

Figure 11.2 shows a HyperChem MD run on a single molecule of adrenaline, to illustrate some of the points.

The total energy, the kinetic energy and the potential energy are plotted. The temperature was set at 300 K, and you can see the heating phase where the total energy rises to a constant value. The kinetic and potential energies vary during the MD 'experiment', but their sum is constant.

MD therefore models the evolution of a system over time, producing a trajectory of atomic positions and velocities.

The final adrenaline structure is quite different to the starting one, and a MM optimization leads to a new equilibrium geometry. MD runs are often used to find alternative starting geometries in MM calculations, because the MD calculation can surmount energy barriers and so be used to explore conformational space.

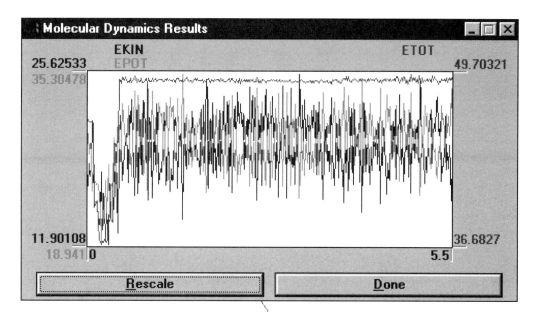

Figure 11.2   A molecular dynamics run on adrenaline at 300 K

11.1.4   Periodic Boundary Conditions

I gave the very simple example of a MD simulation on gas-phase adrenaline at 300 K in order to focus our ideas. Professionals use MD to model liquids and interfaces rather than isolated molecules in the gas phase. They typically take thousands of molecules into account in their calculations.

Think now of a MD simulation of liquid water. We need to choose a representative number of molecules, temperature and volume of space. These variables are interlinked and a suitable 'box' is shown in Figure 11.3.

This cubic box of 216 H_2O molecules has a side of 18.70136 Å and so corresponds to a density of 987.93 kg m^{-3}.

There are obviously very many molecules near the boundaries of the box, and they will experience very different forces from the molecules near the center of the box. The way round this problem is to introduce the same periodic boundary conditions that were mentioned in an earlier chapter. We surround the box with an infinite array of similar boxes and as a particle moves around in the box under study, so do its periodic images in each of the other boxes. If a particle leaves the box then its image appears from one of the other periodic boxes, as discussed earlier. That way the density is kept constant.

In a periodic system, we should include not only all forces between particle A and all other particles in the box, but also the forces between particle A and all other particles in all of the other periodic boxes. This is an infinite number of terms, and poses a problem. For potentials that are essentially short range, the summation can be restricted by the 'minimum image convention'. We take particle A to be the center of coordinates, and draw a further box round A of the same size as the periodic box. Only the forces from particles within this box are considered in the calculation. A further approximation is to take an additional spherical cut-off around the particle.

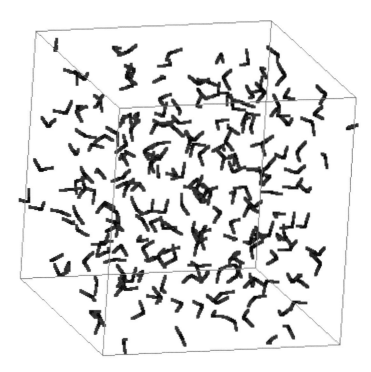

Figure 11.3   A box of water molecules

Long-range potentials such as the potentials between ions have to be dealt with explicitly by summing the series in much the same way as discussed in Chapter 4.

11.2   THE MONTE CARLO METHOD

The Monte Carlo method was developed at the end of the Second World War in order to study the diffusion of neutrons in fissionable material. The name Monte Carlo was allegedly chosen because of the extensive use of random numbers in the calculation. As is often the case in science, there was nothing new about these calculations. According to popular legend, the eighteenth century French naturalist Búffon discovered that if a needle of length l were dropped at random onto a set of parallel lines with spacing d ($>l$), then the probability of the needle crossing the line is $2l/\pi d$.

As you will know, π is an irrational number given by the ratio of the circumference of a circle to its diameter, or alternatively the ratio of the area of a circle to the square of its diameter. Consider then Figure 11.4, which shows the positive quadrant of a circle of radius R embedded in a square of side R.

The area of the quadrant is $\frac{1}{4}\pi R^2$ and the area of the square is R^2. The ratio of the areas of the quadrant to the square is $\frac{1}{4}\pi R^2$ divided by R^2, which gives $\pi/4$.

The Monte Carlo method for estimating the value of π is to choose points at random within this figure. If the chosen point falls inside the circle quadrant, it counts towards the area of the circle. Otherwise, the point counts towards the area of the square.

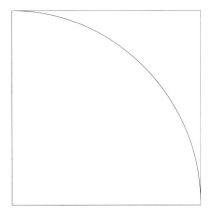

R

Figure 11.4   Monte Carlo calculation

Table 11.1 gives a simple BASIC program to do the calculation, including the selection of random numbers for the x and y coordinates of the point chosen.

You might have to modify the statements

XI = RND
YI = RND

Table 11.1   BASIC program for the Monte Carlo calculation

```
REM   MONTE CARLO CALCULATION OF PI
REM
DIM NTRIES(10)
DATA 10,100,1000,10000,100000,1000000
NGOS = 6
RESTORE
CLS
FOR I = 1 TO NGOS
READ NTRIES(I)
NEXT I
FOR IGOS = 1 TO NGOS
NTRY = NTRIES(IGOS)
QUADRANT = 0
FOR I = 1 TO NTRY
XI = RND
YI = RND
DIST = SQR(XI * XI + YI * YI)
IF DIST  < 1 THEN QUADRANT = QUADRANT + 1
1000 NEXT I
NEWPI = 4 * QUADRANT / NTRY
PRINT
PRINT "VALUE OF PI AFTER"; NTRY; "TRIES IS"; NEWPI
NEXT IGOS
END
```

Table 11.2   Results of the Monte Carlo calculation

NTRIES	10	100	1000	10 000	100 000	1 000 000
NEWPI	2.4	3.32	3.088	3.1548	3.1382	3.1437

which return random numbers in the range $[0, 1]$ depending on your version of BASIC, but otherwise you will find results for NEWPI similar to the following ones in Table 11.2.

Atoms and molecules in fluids are in ceaseless random motion and it is possible to set up models which imitate the motion for systems containing thousands of particles. In the Monte Carlo method, the thermodynamic properties of the model system are calculated according to the methods of statistical thermodynamics.

Refer back now to Figure 11.3, which shows a number of water molecules in a sample box. For the minute, ignore molecules in the other surrounding periodic boxes. To imitate the motion of the water molecules in this box, they are programmed to move every time step in accordance with certain rules to be discussed, thereby generating new configurations. Once enough moves have been made, the thermodynamic properties can be found by averaging over the configurations.

Taking a Cartesian coordinate system for the sake of argument, we select three random numbers and use these to increment the x, y and z coordinates of each atom in turn. The total potential energy U is calculated, together with the Boltzmann factor $\exp(-U/k_\mathrm{B}T)$ for the temperature of interest. From that point on, we could calculate the thermodynamic properties according to the formulae in Chapter 2. There is however a difficulty with this procedure, especially when applied to a fluid.

Because of the close packing of atoms in a fluid, an arbitrary displacement of coordinates will very often bring one atom very close to one or more of its neighbors, and in this case the potential energy can become very large indeed. The Boltzmann factor then becomes zero and this can be the case for a very high proportion of all the new configurations generated. The method is then essentially useless for the calculation of sample properties, since they involve an integral over the Boltzmann factor. In Metropolis' modification to the Monte Carlo method, we get round this problem by allowing atoms the opportunity to 'bounce back' to their starting position if the potential energy rise is unfavorable. We start with the atoms at fixed positions and calculate the possible changes in their coordinates as described above. We also calculate the predicted change in the potential energy, ΔU. If ΔU is negative then the predicted (new) configuration is accepted. If ΔU is positive, then we first calculate the Boltzmann factor $\exp(-\Delta U/k_\mathrm{B}T)$, a number that must lie between 0 and 1. We then select a random number in the range 0 to 1, and compare this to the Boltzmann factor. If the Boltzmann factor is greater than the random number then the change is allowed. If the Boltzmann factor is smaller than the random number then the change is not allowed and so the atoms bounce back to their original position.

The periodic box is used in Monte Carlo calculations just as in the MD case.

12   The Ideal Monatomic Gas

In this chapter I want to show how to go about relating microscopic properties to macroscopic ones. In particular, I will show how to derive a simple equation of state from a knowledge of the properties of individual atoms and molecules. We will consider both classical and quantum mechanical models, and in this particular case they both give the same answer; the ideal equation of state.

Consider then a collection of N 'ideal-gas' particles contained in a simple three-dimensional cubic box of side L, of the kind discussed in Chapter 5 (Figure 12.1). The only thing we need know about the particles is that they have a translational energy. They have no internal structure and no internal energy levels, and they do not interact with each other.

I have labelled the axes x, y, z in the usual sense. The box is thermally insulated so that no energy can enter or escape. In this chapter the symbol U will be used for internal energy, in the thermodynamic sense of Chapter 2. This unfortunately clashes with the use of U for the mutual potential energy in other chapters, but it should not cause confusion.

12.1   THE CLASSICAL TREATMENT

In the simple classical treatment, we focus attention on the individual particles and the fact that they have translational energy. This means that in the course of time many of them will collide with the walls of the container and hence exert a force on the walls of the container. Pressure is force per area and I want to show that the product of pressure and volume is a constant at a constant temperature.

Each atom is assumed to be moving in a random direction with a velocity vector

$$\mathbf{u} = u_x \mathbf{e}_x + u_y \mathbf{e}_y + u_z \mathbf{e}_z$$

The magnitude of this vector is the speed, which I will write as u, or in terms of the components

$$u = \sqrt{(u_x^2 + u_y^2 + u_z^2)}$$

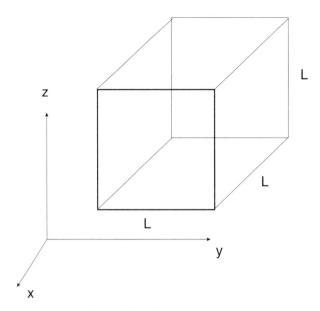

Figure 12.1   Three-dimensional box

As explained in Chapter 8, the particles will not necessarily have the same speed. I gave you a precise mathematical description of the distribution of these speeds, but for this chapter I am going to assume that all particles have the same speed u. This is best identified with the root mean square speed.

Velocity is a vector quantity, and we can treat the components of the vector independently. Although the directions of the particle velocities will be different, we might expect that on average the components of the vector will be the same along each of the directions specified as $+x$, $-x$, $+y$, $-y$, $+z$ and $-z$. This is not as great an assumption as you might imagine, it relies on the fact that space is isotropic and that we are dealing with a very large number of particles.

What I want to do is to calculate the pressure on one of the walls of the box. Pressure is force divided by area, and according to Newton's second law, force is also given by the rate of change of momentum.

The calculation falls neatly into two parts. First of all, I want to calculate the rate of change of momentum for a particle striking a wall of the box. This gives the force exerted per collision. I then need to find the number of particles striking a given wall per unit time. A product of these two factors gives the pressure.

(i) Let us focus attention on those particles moving in the $+x$ direction. I will take the speed in this direction to be u_x and so the linear momentum is $p_x = mu_x$. Every time a particle strikes the wall, it changes its x-momentum by $2\,mu_x$. For a time interval Δt the rate of change of momentum is therefore $2\,mu_x/\Delta t$.

(ii) The number of particles colliding with the wall per unit area and per unit time can be found as follows. Consider a single particle traveling along the $+x$ axis. If it is within a distance $u_x\Delta t$ of the box wall, then it will strike the wall within the time interval Δt, providing that it is traveling in the $+x$ direction. If the particle is traveling in the $-x$ direction then it will not reach the wall in the time interval. Because space is isotropic, we can assume that exactly as many particles are moving in the $+x$ direction as are moving in the $-x$ direction.

The wall of the box has area $A = L^2$ and so all particles traveling along the $+x$ axis and lying in a volume $Au_x\Delta t$ will reach the wall. If there are N particles in total then there are N/V particles per unit volume V and so the number colliding with the wall will be

$$\frac{1}{2}\frac{N}{V}Au_x\Delta t$$

(the factor $\frac{1}{2}$ arising because as many particles will be traveling in the $+x$ direction as in the $-x$ direction, as discussed above).

The pressure exerted on a single wall will therefore be

$$p = \frac{\text{force}}{L^2}$$

which is

$$p = \frac{1}{L^2}\frac{2mu_x}{\Delta t} \times \frac{1}{2}\frac{N}{V}L^2 u_x\Delta t$$

from our arguments above. Hence

$$pV = mNu_x^2$$

Now

$$u^2 = (u_x^2 + u_y^2 + u_z^2)$$

and since space is isotropic, the values of these square values should also be equal. Thus if we call v^2 the mean square speed, $u_x^2 = \frac{1}{3}u^2$ and so

$$pV = \frac{1}{3}mNu^2$$

As I mentioned earlier, not all particles will be traveling at the same speed. There will be a distribution of speeds, with a certain mean value just like every other distribution of anything. If we denote the mean value of the square of the speeds $\langle u^2 \rangle$ then the correct expression is

$$pV = \frac{1}{3}mN\langle u^2\rangle$$

The kinetic energy of a single particle is $\varepsilon_{\text{kin}} = \frac{1}{2}mu^2$ and so we can write the result as

$$pV = \frac{2}{3}N\langle\varepsilon_{\text{kin}}\rangle$$

Because the only energy being considered is the kinetic energy of the particles, the internal energy of the box U is $N\langle\varepsilon_{\text{kin}}\rangle$.

$$pV = \frac{2}{3}U$$

According to the classical equipartition of energy principle, the internal energy of the particles is $\frac{3}{2}Nk_{B}RT$, provided the particles are indeed monatomic. So finally we have

$$pV = nRT$$

which is of course the ideal equation of state.

12.2   THE QUANTUM TREATMENT

In Chapter 5, I discussed the quantum mechanical description of a particle in a selection of potential wells. In the particular case of a three-dimensional well of side L the following results were derived for the energies and wavefunctions;

$$\psi_{n,k,l} = \left(\frac{2}{L}\right)^{3/2} \sin\left(\frac{n\pi}{L}x\right) \sin\left(\frac{k\pi}{L}y\right) \sin\left(\frac{l\pi}{L}z\right)$$

$$\varepsilon_{n,k,l} = (n^2 + k^2 + l^2)\frac{h^2}{8mL^2}$$

where n, k and l are quantum numbers each of which can take values of $1, 2, 3, \ldots$.

What I want to do in this section is to show how we can recover an equation of state from the quantum mechanical treatment of such a system.

In a given box of particles at (say) 300 K, there will be a distribution of particles amongst the available energy levels. Some particles will have high quantum numbers and some will have small ones. Not only that, but as the particles collide they will exchange energy and so they will change their quantum numbers. For the minute we just need to be aware that the particles can have different quantum numbers, subject only to the restriction that the total energy of the box remains constant.

Suppose that, when we make a measurement we find the following arrangement particle 1 has quantum numbers n_1, k_1 and l_1 and hence energy

$$\varepsilon_1 = (n_1^2 + k_1^2 + l_1^2)h^2/8mL^2$$

particle 2 has quantum numbers n_2, k_2 and l_2 and hence energy

$$\varepsilon_2 = (n_2^2 + k_2^2 + l_2^2)h^2/8mL^2$$

and so on until we reach the final particle N which has quantum numbers n_N, k_N and l_N and hence energy

$$\varepsilon_N = (n_N^2 + k_N^2 + l_N^2)h^2/8mL^2$$

If I write the total kinetic energy of all the particles in the box as E_{kin} then

$$E_{kin} = \varepsilon_1 + \varepsilon_2 + \cdots + \varepsilon_N$$

which will be given by

$$E_{kin} = \frac{h^2}{8mL^2} \times (\text{a function of the three quantum numbers})$$

I will write this for brevity as

$$E_{kin} = \frac{h^2}{8mL^2} A$$

I do not need to know A. E_{kin} is actually the internal energy, in the thermodynamic sense, since the potential energy is zero, so I will now write it as U.

$$U = \frac{h^2}{8mL^2} A$$

Also, since the box is a cube of side L then its volume $V = L^3$. I can substitute for L in the expression for U to get

$$U = \frac{h^2}{8m} V^{-2/3} A$$

Now we know from the discussion of thermodynamics in Chapter 2 that $\partial U / \partial V = -p$ and so the particle in a box model gives a relationship between the pressure and volume

$$p = \frac{2}{3} \frac{h^2}{8m} V^{-5/3} A$$

which can be rearranged together with the expression for U to give

$$pV = \frac{2}{3} U$$

that is to say, the pressure times the volume is a constant at a given temperature. Not only that, but if the internal energy of the sample (of amount n) really is given by the equipartition of energy expression $\frac{3}{2} nRT$ then we recover the ideal gas law!

We have now derived an equation of state for an ideal monatomic gas using a quantum mechanical description of the gas. Although I chose to use a cubic box for my derivation, the law is equally valid for other containers.

13   Quantum Gases

The properties of a system in equilibrium can be determined in principle by counting the number of states accessible to the system under different conditions of temperature and pressure.

I am going to illustrate these ideas using three different collections of particles in a three dimensional potential well. First of all, we will consider a collection of atoms that obey just one of the rules of quantum mechanics; that their energy in the potential well is quantized. The atoms themselves have mass but no interesting features such as spin, and so on. They are assumed to be distinguishable. Then we will consider a collection of photons in the potential well, and finally a collection of electrons in a metallic lattice.

For this chapter, I will assume that the particles do not interact with each other, or if they do interact then the interaction is a very weak one or that it can be averaged out in some way. That is why I have referred to a 'gas'. Coulomb's law reminds us that a pair of electrons do in reality interact very strongly with each other, but I will leave the details of that interaction for other chapters.

Earlier, I showed you how to solve the time-independent Schrödinger equation for particles in simple potential wells. In particular, we derived the following wavefunctions and energy levels in the special case of a one-dimensional potential well of length L, with zero potential inside the so-called 'box' and infinite potential elsewhere.

$$\psi_n = \sqrt{\frac{2}{L}} \sin \left(\frac{n\pi}{L} x \right)$$

$$\varepsilon_n = \frac{n^2 h^2}{8mL^2}$$

(13.1)

There is a single quantum number n, which can take values of $1, 2, \ldots$. The energy depends only on n^2 and there are no degeneracies. Each quantum state belongs to a unique energy level. The separation between the individual energy levels increases as n increases.

For a three-dimensional cubic potential well, the corresponding energy levels and wave-functions become

$$\psi_{n,k,l} = \left(\frac{2}{L}\right)^{3/2} \sin\left(\frac{n\pi}{L}x\right) \sin\left(\frac{k\pi}{L}y\right) \sin\left(\frac{l\pi}{L}z\right)$$

$$\varepsilon_{n,k,l} = (n^2 + k^2 + l^2)\frac{h^2}{8mL^2}$$

(13.2)

There are now three quantum numbers that I write n, k and l, each of which can take values of $1, 2, \ldots$. An energy level diagram is shown in Figure 13.1, compared to the one-dimensional case. Many of the energy levels are now degenerate, and many given energy levels each correspond to a number of quantum states.

The energy level pattern for a three-dimensional well is quite different from that for a one-dimensional well for the following reasons.

1. Some quantum states are 'naturally' degenerate simply because the energy depends on the quantum numbers n, k and l in a symmetric way. So for example, all six quantum states $\psi_{1,2,3}$, $\psi_{1,3,2}$, $\psi_{2,1,3}$, $\psi_{2,3,1}$, $\psi_{3,1,2}$ and $\psi_{3,2,1}$ are degenerate and have an energy of $14h^2/(8mL^2)$.

2. There are also many 'accidental' degeneracies; for example, $\psi_{1,2,6}$, $\psi_{1,6,2}$, $\psi_{2,1,6}$, $\psi_{2,6,1}$, $\psi_{6,1,2}$, $\psi_{6,2,1}$, $\psi_{4,4,3}$ $\psi_{4,3,4}$ and $\psi_{3,4,4}$ all have energies of $41h^2/(8mL^2)$. This is due to the fact that $1^2 + 2^2 + 6^2$ and $4^2 + 4^2 + 3^2$ are both equal to 41.

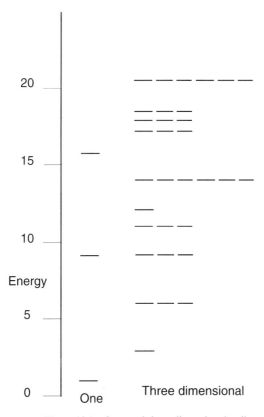

Figure 13.1 One- and three-dimensional wells

Table 13.1   Calculating of density states

Number of states with energies lying between	$D(\varepsilon)\Delta\varepsilon$
(100 and 110) $h^2/8mL^2$	85
1000 and 1010) $h^2/8mL^2$	246
(10 000 and 10 010) $h^2/8mL^2$	1029

The number of natural and accidental degeneracies rises steeply with the energy.

3. Most important of all, as the energy increases, the number of quantum states of near equal energy is seen to increase. The quantum states also crowd together as the energy increases. This crowding together does not happen in the one-dimensional case. Imagine now a macroscopic cubic 'box' of dimension 1 dm^3 in which we have ^{20}Ne atoms at 300 K. According to the classical equipartition of energy principle, each atom has an average energy of $\frac{3}{2}k_BT$, and if we equate the classical average to the quantum mechanical formula for an atom with 'average' quantum numbers

$$(n^2 + k^2 + l^2)\frac{h^2}{8mL^2} = \frac{3}{2}k_BT$$

then direct substitution shows that we can expect to find quantum numbers as high as 10^9.

For such high quantum numbers, the energy levels crowd together so much that they essentially form a continuum. Energy is certainly quantized, but the separation between the energy levels is minute in comparison to k_BT. Also, the number of individual quantum states increases markedly with the quantum numbers (i.e. with the energy). There is so little difference in energy between the neighboring energy levels compared with k_BT that we can treat the energy (and hence the quantum numbers) as being continuous. This is called the *continuum approximation*.

What matters in many experiments is not the precise details of the energy level diagram but a quantity called the *density of states* $D(\varepsilon)$, defined as follows.

Number of states between ε and $\varepsilon + \Delta\varepsilon = D(\varepsilon)\Delta\varepsilon$

Direct calculation gives the results shown in Table 13.1 for the three-dimensional potential well.

It can be seen from this simple calculation that the number of quantum states in a small energy interval rises dramatically with the energy.

13.1   RAYLEIGH COUNTING

Next I want to prove that the density of quantum states (as distinct from energy levels) for a three-dimensional potential well is proportional to the square root of the energy. I can do this using a process called *Rayleigh counting*.

First of all, I plot out the three quantum numbers n, k and l as Cartesian coordinates (Figure 13.2). The points along each axis are therefore at coordinates $1, 2, \ldots$.

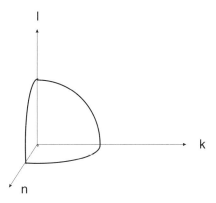

Figure 13.2   Rayleigh counting 1

For large quantum numbers, there will be very many possible combinations of the three quantum numbers n, k and l that correspond to the same energy.

$$\varepsilon_{n,k,l} = (n^2 + k^2 + l^2)\frac{h^2}{8mL^2} \qquad (13.3)$$

In other words, very many quantum states will be degenerate.

According to Pythagoras' theorem, each combination of n, k and l will lie on the surface of a sphere drawn at the coordinate origin shown above, and with radius r given by

$$r^2 = n^2 + k^2 + l^2$$

This r is not the same as L, the fixed dimension of the cubic box.

Part of this sphere in Figure 13.3. The volume of a sphere with radius r is $4/3\pi r^3$.

We only need consider the octant shown because the allowed quantum numbers are all positive and so all the quantum states with equal energy lie on the surface of the sphere octant. In other words, all points on the sphere octant correspond to combinations of the three quantum numbers which give the same energy ε.

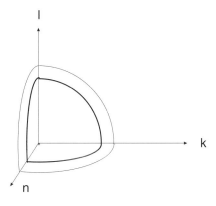

Figure 13.3   Rayleigh counting 2

All points within the sphere octant correspond to combinations of the three quantum numbers that give an energy less than or equal to ε.

The number of states with energy $\leq \varepsilon$ is therefore the volume of the octant

$$\frac{1}{8} \frac{4\pi}{3} (n^2 + k^2 + l^3)^{3/2}$$

or, in terms of the energy ε

$$\frac{1}{8} \frac{4\pi}{3} \left(\frac{8mL^2}{h^2}\right)^{3/2} \varepsilon^{3/2}$$

I now draw a second sphere of radius $r + \Delta r$. The inner sphere in Figure 13.4 has radius r, and the outer sphere has radius $r + \Delta r$. This outer sphere contains all quantum states having energy $\leq \varepsilon + \Delta \varepsilon$, and the volume of this octant is

$$\frac{1}{8} \frac{4\pi}{3} \left(\frac{8mL^2}{h^2}\right)^{3/2} (\varepsilon + \Delta \varepsilon)^{3/2}$$

The number of states having energy between ε and $\varepsilon + \Delta \varepsilon$ is given by the volume enclosed between these two sphere octants

$$\frac{1}{8} \frac{4\pi}{3} \left(\frac{8mL^2}{h^2}\right)^{3/2} (\varepsilon + \Delta \varepsilon)^{3/2} - \frac{1}{8} \frac{4\pi}{3} \left(\frac{8mL^2}{h^2}\right)^{3/2} \varepsilon^{3/2}$$

Use of the binomial theorem shows this to be

$$\frac{1}{8} 2\pi \left(\frac{8mL^2}{h^2}\right)^{3/2} \varepsilon^{1/2} \Delta \varepsilon \tag{13.4}$$

which is seen to be $D(\varepsilon)\Delta\varepsilon$.

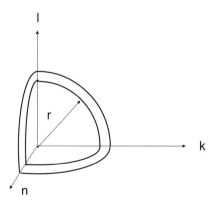

Figure 13.4   Rayleigh counting 3

I will now introduce a proportionality constant B so that

$$D(\varepsilon) = B\varepsilon^{1/2} \tag{13.5}$$

13.2   THE MAXWELL–BOLTZMANN DISTRIBUTION OF ATOMIC KINETIC ENERGIES

It is possible to perform experiments that measure the kinetic energies of atoms in gases. The important quantities are the mean and the standard deviation; there are so many atoms present in a macroscopic sample that we cannot possibly attempt to follow the trajectories of any individual particle and make predictions about the bulk from a knowledge of the properties of an individual.

When the experiments are carried out, the characteristic spread of translational energies shown in Figure 13.5 results. I have followed convention and plotted the quantity $(1/N)(\mathrm{d}N/\mathrm{d}\varepsilon)$ against ε, where N is the number of atoms in the sample.

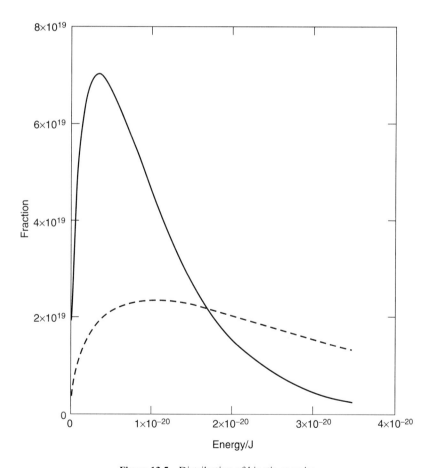

Figure 13.5   Distribution of kinetic energies

The upper curve relates to a temperature of 500 K and the lower one relates to 1500 K. The peak moves to higher energy as the temperature increases, but the area under the curve remains unchanged. The curves are independent of the identity of the atomic gas.

The quantity

$$g(\varepsilon) = \frac{dN}{d\varepsilon} \qquad (13.6)$$

is called the *distribution function* of the translational energy; it gives the number dN of atoms whose energies lie between ε and $\varepsilon + \Delta\varepsilon$. The related quantity

$$\frac{1}{N}\frac{dN}{d\varepsilon} \qquad (13.7)$$

gives the fraction of atoms whose energies lie between ε and $\varepsilon + \Delta\varepsilon$. The number of atoms whose energies lie between ε_A and ε_B is therefore

$$\int_{\varepsilon_A}^{\varepsilon_B} g(\varepsilon)\, d\varepsilon \qquad (13.8)$$

and the total number of atoms N is related to $g(\varepsilon)$ by

$$N = \int_0^\infty g(\varepsilon)\, d\varepsilon \qquad (13.9)$$

For the record, the total translational energy (i.e. the internal energy U) is given by

$$U = \int_0^\infty \varepsilon g(\varepsilon)\, d\varepsilon \qquad (13.10)$$

I will give you a very simple explanation, based on the density of states $D(\varepsilon)$ discussed above.

As we noted earlier, the spacing between the translational energy levels for an atom in a macroscopic potential well is tiny. It is so tiny that it is negligible in comparison to $k_B T$ at room temperature, and so tiny that we can make the continuum approximation. My explanation will only work correctly for atoms at ordinary temperatures. It will not work at low temperatures because the continuum approximation will not be valid, and it will not work for molecules because they have vibrational and rotational energies in addition to the translational ones.

Apart from the density of quantum states $D(\varepsilon)$, we have to consider the probability that a given quantum state will be occupied at a certain temperature. This is given by the Maxwell–Boltzmann formula

$$p(\varepsilon) = AN \exp\left(-\frac{\varepsilon}{k_B T}\right)$$

where A is a constant and N the number of atoms.

Thus, combining the density of states and the probabilities together we predict that

$$g(\varepsilon) = AN \exp\left(-\frac{\varepsilon}{k_B T}\right) B\varepsilon^{1/2}$$

It is convenient to eliminate the constants A and B in terms of N, since

$$N = \int_0^\infty g(\varepsilon)\, d\varepsilon$$

We find that

$$g(\varepsilon)\Delta\varepsilon = \frac{2N}{\sqrt{\pi}\,(k_B T)^{3/2}}\,\varepsilon^{1/2}\,\exp\left(-\frac{\varepsilon}{k_B T}\right)\Delta\varepsilon \qquad (13.11)$$

which gives exact agreement (Figure 13.5) with experiment at high temperatures. The lower curve corresponds to the higher temperature.

It can be shown by direct differentiation that the peak occurs at energy $\varepsilon = \frac{1}{2}k_B T$ and it can also be shown from integration of the above equations that the average energy per atom is $\frac{3}{2}k_B T$, in accord with the equipartition of energy principle.

I have derived the result on the basis of quantum mechanics and the Maxwell–Boltzmann distribution.

13.3 BLACK BODY RADIATION

In Chapter 5 I mentioned the black body radiation problem, which is a study of the electro-magnetic radiation emitted by a cavity. A study of the black body phenomenon was a mile-stone in the path to modern quantum mechanics.

The curves in Figure 13.6 relate to the energy emitted per wavelength at temperatures of

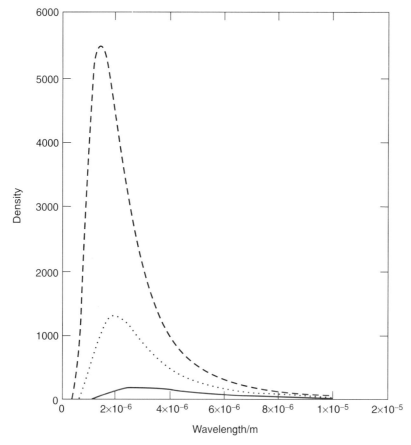

Figure 13.6 Black body radiation

1000 K, 1500 K and 2000 K. The quantity plotted is $dU/d\lambda$. The top curve relates to the highest temperature, and the bottom curve relates to the lowest temperature.

It is seen that

- each of the curves has a peak at a certain wavelength λ_{max}
- λ_{max} moves to shorter wavelength as the temperature increases

The black body curves are totally different to the atomic translational energy curves shown earlier in this chapter. They have a different shape altogether.

Analysis of the experimental data confirms that there is a quantitative relationship between the temperature and the wavelength maximum

$$T\lambda_{max} = A \tag{13.12}$$

where A is a constant, and this is known as Wien's displacement law. The experimental value of the constant A is 2.9×10^{-3} m K and so at 1000 K the maximum wavelength is about 2900 nm.

Stefan also noticed that the energy U obeys

$$U = aT^4 \tag{13.13}$$

where a is the Stefan–Boltzmann constant with a value of 5.67×10^{-8} W m^{-2} K^{-4}. Rayleigh and Jeans treated the electromagnetic radiation as a collection of classical oscillators. They then applied the equipartition of energy principle to each oscillator, to give the Rayleigh–Jeans law

$$\frac{1}{V}\frac{dU}{d\lambda} = \left(\frac{8\pi k_B T}{\lambda^4}\right) \tag{13.14}$$

The quantity on the left-hand side is the energy density per wavelength (hence the $1/V$). This expression does not have a maximum, it simply increases as the wavelength gets smaller. The fact that their predicted curve did not have a maximum became known as the *ultraviolet catastrophe*.

Planck studied the problem, and was able to deduce an expression for the energy density as a function of wavelength which fitted the experimental data very closely

$$\frac{1}{V}\frac{dU}{d\lambda} = \left(\frac{8\pi h c_0}{\lambda^5}\right)\left[\frac{\exp(-hc_0/\lambda k_B T)}{1 - \exp(-hc_0/\lambda k_B T)}\right] \tag{13.15}$$

It is profitable to think of the problem in a slightly different way; we know that the cavity contains photons, and that the energy of a photon is hc_0/λ. What is actually measured is a distribution function $g(\varepsilon)$, such that the number of photons with energies between ε and $\varepsilon + \Delta\varepsilon$ is $g(\varepsilon)\Delta\varepsilon$. Rearrangement of the equation above gives

$$g(\varepsilon) = \frac{8\pi V}{h^3 c_0^3}\varepsilon^2\left[\frac{\exp(-\varepsilon/k_B T)}{1 - \exp(-\varepsilon/k_B T)}\right]$$

or, on multiplying through top and bottom by the exponential $\exp(\varepsilon/kT)$

$$g(\varepsilon) = \frac{8\pi V}{h^3 c_0^3}\varepsilon^2\left[\frac{1}{\exp(\varepsilon/k_B T) - 1}\right] \tag{13.16}$$

If we rewrite this expression as

g(ε) = (density of states) × (probability that a given state is occupied)

we see that the density of states now depends on ε^2.

Not only that but the probability factor $p(\varepsilon)$ is no longer equal to the Boltzmann factor. It is equal to

$$p(\varepsilon) = \left(\frac{1}{\exp\left(\dfrac{\varepsilon}{k_{\mathrm{B}}T}\right) - 1} \right) \tag{13.17}$$

rather than

$$p(\varepsilon) = AN \exp\left(-\frac{\varepsilon}{k_{\mathrm{B}}T} \right)$$

This probability is often referred to as the Bose factor (after S. N. Bose).

I can explain the density of states quite easily. Photons have zero mass, and so the energy expression for a particle in a three-dimensional potential well

$$\varepsilon_{n,k,l} = (n^2 + k^2 + l^2) \frac{h^2}{8mL^2}$$

is clearly not appropriate. The problem is that the version of quantum mechanics I used in order to derive this result does not apply to photons, which are relativistic particles. To cut a long story short, the energy level formula for photons is given by

$$\varepsilon_{n,k,l} = (n^2 + k^2 + l^2)^{1/2} \frac{hc_0}{2L}$$

There are still three quantum numbers, but the formula does not have a mass. I can give a rough and ready justification of the formula as follows.

The square of the linear momentum of a 'particle' in the potential well is $p^2 = m^2v^2$ which is $2m\varepsilon$. This gives

$$2m\varepsilon = \frac{h^2}{4L^2}(n^2 + k^2 + l^2)$$

The square of the linear momentum for a photon is given by $p^2 = \varepsilon^2/c_0^2$ which is

$$\left(\frac{\varepsilon}{c_0} \right)^2 = \frac{h^2}{4L^2}(n^2 + k^2 + l^2)$$

As the quantum numbers increase, the quantum states become more numerous. There is a difference between the particle in a potential well model and the photon in a potential well model; the quantum states for the photon in the potential well cram together more quickly than those for the atom in the potential well.

Once again, the density of states can be quantified by taking a narrow range of energies $\Delta\varepsilon$ and counting the states that lie between ε and $\varepsilon + \Delta\varepsilon$. This of course is defined as $D(\varepsilon)\Delta\varepsilon$ where $D(\varepsilon)$ is the density of states.

Rayleigh counting shows that, for a photon in a three dimensional potential well,

$$D(\varepsilon) \propto \varepsilon^2$$

A detailed calculation including the proportionality constant gives

$$D(\varepsilon) = \frac{8\pi V}{h^3 c_0^3} \varepsilon^2 \tag{13.18}$$

where V is the volume of the potential well.

Sometimes I want to calculate the number N of photons in the potential well. There is a slight twist to the argument because photons have a spin of $\frac{1}{2}$ and each spatial quantum state can therefore hold a maximum of two photons.

So far I have explained the density of states, but not the probability factor. In fact, photons do not obey the Maxwell–Boltzmann probability law. I will explain why shortly.

13.4   MODELING METALS

There are three types of material that interest people who study electrical properties; conductors, insulators and semiconductors. Conductors are materials like metals that conduct electricity. Insulators are to be thought of as materials that do not usually conduct electricity, and semiconductors show properties somewhere in between. We think of metals as rigid arrays of metallic cations surrounded with mobile electrons, as shown in the two-dimensional representation in Figure 13.7.

The cations are the larger circles, and the electrons are the smaller ones in Figure 13.7.

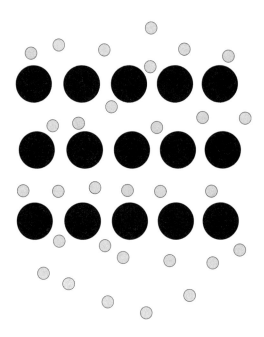

Figure 13.7   A metallic conductor

Metals are good conductors of electricity. They have smaller binding energies than ionic solids, and they also have very high thermal conductivities. The first attempt to explain these properties was the free electron model of metals developed independently by Drude and by Lorentz. According to the free electron theory, each metallic atom gives up at least one electron to the body of the metal. An alkali metal would give up one electron, an alkaline earth metal would give up two electrons, and so on. These conduction electrons move freely, like the particles in an ideal gas, about the lattice points at which the positive ions are fixed.

The conduction electrons are in constant motion and so they collide with each other and with the rigid cation metallic frame.

If we apply an electric field to these metallic electrons, then they move on average to points of lower electric potential and so an electric current can be generated. The mechanism of electrical conduction will be discussed in a later chapter, where I will have a great deal to say about the Drude model.

The conduction electrons in a metal have a wide spread of energies, and a fundamental question for this chapter is 'What is the distribution of kinetic energies of the electrons in a metallic conductor?'

There is no experimental technique that can measure this spread to the same accuracy as can be done with photons in the black body experiment or with molecules in the kinetic theory experiments. However, a variation on the theme of the photoelectric experiment (called high-energy photoemission) can be used to probe electron kinetic energies in a metal.

Experimentally it turns out that the distribution law is different from the two discussed above. First of all, the density of states turns out to be proportional to $\varepsilon^{1/2}$, as in the atomic translational energy case. The probability factor $p(\varepsilon)$ turns out to be quite different

$$p(\varepsilon) = \frac{1}{\exp[(\varepsilon - \varepsilon_F)/k_B T] + 1} \tag{13.19}$$

where ε_F is a parameter called the Fermi energy (or the Fermi level). As can be seen from Figures 13.8 and 13.9, it is quite different from the Maxwell–Boltzmann expression $p(\varepsilon) = AN \exp(-\varepsilon/k_B T)$.

Figure 13.8 shows the Fermi probability for the case of a Fermi energy of 2.5 eV and temperatures of 600 K and 6000 K. The corresponding Maxwell–Boltzmann probabilities are shown in Figure 13.9.

The experimental results for the metal can be summarized as

$$g(\varepsilon) = B\varepsilon^{1/2} \frac{1}{\exp[(\varepsilon - \varepsilon_F)/k_B T] + 1} \tag{13.20}$$

I cannot explain the probability on the basis of the Drude classical electron gas model; a quantum model is needed. In Pauli's quantum model, the conduction electrons are still treated as a gas of non-interacting particles, but the gas is analyzed more rigorously according to the rules of quantum mechanics.

Pauli assumed that each quantum state could contain a maximum of two electrons. This means that the energy level diagram for a one-dimensional potential well containing six particles in their lowest energy arrangement should be drawn as in Figure 13.10 but certainly not where I have allocated four electrons to the $n = 1$ level (Figure 13.11).

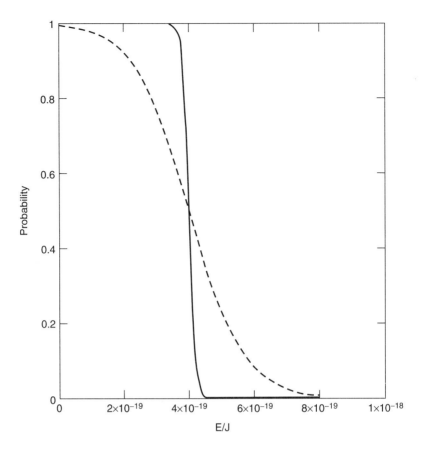

Figure 13.8 The Fermi probability

If the diagram in Figure 13.10 is correct, then I can explain the Fermi energy as follows.

(i) At $T = 0\,\mathrm{K}$, $\mathrm{p}(\varepsilon) = 1$ for $\varepsilon < \varepsilon_\mathrm{F}$ and $\mathrm{p}(\varepsilon) = 0$ for $\varepsilon > \varepsilon_\mathrm{F}$ and so all the available states are occupied up to the Fermi energy ε_F. The N electrons will fully occupy the lowest $N/2$ energy levels and the electron energy cannot be greater than the Fermi energy. The total number of states available is

$$\int_0^{\varepsilon_\mathrm{F}} \mathrm{g}(\varepsilon)\,\mathrm{d}\varepsilon$$

which must give the total number of electrons in the metal.

(ii) For electron energies much above the Fermi level, so that $\varepsilon - \varepsilon_\mathrm{F} \gg k_\mathrm{B}T$, the probability can be approximated by ignoring the 1 in the denominator to give

$$\mathrm{p}(\varepsilon) \approx \exp\left(-\frac{\varepsilon - \varepsilon_\mathrm{F}}{k_\mathrm{B}T}\right)$$

This expression is very similar to the Maxwell–Boltzmann probability. For sufficiently large energies, the Fermi probability reduces to the Maxwell–Boltzmann distribution.

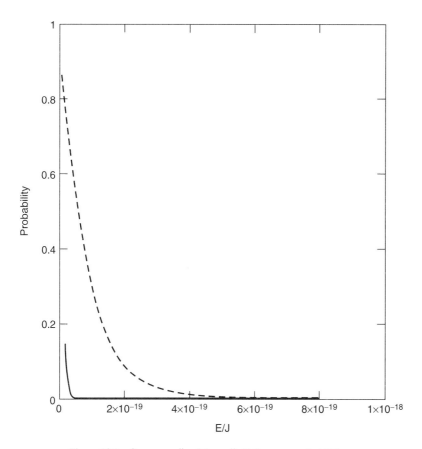

Figure 13.9   Corresponding Maxwell–Boltzmann probabilities

(iii) For electron energies below the Fermi level, the probability can be written

$$p(\varepsilon) \approx 1 - \exp\left(\frac{\varepsilon - \varepsilon_F}{k_B T}\right)$$

That is, the probability of a state being occupied is very close to unity. Similar conclusions follow if we take a three-dimensional box rather than the one-dimensional case.

We have therefore reached a point where we need to think back to the basic assumptions made when we derived the Boltzmann probability distribution.

13.5   INDISTINGUISHABILITY

I must now take time to discuss a fundamental property of microscopic particles. Consider two moving bodies (for example billiard balls) that are on a collision course with each other. In classical mechanics, it is assumed that the two moving bodies can be labelled, and their trajectories followed in detail. So we would call one of the billiard balls A and the other B,

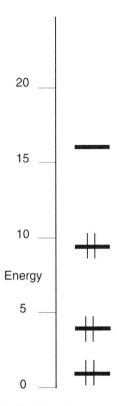

Figure 13.10   An allowed arrangement of electrons

and make deductions based on the fact that (for example) the total momentum is conserved.

$$\frac{\mathrm{d}}{\mathrm{d}t}\left(\mathbf{p}_\mathrm{A} + \mathbf{p}_\mathrm{B}\right) = 0 \tag{13.21}$$

In the case where we have microscopic bodies, it is not possible to attach labels to them. This is a consequence of the Heisenberg Uncertainty Principle. In order to pin down the positions of both particles at a given time, we would have to make an observation on the system and so a photon would have to be involved. In order to make an accurate position measurement, the photon would need to have a small wavelength and hence a large momentum. The consequent disturbance introduced by this photon would thwart all our attempts to identify the two particles at a later time. We take for granted in the macroscopic world that bodies, although superficially indistinguishable because they are mass produced, can in fact be distinguished one from the other. We might for example paint a letter A on one billiard ball, and a letter B on the second billiard ball, and follow them through their trajectories. At the microscopic level we cannot do this.

   A consequence of this fundamental property of microscopic bodies is that experimental properties that depend on the coordinates of individual particles must do so in a symmetric fashion. We cannot single out one particular particle amongst many others for special treatment.

   To give a concrete example of these considerations, let us consider again the one-dimensional potential well problem discussed above. The treatment given was appropriate to a single

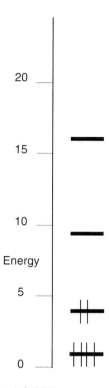

Figure 13.11 A forbidden arrangement of electrons

particle in a constant potential well, where the time-independent Schrödinger equation is

$$-\frac{h^2}{8\pi^2 m}\frac{\mathrm{d}^2\psi}{\mathrm{d}x^2} + (U-\varepsilon)\psi = 0$$

Solutions turned out to be

$$\psi_n = \sqrt{\frac{2}{L}}\sin\left(\frac{n\pi}{L}x\right)$$

$$\varepsilon_n = \frac{n^2 h^2}{8mL^2}$$

where the quantum number n is $1, 2, 3, \ldots$.

Suppose we now consider the case of two similar particles each of mass m in the same potential well. The mutual potential energy U might well now be non zero; for two charged particles such as electrons, U would be strongly dependent on the interelectron separation. For the sake of argument, I am going to assume that the interaction is either zero, or that it somehow averages to a constant (which I will take to be zero or U_0, whichever you like).

The time independent Schrödinger equation is therefore

$$-\frac{h^2}{8\pi^2 m}\left(\frac{\partial^2}{\partial x_A^2} + \frac{\partial^2}{\partial x_B^2}\right)\Psi(x_A, x_B) + (U_0 - \varepsilon)\Psi(x_A, x_B) = 0$$

where I have labelled the particles A and B. The wavefunction $\Psi(x_A, x_B)$ depends on the coordinates of both particles A and B. This labeling of particles is not inconsistent with my discussion above. I just have to be careful to ensure that observables do not appear to favour one of the particles at the expense of the other.

In the case that $U = 0$, the equation becomes

$$-\frac{h^2}{8\pi^2 m}\left(\frac{\partial^2}{\partial x_A^2} + \frac{\partial^2}{\partial x_B^2}\right)\Psi(x_A, x_B) - \varepsilon\Psi(x_A, x_B) = 0$$

and an obvious first step is to ask whether the wavefunction for the two particles $\Psi(x_A, x_B)$ can be written as a product of wavefunctions for each individual particle. That is to say, we wish to investigate whether we can write

$$\Psi(x_A, x_B) = U(x_A)V(x_B)$$

where $U(x_A)$ depends only on the coordinates of particle A, and $V(x_B)$ depends only on the coordinates of particle B.

Substitution and simplification gives

$$-\frac{h^2}{8\pi^2 m}\frac{1}{U}\frac{d^2 U}{dx_A^2} - \frac{h^2}{8\pi^2 m}\frac{1}{V}\frac{d^2 V}{dx_B^2} = \varepsilon$$

The two coordinates are independent of each other, and each of the coordinates can take any value in their ranges. This implies that either term on the left-hand side of the equation must be a constant, and the two constants have to add to the energy ε. In other words

$$-\frac{h^2}{8\pi^2 m}\frac{1}{U}\frac{d^2 U}{dx_A^2} = \varepsilon_A$$

$$-\frac{h^2}{8\pi^2 m}\frac{1}{V}\frac{d^2 V}{dx_B^2} = \varepsilon_B$$

(13.22)

These two equations are instantly recognizable as the equation for a single particle in a potential well, and the time independent Schrödinger equation for the pair of particles can be written as a product of wavefunctions each of which describes a single particle. We refer to models where total wavefunctions are written as products of the wavefunctions for individual particles as *orbital models*. The term *orbital* is widely used in modeling, particularly by chemists when the particles in question are all electrons.

The two-particle wavefunctions depend on two quantum numbers that I will write as n_A and n_B, and for brevity I will write

$$\Psi_{nA,nB}(x_A, x_B) = \psi_{nA}(x_A)\psi_{nB}(x_B)$$

or, to write it out in full

$$\Psi_{nA,nB} = \frac{2}{L}\sin\left(\frac{n_A \pi x_A}{L}\right)\sin\left(\frac{n_B \pi x_B}{L}\right)$$

(13.23)

Wavefunctions by themselves do not have any physical interpretation, but according to the Born interpretation of quantum mechanics, the square of the modulus of a wavefunction has an interpretation as a probability. For a wavefunction that depends on the coordinates of two particles, the interpretation is as follows

$$|\Psi(x_A, x_B)|^2\, dx_A\, dx_B$$

(13.24)

represents the probability that particle A will be found between x_A and $x_A + dx_A$, and that particle B will be found between x_B and $x_B + dx_B$.

So for real wavefunctions, the quantities to concentrate on are the squares of wavefunctions rather than the wavefunctions themselves, since the squares have the physical interpretation.

The lowest energy solution for the two particles in a potential well is

$$\Psi_{1,1} = \frac{2}{L} \sin\left(\frac{\pi x_A}{L}\right) \sin\left(\frac{\pi x_B}{L}\right) \tag{13.25}$$

and this certainly satisfies the requirements of the principle discussed above, because x_A and x_B both appear symmetrically in the probability expression. The next two solutions are

$$\Psi_{1,2} = \frac{2}{L} \sin\left(\frac{\pi x_A}{L}\right) \sin\left(\frac{2\pi x_B}{L}\right)$$

and

$$\Psi_{2,1} = \frac{2}{L} \sin\left(\frac{2\pi x_A}{L}\right) \sin\left(\frac{\pi x_B}{L}\right)$$

and neither of these wavefunctions by themselves satisfy the symmetry requirement because they seem to treat the two particles on an unequal footing.

Now, $\Psi_{1,2}$ and $\Psi_{2,1}$ are degenerate solutions of the time independent Schrödinger equation, and so is any linear combination $c\Psi_{1,2} + d\Psi_{2,1}$. The trick is to arrange the choice of c and d such that the square of the modulus treats each particle on an equal basis. Simple inspection shows that

$$\Psi_+ = \sqrt{\frac{1}{2}}(\Psi_{1,2} + \Psi_{2,1})$$

and

$$\Psi_- = \sqrt{\frac{1}{2}}(\Psi_{1,2} - \Psi_{2,1})$$

satisfy the criterion. These two wavefunctions Ψ_+ and Ψ_- themselves have an important property. If I interchange the names (A and B) of the two particles, the Ψ_+ remains unchanged (and so is referred to as a *symmetric* wavefunction) but Ψ_- changes sign (and so is referred to as an *antisymmetric* wavefunction). The factor $1/\sqrt{2}$ has been added to satisfy the normalization requirement.

13.6 THE EFFECT OF SPIN

Suppose now that the two particles are spin-$\frac{1}{2}$ particles (and so might be electrons). If we denote their spin variables s_A and s_B, and the spin eigenfunctions α and β, then possible spin wavefunctions for the two particles are

$$\alpha(s_A)\alpha(s_B)$$
$$\alpha(s_A)\beta(s_B)$$
$$\beta(s_A)\alpha(s_B)$$
$$\beta(s_A)\alpha(s_B)$$

The same considerations apply to spin wavefunctions as to spatial ones; the first and the fourth are satisfactory because they treat the two particles on an equal footing, but the second and the third by themselves are not satisfactory. These middle two are degenerate eigenfunctions of the spin operator, and so can be combined together without loss of generality to give spin wavefunctions that are still eigenfunctions of the spin operator S_Z, but which also treat the two particles on an equal footing. The four acceptable spin wavefunctions for such spin-$\frac{1}{2}$ particles are

$$\alpha(s_A)\alpha(s_B)$$

$$1/\sqrt{2}[\alpha(s_A)\beta(s_B) + \beta(s_A)\alpha(s_B)]$$

$$\beta(s_A)\beta(s_B)$$

$$1/\sqrt{2}[\alpha(s_A)\beta(s_B) - \beta(s_A)\alpha(s_B)]$$

The first three of these spin wavefunctions are symmetric to exchange of the particle names, and the final one is antisymmetric.

Very often we need to combine spatial and spin wavefunctions. It is a simple matter to show that the following combinations correspond to overall wavefunctions that are symmetric

symmetric space × symmetric spin

antisymmetric space × antisymmetric spin

whilst the following combinations give wavefunctions that are overall antisymmetric to the exchange of particle names

symmetric space × antisymmetric spin

antisymmetric space × symmetric spin

So, to describe the lowest energy state of two spin-$\frac{1}{2}$ particles in the potential well, we have four possible wavefunctions which I can classify as symmetric or antisymmetric (Table 13.2).

In the case of two spin-$\frac{1}{2}$ particles in the potential well with the next highest energy we have eight possible wavefunctions, and once again I can classify them as symmetric or antisymmetric (Table 13.3).

Table 13.2 Symmetric and antisymmetric functions

Space	Spin	Symmetry
$\Psi_{1,1}$	$\alpha(s_A)\alpha(s_B)$	Symmetric
	$1/\sqrt{2}[\alpha(s_A)\beta(s_B) + \beta(s_A)\alpha(s_B)]$	Symmetric
	$\beta(s_A)\beta(s_B)$	Symmetric
	$1/\sqrt{2}(\alpha(s_A)\beta(s_B) - \beta(s_A)\alpha(s_B))$	Antisymmetric

Table 13.3 Symmetric and antisymmetric wavefunctions

Space	Spin	Symmetry
$\Psi_+ = \sqrt{\dfrac{1}{2}}(\Psi_{1,2} + \Psi_{2,1})$	$\alpha(s_A)\alpha(s_B)$	Symmetric
$\Psi_+ = \sqrt{\dfrac{1}{2}}(\Psi_{1,2} + \Psi_{2,1})$	$1/\sqrt{2}[(s_A)\beta(s_B) + \beta(s_A)\alpha(s_B)]$	Symmetric
$\Psi_+ = \sqrt{\dfrac{1}{2}}(\Psi_{1,2} + \Psi_{2,1})$	$(s_A)\beta(s_B)$	Symmetric
$\Psi_+ = \sqrt{\dfrac{1}{2}}(\Psi_{1,2} + \Psi_{2,1})$	$1/\sqrt{2}[\alpha(s_A)\beta(s_B) - \beta(s_A)\alpha(s_B)]$	Antisymmetric
$\Psi_- = \sqrt{\dfrac{1}{2}}(\Psi_{1,2} - \Psi_{2,1})$	$\alpha(s_A)\alpha(s_B)$	Antisymmetric
$\Psi_- = \sqrt{\dfrac{1}{2}}(\Psi_{1,2} - \Psi_{2,1})$	$1/\sqrt{2}[(\alpha(s_A)\beta(s_B) + \beta(s_A)\alpha(s_B)]$	Antisymmetric
$\Psi_- = \sqrt{\dfrac{1}{2}}(\Psi_{1,2} - \Psi_{2,1})$	$\beta(s_A)\beta(s_B)$	Antisymmetric
$\Psi_- = \sqrt{\dfrac{1}{2}}(\Psi_{1,2} - \Psi_{2,1})$	$1/\sqrt{2}[\alpha(s_A)\beta(s_B) - \beta(s_A)\alpha(s_B)]$	Symmetric

13.7 FERMIONS AND BOSONS

It is a fundamental fact of nature that all species of particles fall into two distinct classes, *fermions* and *bosons*. Any particle with half integral spin is a fermion and any particle with integral spin (including zero) is a boson. There are no exceptions to the rule, and composite particles follow the same rule. Thus for example an atom of ^{3}He is composed of an odd number of particles each with spin $\frac{1}{2}$ so that ^{3}He must be a fermion. An atom of ^{4}He has an extra neutron, giving an even number of particles of spin $\frac{1}{2}$ and so it is a boson.

There is one big difference between the wavefunctions of fermions and bosons, and this difference is one that has profound implications. I will state this difference in two ways, both of which are equivalent. First of all, the statement in terms of symmetry and interchange of particle names.

Fermion wavefunctions are antisymmetric to exchange of particle names

Boson wavefunctions are symmetric to exchange of particle names

There are no exceptions.

Looking back at Tables 13.2 and 13.3, it is clear that only certain of the wavefunctions I have written are acceptable for fermions, and only certain other ones are acceptable for bosons.

The orbital model is widely used in modeling and so I must give you an alternative and totally equivalent statement of the above principle in terms of orbitals.

An orbital can be occupied by 0 or 1 fermions of either spin

An orbital can be occupied by any number of bosons

The statement above needs a little explanation. Electrons are fermions, and they have spin 1/2. Apart from a spatial wavefunction, we also have to consider the spin wavefunction. In the orbital approximation, we often think of a *spinorbital* which is a product of a spatial and a spin part, and this is our quantum state.

13.8 ORIGINS OF THE BOSE AND FERMI FACTORS

In Chapter 7 we studied the Boltzmann distribution law which relates the number N_i of particles in a given quantum state of energy ε_i to the total number of particles N and the

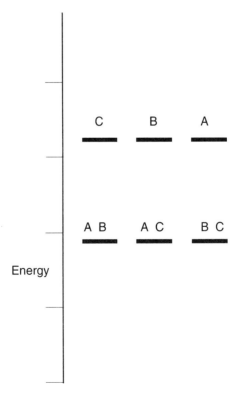

Figure 13.12   The Boltzmann counting rule

temperature T by

$$\frac{N_i}{N} = \frac{\exp\left(-\dfrac{\varepsilon_i}{k_B T}\right)}{\displaystyle\sum_{\text{states}} \exp\left(-\dfrac{\varepsilon_i}{k_B T}\right)}$$

N_i/N is the probability of finding a particle in quantum state ε_i.

In this chapter we have seen that neither the photons in thermal radiation nor the conduction electrons in a metal obey this law. The corresponding probability for photons in quantum state ε_i is given by the Bose factor

$$p_i = \frac{1}{\exp\left(\dfrac{\varepsilon}{k_B T}\right) - 1}$$

whilst the probability for conduction electrons in quantum state ε_i is equal to the Fermi factor

$$p_i = \frac{1}{\exp\left(\dfrac{\varepsilon_i - \varepsilon_F}{k_B T}\right) + 1}$$

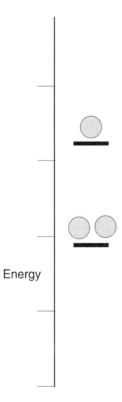

Figure 13.13   Indistinguishable particles

The question then arises as to the origin of the latter two expressions, and under what conditions each of the three is valid.

I 'proved' the Boltzmann distribution in Chapter 7. The key to the proof lies in enumerating the occupancies of quantum states by particles. In the Boltzmann counting rule, the three configurations in Figure 13.12 are all thought to be different; it matters which particle is in which quantum state.

The Boltzmann counting rule (Figure 13.12) does not apply to indistinguishable particles, and we would just consider a single configuration and make no attempt to label the particles.

Finally, since every particle is either a boson or a fermion, we have to take account of whether each individual quantum state is allowed or not; the configuration in Figure 13.13 is allowed for bosons since they can crowd into each quantum state, but is not allowed for fermions since no two fermions can occupy the same quantum level.

A detailed treatment of the counting with the above principles in mind does indeed lead to the Bose and the Fermi probabilities.

In general though, molecular gases under ordinary conditions are found to obey the Boltzmann counting rule irrespective of whether they are fermions or bosons; that is to say, they behave like classical, distinguishable particles. The reason is not hard to find; their de Broglie wavelengths are very much smaller that the characteristic spacing of the problem, in this case the average distance between molecules. It follows that a gas can be treated 'classically', i.e. by the Boltzmann counting rule.

14 Introduction to Statistical Thermodynamics

Molecular structure theory deals with the detailed structure and motion of molecules, whilst thermodynamics deals with their average behavior. For example, the pressure exerted by a gas depends on the rate of collisions of molecules with a wall, yet it is not necessary to know which particular molecule does the colliding. Statistical thermodynamics is based on the principle that thermodynamic quantities are averages of molecular properties, and it sets up a method for the calculation of these properties.

Consider then a system of N particles. Although the total energy can be specified, it is neither possible nor desirable to be definite about the distribution of the energy. The closest we can come is to specify the fraction of particles in an given quantum state, and the specification that

- N_1 particles are in quantum state number 1 with energy ε_1,

- N_2 particles are in quantum state number 2 with energy ε_2 and so on defines the *configuration* of the system.

According to the Boltzmann distribution, the Ns and the εs are related by

$$N_i = \frac{N \exp\left(-\dfrac{\varepsilon_i}{k_B T}\right)}{\displaystyle\sum_{i=1}^{n} \exp\left(-\dfrac{\varepsilon_i}{k_B T}\right)} \tag{14.1}$$

The summation runs over all the quantum states, where n might be infinite, and the formula gives the relative number of particles N_i in the quantum state i whose energy is ε_i. I have assumed that the energies are discrete rather than continuous.

In the case that certain energy levels are degenerate with degeneracy factor g_i, we focus attention on the number of particles N' whose energy is ε_i irrespective of which degenerate quantum state they are in. If there are k possible energy levels, as distinct from n possible quantum states, then the formula above should be rewritten

$$N'_i = \frac{Ng_i \exp\left(-\dfrac{\varepsilon_i}{k_B T}\right)}{\displaystyle\sum_{i=1}^{k} g_i \exp\left(-\dfrac{\varepsilon_i}{k_B T}\right)} \tag{14.2}$$

The summation now refers to the k energy levels; each energy level ε_i has g_i quantum states. Once again, k might be infinite. I have used N_i and N'_i because the numbers we calculate will generally be different, and also to encourage you to keep the difference between energy level and quantum state at the back of your mind.

From now on though, I will use the symbol N_i and the context should make it clear whether we are referring to energy levels or quantum states. Just in case there is a problem, I will write $\sum_{\text{states}}$ when I want to calculate a sum over quantum states, and $\sum_{\text{levels}}$ when I want to calculate a sum over energy levels.

The denominator (which is the same in either expression) plays a special role in our theory, and it is given a special name and symbol. The *partition function* Q (or Z, short for Zustandsumme) is defined as

$$Q = \sum_{\text{states}} \exp\left(-\frac{\varepsilon_i}{k_B T}\right) \tag{14.3}$$

where the sum runs over quantum states, or

$$Q = \sum_{\text{levels}} g_i \exp\left(-\frac{\varepsilon_i}{k_B T}\right) \tag{14.4}$$

where the sum runs over energy levels as discussed above.

In the case that we are dealing with a molecular species, then we refer to the partition function as a *molecular partition function*, and write it as q (or z). I will discuss the molecular partition function in depth in Section 14.2.

In the case where we are dealing with a system of molecules, then I will use the symbol Q (or Z).

14.1 NON-INTERACTING PARTICLES

I showed you in an earlier chapter how to calculate mean values for a system of N non-interacting particles; the mean value of the energy for a two state system is

$$\langle \varepsilon \rangle = \frac{1}{N}\left(N_1 \varepsilon_1 + N_2 \varepsilon_2\right)$$

which works out as

$$E = N\langle \varepsilon \rangle = N \frac{\varepsilon_1 \exp\left(-\dfrac{\varepsilon_1}{k_B T}\right) + \varepsilon_2 \exp\left(-\dfrac{\varepsilon_2}{k_B T}\right)}{Q} \tag{14.5}$$

I can manipulate the sum into an expression involving only Q to

$$E = N \frac{k_B T^2}{Q} \frac{dQ}{dT} \tag{14.6}$$

or

$$E = N k_B T^2 \frac{d \ln Q}{dT}$$

Advanced texts also write such expressions in terms of the variable $\beta = 1/k_B T$, so that we have

$$E = -N \frac{d \ln Q}{d\beta} \tag{14.7}$$

These expressions are perfectly general, and do not depend on my choice of a two-level system.

The internal energy U of the system depends on the volume, and is undefined to within a constant U_0 in classical thermodynamics. I can therefore identify my expression for E with the thermodynamic internal energy, provided that I write

$$U = U_0 + N \frac{k_B T^2}{Q} \left(\frac{\partial Q}{\partial T} \right)_V \tag{14.8}$$

where the differentiation has to be done at constant volume. We can usually take the constant U_0 to be zero. The partition function thus contains all the thermodynamic information needed to describe the internal energy of a system of independent particles at thermal equilibrium.

The reason why partition functions are important is that *all* thermodynamic quantities can be calculated from them. So for example, if we knew the partition functions for all the separate reactants and products in a gaseous chemical reaction then we could calculate the chemical potential of each substance and so evaluate the change in Gibbs energy ΔG at any temperature and pressure.

14.2 THE MOLECULAR PARTITION FUNCTION q

I want to distinguish between the partition function Q for a system (which might comprise N particles) and the partition function q for just one of those particles which is usually taken as an atom or a molecule. As mentioned above, chemists often speak of a *molecular partition function*, written q (or sometimes z). I will use the symbol q, and so

$$q = \sum_{\text{levels}} g_i \exp \left(-\frac{\varepsilon_i}{k_B T} \right)$$

where the sum runs over all possible energy levels of our isolated molecule. Commonly we take the zero of energy to be the lowest value of ε, but this choice can cause problems; for example, I discussed the concept of the zero point energy earlier and sometimes we have to be quite careful in our choice of energy zero.

First of all, it is worth considering the limiting behavior of q. As T approaches $0\,K$, all terms in the summation tend to zero apart from the lowest ε term, which tends to g_1. So $q = g_1$ at $T = 0\,K$.

As T approaches infinity, then every exponential term in the sum contributes a value 1 and the sum is either infinite (if there are an infinite number of quantum states), or becomes equal to the finite number of quantum states (if indeed there are a finite number of such quantum states). The molecular partition function therefore gives an indication of the number of quantum states that are thermally accessible at the temperature of interest.

To a good approximation, molecular energies are given as a sum of translational, rotational, vibrational and electronic contributions

$$\varepsilon_{tot} = \varepsilon_{trans} + \varepsilon_{rot} + \varepsilon_{vib} + \varepsilon_{el}$$

This approximation ignores any nuclear contributions, and any interactions between the translational, rotational, vibrational and electronic motions.

A typical term in the molecular partition function q is thus

$$\exp\left(-\frac{\varepsilon_{trans} + \varepsilon_{rot} + \varepsilon_{vib} + \varepsilon_{el}}{k_B T}\right)$$

or

$$\exp\left(-\frac{\varepsilon_{trans}}{k_B T}\right) \exp\left(-\frac{\varepsilon_{rot}}{k_B T}\right) \exp\left(-\frac{\varepsilon_{vib}}{k_B T}\right) \exp\left(-\frac{\varepsilon_{el}}{k_B T}\right)$$

Thus the sum

$$q_{tot} = \sum_{i=1}^{\infty} \exp\left(-\frac{\varepsilon_{tot}}{k_B T}\right) \tag{14.9}$$

can be written as a product of sums, one for the translational motion, one for the rotational motion, one for the vibrational and one for the electronic;

$$q_{tot} = \sum_{trans} \exp\left(-\frac{\varepsilon_{trans}}{k_B T}\right) \sum_{rot} \exp\left(-\frac{\varepsilon_{rot}}{k_B T}\right) \sum_{vib} \exp\left(-\frac{\varepsilon_{vib}}{k_B T}\right) \sum_{el} \exp\left(-\frac{\varepsilon_{el}}{k_B T}\right) \tag{14.10}$$

where the first sum is over the translational energies and so on. In other words, the molecular partition function can be written

$$q_{tot} = q_{trans} q_{rot} q_{vib} q_{el} \tag{14.11}$$

and each of these can be evaluated separately, as I will now illustrate for a very simple case, namely $^1H^{35}Cl$. Spectroscopic and other data $^1H^{35}Cl$ are summarized in Table 14.1.

14.2.1 q_{trans}

In earlier chapters we considered at some length the energy levels for a particle in a three-dimensional cubic box of side L. The energy turned out to depend on three quantum numbers n, k and l, with energy formula

$$\varepsilon_{n,k,l} = (n^2 + k^2 + l^2) \frac{h^2}{8mL^2} \tag{14.12}$$

The molecular partition function is therefore

$$q_{trans} = \sum_{n=1}^{\infty} \sum_{k=1}^{\infty} \sum_{l=1}^{\infty} \exp\left[-\frac{(n^2 + k^2 + l)h^2}{8mL^2 k_B T}\right]$$

Table 14.1 Spectroscopic data for $^1H^{35}Cl$

$^1H^{35}Cl$

Relative atomic masses
$^1H = 1.008665$
$^{35}Cl = 34.96885$

Reduced mass
$\mu = 1.627\,995 \times 10^{-27}$ kg

Electronic term values
C ($^1\Pi$)   77 575 cm^{-1}
V ($^1\Pi$)   77 293 cm^{-1}
A ($^1\Sigma^+$)   ... Continuous absorption starting at 44 000 cm^{-1}
X ($^1\Sigma^+$)   0

Vibrational wavenumber
$\omega_e = 2989.74$ cm^{-1}

Rotational constant
$B_e = 10.5909$ cm^{-1}

Derived information
Force constant $k_s(4\pi^2\mu\omega_e^2 c_0^2) = 516.3$ N m^{-1}
Bond length $R_e[\sqrt{(h/8\pi^2\mu B_e c_0)}] = 127.46$ pm.

A little consideration will show that this can be written

$$q_{\text{trans}} = \sum_{n=1}^{\infty} \exp\left(-\frac{n^2 h^2}{8mL^2 k_B T}\right) \sum_{k=1}^{\infty} \exp\left(-\frac{k^2 h^2}{8mL^2 k_B T}\right) \sum_{l=1}^{\infty} \exp\left(-\frac{l^2 h^2}{8mL^2 k_B T}\right)$$

Each of the three summations are equal and so we can write

$$q_{\text{trans}} = \left[\sum_{n=1}^{\infty} \exp\left(-\frac{n^2 h^2}{8mL^2 k_B T}\right)\right]^3 \tag{14.13}$$

We found earlier that the spacing between these energy levels is tiny in comparison with $k_B T$ at ordinary temperatures, and so we will now make the continuum approximation (where we replace an infinite sum by an integral and assume that n can take any positive value, not just an integral one) in order to evaluate the molecular partition function.

Thus we substitute

$$\int_0^{\infty} \exp\left(-\frac{n^2 h^2}{8mL^2 k_B T}\right) dn$$

for

$$\sum_{n=1}^{\infty} \exp\left(-\frac{n^2 h^2}{8mL^2 k_B T}\right)$$

The value of the integral can be obtained from standard compilations of integrals, and it is

$$\frac{(2\pi m k_B T)^{1/2} L}{h}$$

The translational molecular partition function q_{trans} is therefore

$$q_{trans} = \frac{(2\pi m k_B T)^{3/2}}{h^3} V \tag{14.14}$$

where $V = L^3$ is the volume of the box. I used a cubic box for the derivation, but the formula for q_{trans} is valid whatever the shape of the container.

Thus for example the translational molecular partition function for $^1H^{35}Cl$ in a container of volume $1 \, dm^3$ at $300 \, K$ is calculated as follows. The mass (not to be confused with the reduced mass) is $5.974\,30 \times 10^{-26} \, kg$ and so

$$q_{trans} = \frac{(2\pi \times 5.97430 \times 10^{-26} \, kg \times 1.38066 \times 10^{-23} \, J\,K^{-1} \times 300 \, K)^{3/2}}{(6.62618 \times 10^{-34} \, J\,s)^3} \, (10^{-3} \, m^3)$$

which gives $q_{trans} = 2.107 \times 10^{29}$. The molecular translational partition function is of course dimensionless, since it is a sum of exponentials.

14.2.2 q_{rot}

I will assume that HCl is a rigid molecule rotating about an axis that passes through its centre of mass and perpendicular to the molecular axis (Figure 14.1).

There are three possible axes of rotation, and by convention they are labelled a, b and c. Figure 14.1 is appropriate for such a motion.

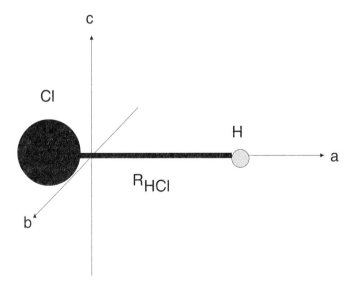

Figure 14.1 A rotating diatomic, HCl

I did not discuss rotational motion in any detail in earlier chapters, but I can give a justification for the energy formula as follows.

The classical rotational energy of a rigid diatomic molecule rotating about such an axis is

$$\varepsilon_{\text{rot}} = \frac{L^2}{2I} \tag{14.15}$$

where I is the moment of inertia about the axis. Once we apply quantum mechanics to such problems, it often turns out that angular momentum is quantized, with the square of $\mathbf{L}$ given by a formula such as $l(l+1)h^2/4\pi^2$ where l is a quantum number. For rotational spectroscopy, the quantum number is written J, and in the case of the simple diatomic motion described above $I = \mu_{\text{AB}}R_{\text{AB}}^2$. Substitution of these quantities into the classical expression yields the correct quantum mechanical energies.

The energy formula is often written

$$\varepsilon_J = hc_0 BJ(J+1) \tag{14.16}$$

where B is called the rotational constant (and has dimensions of length^{-1}), and in the specific case of HCl is

$$B = \frac{h}{8\pi^2 \mu_{\text{HCl}} R_{\text{HCl}}^2 c_0}$$

In this expression, J is the rotational quantum number which can take values of $0, 1, 2, 3, \ldots$. There is a second quantum number written M_J which does not appear in the energy formula. For each value of J, M_J can take values ranging from $-J$ to $+J$ in steps of 1. So if $J = 3$, then possible values of M_J are $-3, -2, -1, 0, 1, 2, 3$.

R_{HCl} is the internuclear separation, and the quantity μ_{HCl} is the reduced mass

$$\frac{1}{\mu_{\text{HCl}}} = \frac{1}{m_{\text{H}}} + \frac{1}{m_{\text{Cl}}}$$

which gives $\mu = 1.627\ 995 \times 10^{-27}$ kg.

Each energy level (characterized by quantum number J) is $2J + 1$-fold degenerate and so the molecular rotational partition function is

$$q_{\text{rot}} = \sum_{J=0}^{\infty} (2J+1)\exp\left[-\frac{hc_0 BJ(J+1)}{k_B T}\right] \tag{14.17}$$

The series does not sum to any closed analytical function. Typically, $B_e = 1059$ m^{-1} for HCl and direct calculation shows that the spacing between rotational energy levels is small in comparison to $k_B T$. Thus I can replace the sum by an integral

$$q_{\text{rot}} = \int_0^{\infty} (2J+1)\exp\left[-\frac{hc_0 BJ(J+1)}{k_B T}\right]\mathrm{d}J$$

which gives directly

$$q_{\text{rot}} = \frac{k_B T}{hc_0 B}$$

or

$$q_{rot} = \frac{8\pi^2 \mu_{AB} R_{AB}^2 k_B T}{h^2}$$

(14.18)

There is however some small print. It turns out that not all of the rotational energy levels are necessarily accessible to the molecule. For an unsymmetrical molecule such as HF, HD or OCS then all the rotational levels are indeed accessible and so the formula is correct. For more symmetrical molecules such as H_2 and CO_2 only half of the rotational energy levels are accessible and so we have to divide the expression for q_{rot} by a *symmetry number* σ (which in this case would be 2).

$$q_{rot} = \frac{8\pi^2 \mu_{AB} R_{AB}^2 k_B T}{\sigma h^2}$$

(14.19)

At first sight, the symmetry number owes its origin to the fact that a homonuclear diatomic produces an identical molecule on rotation by $2\pi/2$ (180°) about the b axis (or indeed any other perpendicular axis through the center of mass). Similar considerations apply for poly-atomic molecules, and although I do not propose to derive the molecular rotational partition function for non-linear molecules, I should state that the symmetry number plays an impor-tant role. H_2O belongs to the C_{2v} point group, and has a C_2 axis of rotation. The symmetry number is therefore 2. NH_3 belongs to the C_{3v} point group, and the symmetry number is 3, and so on.

The symmetry factor σ actually has its origin in quantum mechanics, and it is to do with the exchange of identical particles as discussed earlier. The spin quantum number for electrons is $s = \frac{1}{2}$, giving rise to two possible alignments of the electron spin magnetic dipole moment in an applied magnetic induction. Many nuclei also have an intrinsic angular momentum called *nuclear spin*, but they are not restricted to a quantum number of $\frac{1}{2}$. The nuclear spin quantum number I can take integral or half integral values and it turns out that atoms with integral nuclear spin (including zero) are bosons, whilst atoms with half integral spin are fermions (Table 14.2).

A fermion has to have an antisymmetric total wavefunction, whilst a boson has to have a symmetric total wavefunction.

My initial statement of the overall molecular partition function

$$q_{tot} = q_{trans} q_{rot} q_{vib} q_{el}$$

(14.20)

Table 14.2 Spin data

Nucleus	I	Nucleus	I
^1H	1/2	^2D	1
^{13}C	1/2	^7Li	3/2
^{15}N	1/2	^{14}N	1
^{19}F	1/2	^{17}O	5/2
^{29}Si	1/2	^{23}Na	3/2
^{31}P	1/2	^{33}S	
Nuclei with zero spin			
^{12}C	0	^{16}O	0
^{18}O	0	^{28}Si	0

needs to be amended to take account of the nuclear properties to read

$$q_{tot} = q_{nuc}q_{trans}q_{rot}q_{vib}q_{el} \tag{14.21}$$

and the total wavefunction should be written

$$\Psi_{tot} = \psi_{nuc}\psi_{trans}\psi_{rot}\psi_{vib}\psi_{el} \tag{14.22}$$

Entire textbooks have been written about the energy levels, spectroscopy and statistical thermodynamics of diatomic molecules. They are fascinating molecules and I want to explain some of the consequences of their high symmetry.

Nuclear spin wavefunctions can be either symmetric or antisymmetric to exchange of the names of the nuclei. Rotational states can be either symmetric or antisymmetric; rotational states with even values of J have symmetric wavefunctions whilst those with odd values of J have antisymmetric wavefunctions. Whilst electronic states can be either symmetric or antisymmetric, it is rare to find an antisymmetric ground state. Vibrational states are symmetric, as are translational states.

The properties of Ψ_{tot} to interchange of atom names is therefore determined by the product $\psi_{rot}\psi_{nuc}$.

Hydrogen has a nuclear spin $\frac{1}{2}$ and is a fermion so that the total wavefunction in a diatomic has to be antisymmetric to the exchange of nuclear labels. So rotational states with odd J values have to couple with symmetric nuclear spin wavefunctions and rotational states with even J values have to couple with antisymmetric nuclear spin wavefunctions. In dideuterium the situation is reversed; ^2D has a nuclear spin of 1 and so is a boson. This means that the total wavefunction must be symmetric and thus odd J values couple with the antisymmetric nuclear spin wavefunctions and rotational states with even J quantum numbers must couple with symmetric nuclear spin wavefunctions. This justifies the inclusion of a factor of $\sigma = 2$ in the above expressions; in order to make the total wavefunction either symmetric or antisymmetric we can only take half of the available rotational states.

14.2.3 q_{vib}

Vibrational energy levels have separations that are at least on order of magnitude greater than rotational modes, which in turn are some 20 orders of magnitude greater than translational modes. Consequently the spacing is comparable to k_BT for everyday temperatures. This means that very few vibrational levels will be occupied at ordinary temperatures for ordinary molecules.

If we restrict the discussion to the one-dimensional (ie, diatomic) harmonic oscillator then

$$\varepsilon_\nu = hc_0\omega_e\left(\nu + \frac{1}{2}\right) \tag{14.23}$$

with

$$\omega_e = \frac{1}{2\pi c_0}\sqrt{\frac{k_s}{\mu}}$$

which for HCl is $2989.74\,\text{cm}^{-1}$. It is common but not essential to take the energy zero as that of the energy level $v = 0$ and so q_{vib} can be written

$$q_{\text{vib}} = \sum_{\nu=0}^{\infty} \exp\left(-\frac{hc_0\nu}{k_{\text{B}}T}\right)$$

This simple geometric series can be summed exactly to give

$$q_{\text{vib}} = \frac{1}{\left[1 - \exp\left(-\dfrac{hc_0\omega_{\text{e}}}{k_{\text{B}}T}\right)\right]} \tag{14.24}$$

Nevertheless, it is worth bearing in mind the magnitude of $hc_0\omega_{\text{e}}/k_{\text{B}}T$. In many situations $hc_0\omega_{\text{e}}$ is greater than $k_{\text{B}}T$. In light diatomic molecules, which have high force constants as well as low reduced masses, there is essentially just one state, the vibrational ground state, that is thermally accessible at room temperature and the molecular vibrational partition function rarely has a value that is far from unity. Most molecules have only one accessible state, the vibrational state with $v = 0$, at room temperature.

14.2.4 q_{el}

It is not possible to find an algebraic formula for the electronic molecular partition function. In the case of HCl, the first excited state is possibly $44\,000\,\text{cm}^{-1}$ above the ground state, and the first characterized excited state is $77\,293\,\text{cm}^{-1}$ above the ground state. In either case, the excited state energies are sufficiently large in comparison to $k_{\text{B}}T$ to lead us to the conclusion that

$$q_{\text{el}} = 1 \tag{14.25}$$

and this conclusion is good for molecules. In the case that the ground electronic state is degenerate with a degeneracy factor g_0 then the electronic molecular partition function becomes

$$q_{\text{el}} = g_0$$

In the rare case that a molecule has a very low lying electronic excited state whose energy ε_1 is much less than $k_{\text{B}}T$ then

$$q_{\text{el}} = g_0 + g_1 \exp\left(-\frac{\varepsilon_1}{k_{\text{B}}T}\right) \tag{14.26}$$

provided that the vibrational and rotational partition functions are the same for the two electronic states.

In the case of HCl, the differences in energy between the ground electronic state and the A, V and C states are all much higher than $k_{\text{B}}T$, and the degeneracy of the ground state is 1.

14.3 THE STATISTICAL ENTROPY

In an earlier chapter I discussed the multiplicity function $g(n, t)$ of a configuration, which when divided by the total number of configurations gives the probability that this configuration can be reached, from amongst all other configurations, by purely random means. So we discussed the binary alloy problem, and concluded that the probabilities were related to the binomial coefficients.

In the context of statistical thermodynamics, $g(n, t)$ is called a *statistical weight* for a given configuration, and it is usually given the symbol W (or sometimes Ω).

The expression linking entropy with W

$$S = k_B \ln W \tag{14.27}$$

was first proposed by Boltzmann in 1877, and the legend $S = k \ln W$ is inscribed on his tombstone in Vienna. I can give you a simple-minded justification for Boltzmann's formula as follows.

Consider once again a closed assembly of N non-interacting particles (e.g. an ideal gas) with fixed energy U, where each particle can have energies $\varepsilon_1, \varepsilon_2$, and so on. If there are N_1 particles with energy ε_1, N_2 particles with energy ε_2 and so on then

$$U = U_0 + \sum_{\text{states}} N_i \varepsilon_i \tag{14.28}$$

The constant of integration U_0 can be taken as zero. A small change in U is related to small changes in N_i and ε_i as follows

$$dU = \sum_{\text{states}} N_i \, d\varepsilon_i + \sum_{\text{states 1}} \varepsilon_i \, dN_i$$

The internal energy of an ideal gas does not depend on the volume and so $d\varepsilon_i = 0$. Hence

$$dU = \sum_{\text{states}} \varepsilon_i \, dN_i \tag{14.29}$$

We know from classical thermodynamics that a infinitesimal reversible heat change produces an entropy change of dS, which is related to dU (for a constant volume change) as follows

$$dS = \frac{dU}{T} = \sum_{\text{states}} \frac{\varepsilon_i}{T} \, dN_i \tag{14.30}$$

In Chapter 7 I gave you a derivation of the Boltzmann distribution; we considered the statistical weight

$$W = \frac{N!}{N_1! N_2! \dots N_p!}$$

and I showed how the maximum of W could be calculated by Lagrange's method. This gave a condition for the predominant configuration when

$$\frac{\partial \ln W}{\partial N_i} + \alpha - \beta \varepsilon_i = 0$$

where β turns out to be

$$\beta = \frac{1}{k_\mathrm{B} T}$$

Eliminating β from these equation gives

$$\mathrm{d}S = k_\mathrm{B} \sum_{i=1}^{p} \frac{\partial \ln W}{\partial N_i} \, \mathrm{d}N_i + k_\mathrm{B} \alpha \sum_{i=1}^{p} \mathrm{d}N_i$$

Since the number of particles is constant, the second term is zero and so

$$\mathrm{d}S = k_\mathrm{B} \, \mathrm{d} \ln W \qquad (14.31)$$

which gives support to the Boltzmann expression

$$S = k_\mathrm{B} \ln W \qquad (14.32)$$

14.4   THE CANONICAL PARTITION FUNCTION

The molecular partition function q relates to single molecule energy states, but in thermodynamics we have to take account of the energy states of large groups of molecules, interacting or otherwise. We therefore define a different kind of partition function in terms of the energies of the complete set of molecules.

  The canonical partition function Q is defined as a sum over all the quantum states for the complete set of molecules in the system.

$$Q = \sum_{\text{states}} \exp\left(-\frac{E_i}{k_\mathrm{B} T}\right) \qquad (14.33)$$

where E_i is a quantized energy for the complete set of particles. In the simple case of identical non-interacting particles I can relate Q very simply to the molecular partition function q. There are two cases, depending whether the particles are distinguishable or indistinguishable.

14.5   DISTINGUISHABLE NON-INTERACTING PARTICLES

Consider a crystal of identical particles in which all the particles are localized. If we label these particles A, B, C... then we see that they are distinguishable simply because we can describe them according to their positions in space. Assume that each particle (for example A) has a set of energy levels ε_1^A, ε_2^A and so on. The system partition function Q is given by

$$Q = \sum_i \sum_j \sum_k \ldots \exp\left[-\frac{(\varepsilon_i^\mathrm{A} + \varepsilon_j^\mathrm{B} + \varepsilon_k^\mathrm{C} + \cdots)}{k_\mathrm{B} T}\right]$$

which can be factored into a product of individual partition coefficients for the molecules A, B, C . . .

$$Q = \sum_i \exp\left(-\frac{\varepsilon_i^A}{k_B T}\right) \sum_j \exp\left(-\frac{\varepsilon_j^B}{k_B T}\right) \sum_k \exp\left(-\frac{\varepsilon_k^C}{k_B T}\right) \ldots$$

The molecules are identical, and so all these sums in the product are identical. This means we can write

$$Q = \left[\sum_i \exp\left(-\frac{\varepsilon_i^A}{k_B T}\right)\right]^N \qquad (14.34)$$

which leads to the important result that

$$Q = q^N \qquad (14.35)$$

14.6   INDISTINGUISHABLE NON-INTERACTING PARTICLES

The calculation above is fine for a crystal, where we can label molecules, or for 'classical' particles like motor cars where each can be identified by its chassis number. In the case of molecules in a gas, or the electrons in a metal, the particles are truly indistinguishable and the calculation given above needs a simple modification. We have to allow for the indistinguishability of the molecules. I showed you how to do this in a previous chapter; what is needed is to divide the distinguishable Q by $N!$ Thus for indistinguishable particles

$$Q = \frac{q^N}{N!} \qquad (14.36)$$

Molecular partition functions are tremendously important in science because they contain all the information about a system of independent particles at thermal equilibrium. The assumption that the particles are independent (and so do not interact with each other) allows us to identify (for example) the internal energy of a system of N particles with their mean energy

$$U = U_0 + N\langle\varepsilon\rangle$$

but the assumption is restrictive and we need to allow for the possibility of intermolecular interactions. The crucial new concept is the *ensemble*. In order to set up an ensemble, we take our set of interacting molecules and reproduce it a large number of times keeping every reproduction similar to the original set. This collection of reproductions is called the *ensemble*. The system we want to study is just one member of the collection of the N^* copies.

There are several important ensembles in the theory of statistical thermodynamics, depending on what is kept constant between the copies and what is allowed to vary. We are concerned only with the *canonical ensemble* where the number of particles N, the volume V and the temperature T are kept constant between each member of the ensemble, but energy can be transferred between the members of the ensemble. The total energy of the ensemble is kept constant.

One point of introducing the canonical ensemble is to allow comparison of the time average of any bulk property of any one assembly with the average over all assemblies as the ensemble at a single instant of time.

According to the so-called *ergodic hypothesis* these two are equal.

Figure 14.2 shows a small part of an ensemble of N^* assemblies. Each assembly contains the same number of particles N as the original system (shaded). Each member of the ensemble is at the same temperature T, and occupies the same volume V. All the members of the ensemble are in thermal contact, and energy may flow from one to another.

As time progresses, the members of the ensemble may well have different energies. I will write the possible values of the energies of each assembly E_1^*, E_2^* (not to be confused with the energy of each individual particle that makes up the N total in each assembly), and I expect to find at some instant of time that there are

N_1^* members of the ensemble with energy E_1^*

N_2^* members of the ensemble with energy E_2^*

and so on. A calculation along the lines of that given earlier (with the same premise of constant ensemble energy and volume) gives

$$N_i^* = \frac{N^* \exp\left(-\dfrac{E_i^*}{k_B T}\right)}{\displaystyle\sum_{i=1}^{N^*} \exp\left(-\dfrac{E_i^*}{k_B T}\right)}$$

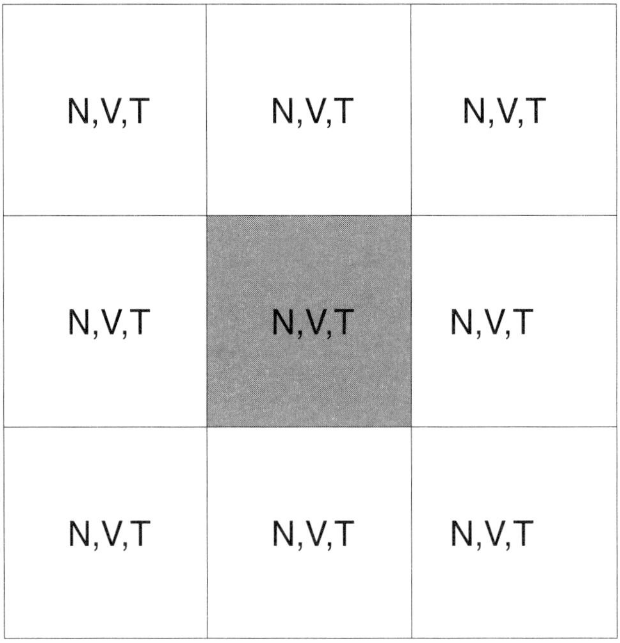

N,V,T	N,V,T	N,V,T
N,V,T	N,V,T	N,V,T
N,V,T	N,V,T	N,V,T

Figure 14.2   A canonical ensemble

This expression shows how the total energy of the ensemble is distributed amongst the $N*$ members of the ensemble, and the denominator

$$Q = \sum_{i=1}^{N*} \exp\left(-\frac{E_i^*}{k_B T}\right) \tag{14.37}$$

is referred to as the *canonical partition function*.

14.7 THERMODYNAMIC QUANTITIES FROM PARTITION FUNCTIONS

14.7.1 The Internal Energy U

The average value of $U - U_0$ for the ensemble of $N*$ members is given by

$$U - U_0 = \frac{\displaystyle\sum_{i=1}^{N*} N_i^* E_i^*}{N*}$$

and according to the ergodic hypothesis, this is equal to the time average of $U - U_0$. Thus

$$U - U_0 = \frac{\displaystyle\sum_{i=1}^{N*} E_i^* \exp\left(-\frac{E_i^*}{k_B T}\right)}{Q}$$

I can tidy up this equation by noting that

$$\left(\frac{\partial Q}{\partial T}\right)_V = \frac{1}{k_B T^2} \sum_{i=1}^{N*} E_i^* \exp\left(-\frac{E_i^*}{k_B T}\right)$$

and so on substitution

$$U - U_0 = \frac{k_B T^2}{Q}\left(\frac{\partial Q}{\partial T}\right)_V \tag{14.38}$$

or, in terms of $\ln Q$ and β

$$U - U_0 = -\left(\frac{\partial \ln Q}{\partial \beta}\right)_V \tag{14.39}$$

These equations are very similar to those relating U to the molecular partition function q derived earlier for the special case of non-interacting particles.

$$U = U_0 + \frac{N k_B T^2}{q}\left(\frac{\partial q}{\partial T}\right)_V$$

There is no multiplying factor of N because we considered all the N particles in the system at once.

The construction of Q from the properties of individual particles is relatively easy provided that the particles do not interact with each other. In the case of non-interacting particles we have $Q = q^N$ (if the particles are distinguishable) or $Q = q^N/N!$ (if the particles are not distinguishable).

The two equations for $U - U_0$ given above are entirely consistent, provided that $Q = q^N$ or $Q = q^N/N!$ as appropriate.

14.7.2 The Entropy S

To obtain the entropy we make use of the standard formula

$$\int_0^T dS = \int_0^T \frac{C_V}{T} \, dT \tag{14.40}$$

which applies for a constant volume change with no intervening phase changes.

Starting from the expression for U

$$U = U_0 + \frac{k_B T^2}{Q} \left(\frac{\partial Q}{\partial T} \right)_V$$

then we have

$$C_V = \frac{\partial}{\partial T} \left[\frac{k_B T^2}{Q} \left(\frac{\partial Q}{\partial T} \right)_V \right]_V$$

After considerable but straightforward manipulation we find

$$S_T = \frac{U - U_0}{T} + k_B \ln Q \tag{14.41}$$

This expression for S_T can also be written in terms of Q alone to give

$$S^T = \frac{k_B T}{Q} \left(\frac{\partial Q}{\partial T} \right)_V + k_B \ln Q \tag{14.42}$$

and it is in fact independent of the assumption that there were no phase changes.

14.7.3 The Sackur–Tetrode Equation

A simple example is afforded by the case of a gas of N non-interacting atomic particles. In the case where these particles are distinguishable, then the canonical partition function Q is related to the molecular partition function q by

$$Q = q^N$$

In the case that the particles are indistinguishable then

$$Q = \frac{q^N}{N!}$$

For an atomic particle in a non-degenerate electronic ground state that is well separated from the higher electronic states, the only contribution to q is the translational partition function

$$q_{\text{trans}} = \frac{(2\pi m k_B T)^{3/2}}{h^3} V$$

and since

$$Q = \frac{q_{\text{trans}}^N}{N!}$$

we have

$$S_T = k_B T \left(\frac{\partial \ln Q}{\partial T}\right)_V + N k_B \ln q_{\text{trans}} - k_B \ln(N!)$$

The third term on the right hand side can be simplified using Stirling's approximation, and the first term on the right hand side can be easily differentiated. Finally, on substitution of the ideal gas equation of state, we obtain the Sackur–Tetrode equation

$$S = nR\left\{\frac{5}{2} + \ln\left[\left(\frac{2\pi m k_B T}{h^2}\right)^{3/2} \frac{k_B T}{p}\right]\right\} \tag{14.43}$$

14.7.4 Residual Entropy

According to the second law of thermodynamics, $\mathrm{d}S = \mathrm{d}q_{\text{rev}}/T$. The absolute entropy of a pure substance at temperature T can therefore be evaluated to give

$$S_T - S_0 = \int_{0\,\text{K}}^{T} \frac{C_p}{T} \, \mathrm{d}T \tag{14.44}$$

with the assumption that there are no phase changes between 0 and T.

According to the third law of thermodynamics, the constant of integration S_0 is zero for a perfectly crystalline material. To this expression we might need to add entropies for phase changes, and these are given typically by $\Delta H_{\text{melt}}/T_{\text{melt}}$ (which relates to the change in entropy for the change from a solid to a liquid at the melting point) depending on the temperature of interest and the pressure.

Absolute entropies can therefore be determined from thermodynamic measurements of heat capacities and from experimental determinations of the enthalpies and temperatures of phase transitions. They can be corrected for pressure and for the number of particles present in order to give standard molar entropies S_m^0, which usually relate to a standard state of 1 atm pressure.

Absolute entropies can also be evaluated from the canonical partition function

$$S_T = \frac{k_B T}{Q}\left(\frac{\partial Q}{\partial T}\right)_V + k_B \ln Q \tag{14.45}$$

Entropies evaluated in this way are often called *spectroscopic entropies* because they have to be evaluated using particle energy levels, which can be determined from spectroscopic studies. They can also be found from molecular structure calculations, but the name 'spectroscopic entropy' seems to have stuck for all studies that are non-calorimetric.

The agreement between entropies measured thermochemically and entropies determined from spectroscopic studies is usually extremely good, but in a few instances there is a discrepancy. This discrepancy is called the *residual entropy*, and it can almost always be traced back to a deviation from the Third Law. A couple of examples will illustrate the principle.

(i) Solid CO. The residual molar entropy is found to be $4.6\,\mathrm{J\,K^{-1}\,mol^{-1}}$. The CO molecule has a very small electric dipole moment, and so the CO molecule can presumably occupy any site on the crystal lattice with orientation CO or OC. If there are N lattice sites and just two possible orientations, then $W = 2^N$ and

$$S_{\text{resid}} = Nk_B \ln 2$$

which gives a residual molar entropy of $R \ln 2 = 5.8\,\mathrm{J\,K^{-1}\,mol^{-1}}$, in agreement with experiment.

(ii) Solid CH_3D. The residual molar entropy is found to be $11.56\,\mathrm{J\,K^{-1}\,mol^{-1}}$. There are four equivalent tetrahedral positions for the deuterium atom in solid CH_3D and so we would predict that $W = 4^N$. This suggests a residual molar entropy of $R \ln 4 = 11.52\,\mathrm{J\,K^{-1}\,mol^{-1}}$ in very good agreement with experiment.

14.7.5   Other Thermodynamic Quantities

Once we have established expressions for U and S in terms of Q, it is a straightforward matter to find corresponding expressions for the three remaining thermodynamic functions H, A and G. These turn out to be

$$A = A_0 - k_B T \ln Q$$

$$H = H_0 + \frac{k_B T^2}{Q}\left(\frac{\partial Q}{\partial T}\right)_V + Nk_B T \tag{14.46}$$

$$G = G_0 - k_B T \ln Q + \frac{Nk_B V}{Q}\left(\frac{\partial Q}{\partial V}\right)_T$$

Here A_0, H_0 and G_0 are the values of A, H and G at $0\,\mathrm{K}$.

Molar quantities can be obtained by substituting $N = nL$ in the above equations, where n is the amount of substance present and L is the Avogadro constant.

14.8   EQUATION OF STATE FOR GAS REACTIONS

A simple route to an equation of state is provided by the thermodynamic identity

$$p = -\left(\frac{\partial A}{\partial V}\right)_T \tag{14.47}$$

If we start from the expression for the Helmholtz energy given above

$$A = A_0 - k_B T \ln Q$$

then we find (after some manipulation) that

$$p = \frac{k_B T}{Q}\left(\frac{\partial Q}{\partial V}\right)_T \tag{14.48}$$

Statistical thermodynamicists refer to this equation as the equation of state.

14.9 EQUILIBRIUM CONSTANTS

In Chapter 2, I showed you how we go about calculating equilibrium constants for ideal gas reactions such as

$$A + B \;\leftrightarrow\; C + D$$

in terms of the classical thermodynamic functions. In particular, I gave you the equation

$$RT \log_e(K_p) = -\Delta G^0_m \tag{14.49}$$

relating the pressure equilibrium constant to the change in the standard Gibbs free energy.
   For a more general equation

$$\nu_A A + \nu_B B + \cdots \;\leftrightarrow\; \nu_C C + \nu_D D + \cdots$$

where the νs are stoichiometry numbers, we have a similar expression

$$RT \log_e(K_p/\text{unit}) = -\Delta G^0_m$$

where the pressure equilibrium constant is now no longer dimensionless

$$K_p = \frac{p_C^{\nu C} p_D^{\nu D} \cdots}{p_A^{\nu A} p_B^{\nu D} \cdots}$$

and I have divided by its unit in order to make the logarithm argument dimensionless.
   Starting with the two statistical equations (from above)

$$p = \frac{k_B T}{Q}\left(\frac{\partial Q}{\partial V}\right)_T$$

and

$$G = G_0 - k_B T \ln Q + \frac{N k_B V}{Q}\left(\frac{\partial Q}{\partial V}\right)_T$$

we have, for an ideal gas where $pV = N k_B T$

$$G = G_0 - k_B T \ln Q + N k_B T \tag{14.50}$$

For a gas comprising N indistinguishable particles

$$Q = \frac{q^N}{N!}$$

and we find, on using Stirling's approximation for the factorial term,

$$G = G_0 - Nk_{\mathrm{B}}T \, \ln\left(\frac{q}{N}\right)$$

If $N = nL$ (where L is the Avogadro constant and n the amount) then we have finally

$$G_{\mathrm{m}} = G_{\mathrm{m},0} - nRT \ln\left(\frac{q}{nL}\right)$$

If we now specify a standard pressure p^0, which we might take for example as 1 atm,

$$G_{\mathrm{m}}^0 = G_{\mathrm{m},0}^0 - nRT \ln\left(\frac{q^0}{nL}\right) \tag{14.51}$$

where q^0 is the value of the molecular partition function under standard conditions. This equation is often written in terms of the molar molecular partition function, $q_{\mathrm{m}} = q/n$.

Only the translational molecular partition function depends on the pressure (and on the volume, through the ideal equation of state).

The only form of energy that does not drop at $0\,\mathrm{K}$ to a common value, which we conventionally take to be zero, is the vibrational energy. At $0\,\mathrm{K}$, this energy falls to the zero point energy whose value varies from species to species. The Gibbs energy change for the reaction at $0\,\mathrm{K}$ is equal to the internal energy change, which is therefore the difference between the zero point energies of the products and reactants. By convention this is given the symbol ΔE_0.

The equilibrium constant for the gas phase equation

$$A + B \;\leftrightarrow\; C + D$$

can finally be written

$$-\Delta G_m = RT\left[-\frac{\Delta E_0}{RT} + \ln\left(\frac{q_{\mathrm{C,m}} q_{\mathrm{D,m}}}{q_{\mathrm{A,m}} q_{\mathrm{B,m}}}\right)\right] \tag{14.52}$$

and this significant result enables us to calculate equilibrium constants for very simple gas-phase reactions from a knowledge of the partition functions of reactants and products, together with the change in the zero point energy.

15   Modeling Atoms

15.1   THE OLD QUANTUM THEORY

The simplest possible atom is the hydrogen atom, which consists of a single electron and a proton. Many other one-electron atoms (such as He^+) are known to be stable, and have been extensively studied.

Experimentally, it turns out that a hydrogen atom can emit (or absorb) electromagnetic radiation having only very specific frequencies. For example, Balmer discovered the famous series of lines in the visible region of the electromagnetic spectrum with wavelengths $\lambda = 656$, 486, 434 and 410 nm that fitted the equation

$$\frac{1}{\lambda} = R_H \left(\frac{1}{2^2} - \frac{1}{n^2} \right) \quad \text{where } n = 3, 4, 5, 6 \ldots \tag{15.1}$$

The proportionality constant $R_H = 109\,677$ cm^{-1} is called the Rydberg constant; do not get R_H mixed up with the gas constant R. For the He^+ atom, the same formula applies but with a proportionality constant of (very nearly) $\frac{1}{4} R_H$, for Li^{2+} the proportionality constant is (very nearly) $\frac{1}{9} R_H$ and so on.

As time went on, other series of hydrogen atom lines were discovered in different regions of the electromagnetic spectrum. Their wavelengths were found to fit the general expression

$$\frac{1}{\lambda} = R_H \left(\frac{1}{n_1^2} - \frac{1}{n_2^2} \right) \quad \text{where } n_2 > n_1 = 1, 2, 3, 4, 5, 6, \ldots \tag{15.2}$$

For each value of the smaller quantum number n_1, the series of lines is named after the person who discovered it as shown in Table 15.1.

One of the greatest problems in physics at the turn of the nineteenth century was that of the structure, stability and electronic spectrum of such one-electron atoms.

Neils Bohr pictured a negatively charged electron orbiting the positively charged nucleus in a number of permitted circular orbits. For each orbit, the angular momentum was quantized in that it could only have a magnitude that was an integral multiple of $h/(2\pi)$. If we balance out the centrifugal force for an electron moving in a circular orbit of radius r with the electrostatic

Table 15.1 Names given to spectral series

Lower state quantum number n_1	H-atom spectral series
1	Lyman
2	Balmer
3	Paschen
4	Brackett
5	Pfund

attraction between the electron and the nucleus, and substitute in Bohr's quantum formula, we arrive at the correct expression for the energy levels in any one-electron atom.

The electron can jump only from orbit to orbit, and when it does so it emits or absorbs radiation with a wavelength calculated from Bohr's theory. The calculated wavelengths agreed well with those measured to the experimental precision of the day. We now know that the spectra show fine structure, and subsequent measurements of the spectrum of atomic hydrogen have been made with great precision. The frequency of one particular line is known to an accuracy of one part in 10^{11}.

Sommerfeld extended Bohr's ideas by allowing elliptic orbits, and we talk about the Bohr–Sommerfeld model of the atom. The treatment is intrinsically flawed, but you should have it brought to your attention. People call it the 'Old Quantum Theory'.

This model posed a problem as follows. Electrons are negatively charged, whilst protons carry an equal and opposite positive charge. According to the Old Quantum Theory, the electron was allowed to orbit the nucleus in one of a possible number of circles of fixed radius (or ellipses) with certain corresponding angular frequencies ω. According to classical electromagnetic theory, an orbiting negatively charged electron whose angular frequency is ω should emit radiation with the same frequency ω. This is in complete disagreement with experiment.

Classical electromagnetic theory has stood the test of time in many respects, but it can not describe many phenomena at the atomic and molecular level. The fact is that electrons can in some sense 'orbit' a nucleus without emitting radiation. Classical electromagnetic theory cannot explain this, and so we have to turn to the world of quantum mechanics for an explanation.

15.2 THE ONE-ELECTRON ATOM

So, back to the *one-electron atom* so-called because our treatment is going to be valid for H, He$^+$, Li^{2+}, and so on. All that is important is that we have a single electron and a positively charged nucleus, which could be H$^+$, He^{2+} ... and so on. The mutual potential energy U between an electron of charge $-e$ and a nucleus of charge $+Ze$ distant r is

$$U = \frac{1}{4\pi\varepsilon_0} \frac{(-e)(+Ze)}{r} \tag{15.3}$$

The potential is time-independent, and so we interest ourselves in solution of the time-independent three-dimensional Schrödinger equation.

$$-\frac{h^2}{8\pi^2\mu} \left[\frac{\partial^2 \psi(\mathbf{r})}{\partial x^2} + \frac{\partial^2 \psi(\mathbf{r})}{\partial y^2} + \frac{\partial^2 \psi(\mathbf{r})}{\partial z^2} \right] + U(\mathbf{r})\psi(\mathbf{r}) = \varepsilon\psi(\mathbf{r}) \tag{15.4}$$

where μ is the reduced mass of the electron-nucleus pair, $U(r)$ the electrostatic mutual potential energy, ε the total energy and $\psi(\mathbf{r})$ the *electronic* wavefunction. $\psi(\mathbf{r})$ depends only on the position of the electron in space with respect to the center of mass, and it does not depend on time. The position of the center of mass is almost exactly equal to the proton position, because of the large difference between the proton and electron masses. For the same reason, the reduced mass μ of the electron–proton pair is almost exactly equal to the electron mass.

Be careful to distinguish between the vector $\mathbf{r}$, which denotes a position in space of the electron relative to the center of mass, and the scalar $\mathbf{r}$, which is the scalar distance between these two particles.

Notice that the potential U is *spherically symmetric*. For such problems, it is usually beneficial to recast the equation in spherical polar coordinates. When we do this, we have

$$-\frac{h^2}{8\pi^2\mu}\left[\frac{1}{r^2}\frac{\partial}{\partial r}\left(r^2\frac{\partial\psi}{\partial r}\right) + \frac{1}{r^2\sin\theta}\frac{\partial\psi}{\partial\theta} + \frac{1}{r^2\sin^2\theta}\frac{\partial^2\psi}{\partial\phi^2}\right] + U\psi = \varepsilon\psi \tag{15.5}$$

This can also be written in terms of the angular momentum operator $\hat{l}^2$ discussed in Chapter 5 as

$$-\frac{h^2}{8\pi^2\mu}\frac{1}{r^2}\frac{\partial}{\partial r}\left(r^2\frac{\partial\psi}{\partial r}\right) + \frac{1}{2\mu}\frac{\hat{l}^2}{r^2}\psi + U\psi = \varepsilon\psi \tag{15.6}$$

Both the angular momentum operator $\hat{l}^2$ and the operator for the z component of this operator $\hat{l}_z$ commute with the Hamiltonian operator, and so it should come as no surprise to find that the eigenfunctions and eigenvalues of the Hamiltonian can be written

$$\psi_{n,l,m}(r,\theta,\phi) = NR_{n,l}(r)Y_{l,m}(\theta,\phi)$$

where the $Y_{l,m}(\theta,\phi)$ are eigenfunctions of angular momentum, as discussed in Chapter 5. Separation of variables shows that the radial function $R_{n,l}(r)$ satisfies an equation

$$-\frac{h^2}{8\pi^2\mu}\frac{d}{dr}\left(r^2\frac{dR}{dr}\right) + \left[\frac{h^2l(l+1)}{16\mu r^2} + U\right]R = \varepsilon R \tag{15.7}$$

called the radial equation. It is usually written more compactly in terms of $S(r) = rR(r)$ to give

$$-\frac{h^2}{8\pi^2\mu}\frac{d}{dr}\left(\frac{dS}{dr}\right) + \left(\frac{h^2l(l+1)}{16\mu r^2} + U\right)S = \varepsilon S \tag{15.8}$$

The term in $l(l+1)$, which is called the centrifugal potential, adds to the Coulomb potential to give an effective potential.

Solution of the radial equation can be done by expanding $S(r)$ as a power series in r, and requiring that the wavefunction vanishes as r gets large. These functions represent the bound states, and have energy $\varepsilon < 0$.

15.2.1   General Features of the Bound States

- Energy quantized, and it depends on a single quantum number that we call n, the principal quantum number. n can take values from 1 to infinity, in steps of 1.

$$\varepsilon_n = -\left(\frac{Z^2\mu e^4}{8\varepsilon_0^2 h^2}\right)\frac{1}{n^2}$$

- There are actually three quantum numbers, that we write n, l and m_l but the energy of a free atom only depends on n. The other two quantum numbers are given the symbols l and m_l, and they are called the azimuthal and magnetic quantum numbers, respectively. They are particularly important when the atom experiences an external electric (the Stark effect) or magnetic (the Zeeman effect) field. The quantum numbers l and m_l are just the eigenvalues of angular momentum discussed in Chapter 5. The angular momentum is quantized.

- The energy is negative for these *bound* states. There are also solutions of the time independent Schrödinger equation that correspond to positive energies and these are called the *unbound* states.

- The energy of a H atom is less than that of proton plus electron at infinity so that a H atom is predicted to be stable.

15.3   NORMALIZED HYDROGEN ATOM WAVEFUNCTIONS

For a problem with spherical symmetry, the spherical polar coordinates r, θ and ϕ are the obvious choice. By convention though, we use the dimensionless variable $\rho = Zr/a_0$ rather than r. Here Z is the nuclear charge number and a_0 the first bohr radius (approximately 52.9 pm). So for He$^+$ you would substitute $Z = 2$ into the equations shown in Table 15.2.

15.3.1   Visualization

How do we picture these atomic orbitals? Let us concentrate on the lowest energy solution, the so-called 1s atomic orbital. From Table 15.2, you will see that it has the mathematical form

$$\psi_{1s} = \frac{1}{\sqrt{\pi}} \left(\frac{Z}{a_0}\right)^{3/2} \exp(-\rho)$$

If we take a H atom where $Z = 1$, then

$$\psi_{1s} = \frac{1}{\sqrt{\pi}} \left(\frac{1}{a_0}\right)^{3/2} \exp\left(-\frac{r}{a_0}\right)$$

and a plot of $a_0^{3/2}\psi_{1s}$ against r/a_0 gives the curve shown in Figure 15.1.

The dimension of ψ is (length)$^{-3/2}$. Figure 15.1 shows the variation of the 1s atomic orbital with distance from the nucleus.

According to the Born interpretation of quantum mechanics, we should concentrate on the square of the wavefunction rather than the wavefunction itself, because the probability of finding a particle in the volume element $d\tau$ is $\psi^2 \, d\tau$. Figure 15.2 shows $a_0^3 \psi^2$ as a function of r/a_0.

The only features I would draw your attention to are that the wavefunction falls off to zero as the distance between the electron and the nucleus increases, and that the wavefunction has a 'cusp' at the nuclear position; the wavefunction is not flat at $r = 0$, it still has a gradient.

According to the Born interpretation of quantum mechanics, the probability of finding a particle between x and $x + \Delta x$ is $\psi^2 \Delta x$ when the wavefunction only depends on the variable x.

Table 15.2 Hydrogenic wavefunctions

n, l, m	Symbol	Expression
1, 0, 0,	1s	$\dfrac{1}{\sqrt{\pi}}\left(\dfrac{Z}{a_0}\right)^{3/2}\exp(-\rho)$
2, 0, 0	2s	$\dfrac{1}{4\sqrt{2\pi}}\left(\dfrac{Z}{a_0}\right)^{3/2}(2-\rho)\exp\left(-\dfrac{\rho}{2}\right)$
2, 1, 0	$2p_z$	$\dfrac{1}{4\sqrt{2\pi}}\left(\dfrac{Z}{a_0}\right)^{3/2}\rho\cos(\theta)\exp\left(-\dfrac{\rho}{2}\right)$
2, 1, ±1	$2p_x$	$\dfrac{1}{4\sqrt{2\pi}}\left(\dfrac{Z}{a_0}\right)^{3/2}\rho\sin(\theta)\cos(\phi)\exp\left(-\dfrac{\rho}{2}\right)$
	$2p_y$	$\dfrac{1}{4\sqrt{2\pi}}\left(\dfrac{Z}{a_0}\right)^{3/2}\rho\sin(\theta)\sin(\phi)\exp\left(-\dfrac{\rho}{2}\right)$
3, 0, 0	3s	$\dfrac{2}{81\sqrt{3\pi}}\left(\dfrac{Z}{a_0}\right)^{3/2}(27-18\rho+2\rho^2)\exp\left(-\dfrac{\rho}{3}\right)$
3, 1, 0,	$3p_z$	$\dfrac{2}{81\sqrt{\pi}}\left(\dfrac{Z}{a_0}\right)^{3/2}(6\rho-\rho^2)\cos(\theta)\exp\left(-\dfrac{\rho}{3}\right)$
3, 1, ±1	$3p_x$	$\dfrac{2}{81\sqrt{\pi}}\left(\dfrac{Z}{a_0}\right)^{3/2}(6\rho-\rho^2)\sin(\theta)\cos(\phi)\exp\left(-\dfrac{\rho}{3}\right)$
	$3p_y$	$\dfrac{2}{81\sqrt{\pi}}\left(\dfrac{Z}{a_0}\right)^{3/2}(6\rho-\rho^2)\sin(\theta)\sin(\phi)\exp\left(-\dfrac{\rho}{3}\right)$
3, 2, 0	$3d_z^2$	$\dfrac{2}{81\sqrt{6\pi}}\left(\dfrac{Z}{a_0}\right)^{3/2}\rho^3[3\cos^2(\theta)-1]\exp\left(-\dfrac{\rho}{3}\right)$
3, 2, ±1	$3d_{xz}$	$\dfrac{\sqrt{2}}{81\sqrt{\pi}}\left(\dfrac{Z}{a_0}\right)^{3/2}\rho^2\sin(\theta)\cos(\theta)\cos(\phi)\exp\left(-\dfrac{\rho}{3}\right)$
	$3d_{yz}$	$\dfrac{\sqrt{2}}{81\sqrt{\pi}}\left(\dfrac{Z}{a_0}\right)^{3/2}\rho^2\sin(\theta)\cos(\theta)\sin(\phi)\exp\left(-\dfrac{\rho}{3}\right)$
3, 2, ±2	$3d_{x^2-y^2}$	$\dfrac{1}{81\sqrt{2\pi}}\left(\dfrac{Z}{a_0}\right)^{3/2}\rho^2\sin^2(\theta)\cos(2\phi)\exp\left(-\dfrac{\rho}{3}\right)$
	$3d_{xy}$	$\dfrac{1}{81\sqrt{2\pi}}\left(\dfrac{Z}{a_0}\right)^{3/2}\rho^2\sin^2(\theta)\sin(2\phi)\exp\left(-\dfrac{\rho}{3}\right)$

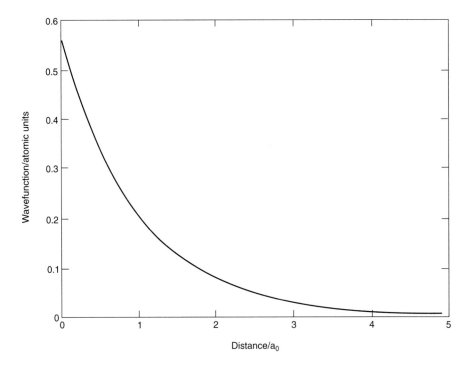

Figure 15.1   A hydrogen 1s orbital

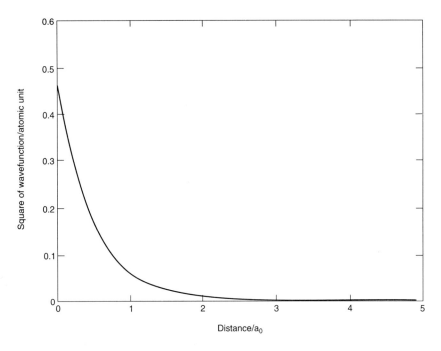

Figure 15.2   Square of the 1s atomic orbital

When the wavefunction $\psi(x, y)$ depends on x and y, then the probability that we will find the particle between x and $x + \Delta x$ and y and $y + \Delta y$ is $\psi^2(x, y)\Delta x \Delta y$. In three dimensions where the wavefunction $\psi(x, y, z)$ depends on x, y and z then the probability that we will find a particle between x and $x + \Delta x$, y and $y + \Delta y$ and z and $z + \Delta z$ is $\psi^2(x, y)\Delta x \Delta y \Delta z$.

An atom is a spherical entity, and we might ask the question 'what is the probability that we would find an electron anywhere in space at a distance r from the nucleus?' rather than asking about the occupancy of various equivalent regions of space around the atom that are just related by symmetry. This probability is equivalent to the chance that we will find the electron anywhere between two spheres of radius r and $r + \Delta r$, centred on the nucleus. The volume of a sphere of radius r is $\frac{4}{3}\pi r^3$ then the volume between two such spheres of radii r and $r + \Delta r$ is $4\pi r^2 \Delta r$.

For a wavefunction with spherical symmetry, the quantity $4\pi r^2 \psi^2$ is important. It is called the *radial distribution function*. It gives the probability of finding the particle anywhere at a distance r from the nucleus.

Figure 15.3 is a plot of this radial distribution function for a H atom.

Note that it has a maximum. A careful examination of the curve shows that the maximum occurs when $r = a_0$, the first Bohr radius.

There is a very clear distinction to be made between the Old Quantum Theory and the present treatment. In the Bohr theory, the electron was constrained to orbit a nucleus, and a balance of the classical electrostatic and the centripetal acceleration forces, together with Bohr's hypothesis, suggested that only certain orbits were permitted. The lowest energy orbit had radius a_0, hence the name of the 'first Bohr orbit'.

There are several ways in which we can visualize the one-electron atom solutions. Two have been shown above. Two other popular ways of visualization are as follows.

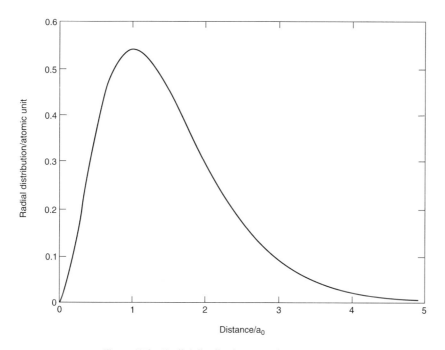

Figure 15.3 Radial distribution curve for a 1s orbital

The contour diagram in Figure 15.4 should be familiar to all hikers and map readers. It gives a two-dimensional representation of the wavefunction or its square at all positions in a plane.

Let us look now at the 2s orbital. It is given by

$$\frac{1}{4\sqrt{2\pi}} \left(\frac{Z}{a_0}\right)^{3/2} (2 - \rho) \exp\left(-\frac{\rho}{2}\right)$$

and so once again it is spherically symmetrical and depends only on the distance between the nucleus and the electron. There is a difference from the 1s orbital in that the function changes sign at $\rho = 2$. Of interest is the radial distribution curve shown in Figure 15.5.

Notice that the curve (and hence the probability) goes to 0 at $r = 0$, at infinity and at $r = 2a_0$ where the 2s wavefunction has a node.

For $l = 1$, there are three possible values of m, $m = -1, 0$ and $+1$. These may be designated p_{+1}, p_0 and p_{-1} states. The angular factors associated with these states are (apart from numerical factors)

p_{+1} is proportional to $\sin \theta \exp(j\phi)$

p_0 is proportional to $\cos \theta$

p_{-1} is proportional to $\sin \theta \exp(-j\phi)$

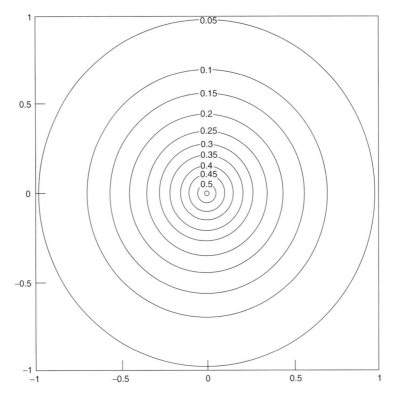

Figure 15.4 Contour diagram for a 1s orbital

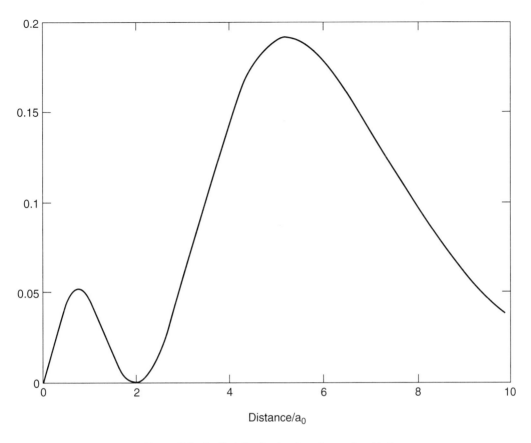

Figure 15.5   Radical distribution for a 2s atomic orbital

where j is the square root of -1. For most purposes in chemistry it is convenient to replace these with the so-called *real equivalents*, shown in Table 15.2

$$p_x = \frac{p_{+1} + p_{-1}}{\sqrt{2}}$$

$$p_z = p_0 \qquad\qquad\qquad (15.9)$$

$$p_y = -j\left(\frac{p_{+1} - p_{-1}}{\sqrt{2}}\right)$$

The names p_x, p_y and p_z indicate that the angular parts of these wavefunctions have their maximum values in the x, y and z directions respectively.

Figure 15.6 is a two-dimensional contour diagram for one of the three equivalent 2p atomic orbitals, in this case the $2p_z$ orbital. The z axis is the vertical axis.

The three p orbitals (p_x, p_y and p_z) are totally equivalent to each other in that they can be rotated into each other by rotating the coordinate axes (or by rotating the three orbitals and keeping the axes fixed).

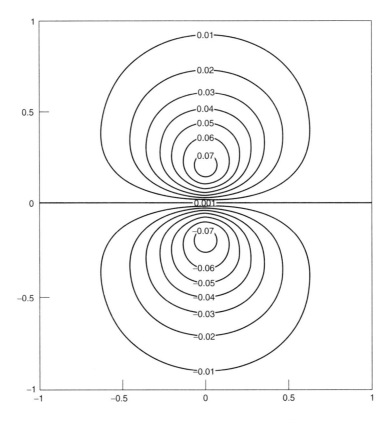

Figure 15.6 Contour plot of a 2p atomic orbital

The d orbitals ($l = 2$, $m_l = -2, -1, 0, 1, 2$) are not quite so straightforward. In order to get real combinations we take

$$\mathrm{d}_{z^2} = \mathrm{d}_0$$

$$\mathrm{d}_{xz} = \frac{\mathrm{d}_{+1} + \mathrm{d}_{-1}}{\sqrt{2}}$$

$$\mathrm{d}_{yz} = -j\,\frac{\mathrm{d}_{+1} - \mathrm{d}_{-1}}{\sqrt{2}} \qquad (15.10)$$

$$\mathrm{d}_{x^2-y^2} = \frac{\mathrm{d}_{+2} + \mathrm{d}_{-2}}{\sqrt{2}}$$

$$\mathrm{d}_{xy} = -j\,\frac{\mathrm{d}_{+2} - \mathrm{d}_{-2}}{\sqrt{2}}$$

The $3\mathrm{d}_{z^2}$ orbital has the contour diagram shown in Figure 15.7.

Remember once again that these diagrams are two-dimensional representations of three-dimensional bodies. Mentally rotate the figure about the z axis in order to come up with a three-dimensional body. Notice the change in sign (the zero contour).

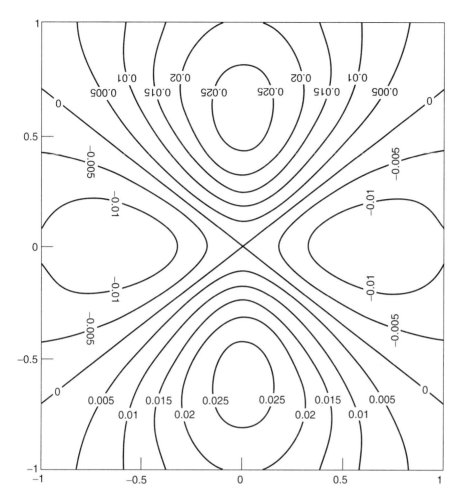

Figure 15.7 Cotour diagram for a 3d$_z$ atomic orbital

The 3d$_{xy}$, 3d$_{xz}$ and 3d$_{yz}$ are all equivalent, and they all have their maxima between the axes referred to. For example, the 3d$_{xz}$ hydrogenic orbital has the diagram contour diagram shown in Figure 15.8.

Notice the change of sign between the four sectors of the diagram, which is consistent with the sign of the x and the y axes.

If we examine the behavior at small values of r of the one-electron 3s, 3p and 3d atomic orbitals then apart from the normalization constant, the common exponential factor and the 'angular' factor, the 3s orbital is roughly constant (for small r), the 3p orbital has a factor of r and the 3d orbital has a factor of r^2. These radial wavefunctions are plotted in Figure 15.9 to illustrate the point. The solid curve is the 3s orbital, the dotted curve is the 3p orbital and the dashed curve is the 3d orbital.

Another way is to investigate the probability of finding the electron between r and $r + dr$, regardless of angle. Figure 15.10 gives the radial distribution functions for the same wavefunctions.

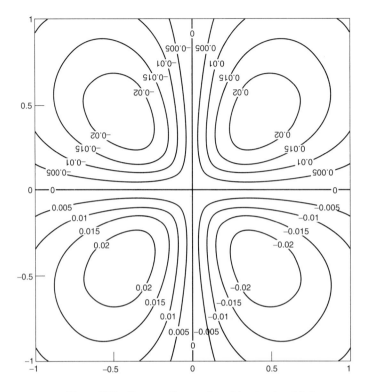

Figure 15.8   Contour diagram for a $3d_{xz}$ atomic orbital

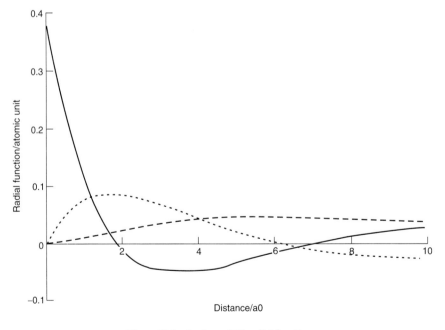

Figure 15.9   3s, 3p and 3d radial functions

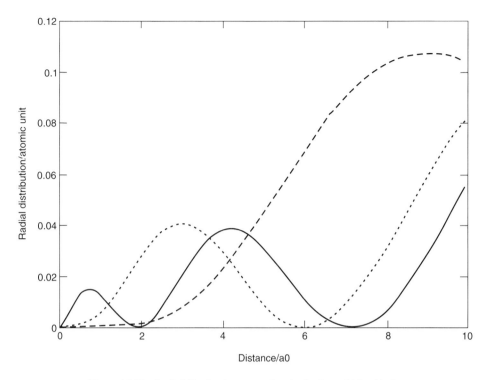

Figure 15.10   Radial distribution curves for the 3s, 3p and 3d orbitals

15.4   THE PHYSICAL SIGNIFICANCE OF *l* AND m_l; THE STERN–GERLACH EXPERIMENT

For bodies moving in three-dimensional space, the *angular momentum* is a profoundly important quantity. I have already mentioned the quantization of energy for particles in bound states, and the quantization of electron orbital angular momentum in a one-electron atom.

In Chapter 8, I mentioned that a particle with mass M and charge Q, and non-zero angular momentum **l** was a magnetic dipole moment

$$\mathbf{p}_\mathrm{m} = \frac{Q}{2M}\,\mathbf{l} \tag{15.11}$$

and so you might expect that measurements on the magnetic moments of atoms would lead to direct evidence for the quantization of angular momentum. This idea forms the basis for the Stern–Gerlach experiment.

The key to the Stern–Gerlach experiment is that whilst atomic magnetic moments do not experience a net force in a uniform magnetic field, they do experience such a force in a non-uniform field. Passing a beam of atoms through a non-uniform magnetic field will deflect the beam, just as a beam of charged particles gets deflected by an electrostatic field. This is shown schematically in Figure 15.11. The beam of atoms emerges from an oven and passes through a collimator and thence a Stern–Gerlach magnet. I have shown the beam splitting into two

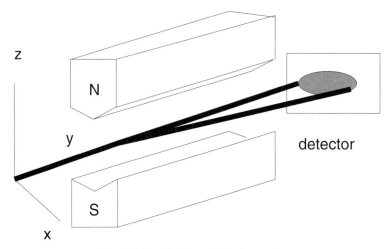

Figure 15.11 The Stern–Gerlach experiment

components. The experiment is actually quite difficult, and very low beam intensities have to be used to minimize the effect of collisions. The magnet is a powerful electromagnet, and the detector is usually a photographic plate.

The important thing is that the magnetic field B is not constant in the z direction and so it has a non-zero differential $\partial B_z/\partial z$. I will just tell you that we can prove from classical electromagnetic theory that there is a force in the z direction on the atom, with magnitude

$$F_z = p_e \frac{\partial B_z}{\partial z} \tag{15.12}$$

It would be difficult to suspend a single atom in between the poles of a magnet and then watch it accelerate, so what we do is to study the deflection of a beam of atoms when they pass through such a magnet. In their first experiment, Stern and Gerlach confirmed the prediction of angular momentum quantization, for they observed a series of bands at the detector rather than a continuous spread of atoms. For angular momentum quantum number l, $2l + 1$ bands are observed.

When they used silver atoms, Stern and Gerlach observed just two bands. This seems to conflict with the discussion above, because two orientations correspond to an angular momentum quantum number of $\frac{1}{2}$, and this does not appear from the Schrödinger treatment. The resolution of this conflict came with the suggestion that the angular momentum observed in the case of Ag atoms was not due to the orbital angular momentum (for which in any case $l = 0$), rather to the intrinsic angular momentum that we now call *spin*.

In fact the formula for the magnetic moment

$$\mathbf{p}_m = \frac{Q}{2M} \mathbf{l}$$

needs modification in the case of electron spin

$$\mathbf{p}_m = g \frac{(-e)}{2m_e} \mathbf{s} \tag{15.13}$$

where

$$s = \sqrt{\frac{1}{2}\left(\frac{1}{2}+1\right)}\frac{h}{2\pi} \qquad (15.14)$$

and the *g factor* is known very accurately to be

$$g = 2.002\ 319\ 304\ 386 \pm (20 \times 10^{-12})$$

The first step to understanding electron spin and the *g* factor came from the work of Paul Dirac in 1928. He developed a *relativistic* version of quantum theory of the electron in which spin and $g = 2$ emerged naturally. The remaining list of decimals is a consequence of a branch of physics called quantum electrodynamics.

15.5   MANY-ELECTRON ATOMS

In a previous section, I showed you how to apply quantum mechanics to model one-electron atoms. What we did was to write down the time-independent Schrödinger equation

$$\left[\frac{\partial^2 \psi(\mathbf{r})}{\partial x^1} + \frac{\partial^2 \psi(\mathbf{r})}{\partial y^2} + \frac{\partial^2 \psi(\mathbf{r})}{\partial z^2}\right] + \frac{8\pi^2 \mu}{h^2}(U - \varepsilon)\psi(\mathbf{r}) = 0$$

and then seek solutions that were compatible with the spatial symmetry of the problem, and that satisfied the boundary conditions. We refer to such solutions as atomic orbitals, since they give information about a single electron.

As shown in the Appendix, we can write the equation in a more compact notation using the ∇^2 operator as

$$\nabla^2 \psi(\mathbf{r}) + \frac{8\pi^2 \mu}{h^2}(U - \varepsilon)\psi(\mathbf{r}) = 0$$

or in the form of an eigenvalue equation as

$$-\frac{h^2}{8\pi^2 \mu}\nabla^2 \psi(\mathbf{r}) + U\psi(\mathbf{r}) = \varepsilon \psi(\mathbf{r})$$

The first term involving the operator ∇^2 represents the kinetic energy of the electron. U is the mutual potential energy of the nucleus (of charge Ze) and the electron (of charge $-e$),

$$U = -\frac{Ze^2}{4\pi\varepsilon_0 r}$$

and so represents the work done in bringing one of those two charged particles from infinity up to the point distance r from the other one.

For a many electron atom of nuclear charge Ze, we have to consider three contributions to the energy: first of all the kinetic energy of each electron (the nucleus is fixed in space for this calculation and so has no kinetic energy); second, the mutual potential energy of the nucleus (charge Ze) and each electron (of charge $-e$); and finally, we have to consider the mutual potential energies of each pair of electrons.

$$\left(\begin{array}{c} \dfrac{-h^2}{8\pi^2 m}\nabla_1^2 - \dfrac{h^2}{8\pi^2 m}\nabla_2^2 - \cdots - \dfrac{h^2}{8\pi^2 m}\nabla_n^2 \\[2mm] -\dfrac{Ze^2}{4\pi\varepsilon_0 r_1} - \dfrac{Ze^2}{4\pi\varepsilon_0 r_2} - \cdots - \dfrac{Ze^2}{4\pi\varepsilon_0 r_n} \\[2mm] +\dfrac{e^2}{4\pi\varepsilon_0 r_{12}} + \dfrac{e^2}{4\pi\varepsilon_0 r_{13}} + \cdots + \dfrac{e^2}{4\pi\varepsilon_0 r_{n,n-1}} \end{array} \right) \Psi(\mathbf{r}_1, \mathbf{r}_2, \ldots, \mathbf{r}_n) = \varepsilon\Psi(\mathbf{r}_1, \mathbf{r}_2, \ldots, \mathbf{r}_n) \quad (15.15)$$

As before, each term involving ∇^2 represents the kinetic energy of each of the electrons. Each of the terms involving a $1/r_i$ represents the electrostatic attraction between the nucleus and the electron. The terms involving the separation between the electrons as typified by $e^2/4\pi\varepsilon_0 r_{12}$ represent the mutual repulsive potential energy of each pair of electrons, and it is these terms that make our life very difficult.

At first sight, we might be tempted to ignore the electron repulsion terms and try the 'separation of variables' technique that has served us so well.

If the electron repulsion terms were not present, then the separation of variables technique would indeed give a solution $\Psi_0(\mathbf{r}_1, \mathbf{r}_2, \ldots, \mathbf{r}_n)$

$$\left(\begin{array}{c} \dfrac{-h^2}{8\pi^2 m}\nabla_1^2 - \dfrac{h^2}{8\pi^2 m}\nabla_2^2 - \cdots - \dfrac{h^2}{8\pi^2 m}\nabla_n^2 \\[2mm] \cdots - \dfrac{Ze^2}{4\pi\varepsilon_0 r_1} - \dfrac{Ze^2}{4\pi\varepsilon_0 r_2} - \cdots - \dfrac{Ze^2}{4\pi\varepsilon_0 r_n} \end{array} \right) \Psi_0(\mathbf{r}_1, \mathbf{r}_2, \ldots, \mathbf{r}_n) = \varepsilon\Psi_0(\mathbf{r}_1, \mathbf{r}_2, \ldots, \mathbf{r}_n)$$

such that $\Psi_0(\mathbf{r}_1, \mathbf{r}_2, \ldots, \mathbf{r}_n)$ was a product of individual one-electron functions (orbitals), $\chi_i(\mathbf{r}_i)$

$$\Psi_0(\mathbf{r}_1, \mathbf{r}_2, \ldots, \mathbf{r}_n) = \chi_1(\mathbf{r}_1)\chi_2(\mathbf{r}_2) \cdots \chi_n(\mathbf{r}_n) \quad (15.16)$$

and ε was the sum of one-electron energies

$$\varepsilon = \varepsilon_1 + \varepsilon_2 + \cdots + \varepsilon_n$$

but such solutions are hopelessly inaccurate. It is the repulsions between the individual electrons that make the problem intractable mathematically yet interesting at the atomic level.

The many-electron atom is an example of a so-called *many body problem*. These problems are not unique to the study of atomic and molecular structure. For example, Newton's theory of planetary motion gives an exact classical solution to the problem of one body orbiting under the gravitational influence of another but real planetary systems are more complicated because they too involve the gravitational interactions of many planets with each other. During the eighteenth and nineteenth centuries, a great deal of effort was spent on trying to solve such problems; all these attempts were unsuccessful and we generally accept today that exact solutions do not exist, even for three interacting bodies. That is not to say that we cannot solve the problems numerically to any given degree of accuracy, but algebraic solutions do not exist.

Given these difficulties it should come as no surprise to learn that exact solution of the Schrödinger equation is impossible, because of the electron–electron repulsion terms. Yet in some ways, the many-electron Schrödinger equation should be easier to solve than most many-body problems, for the following reason. I have stressed to you the importance of the Born interpretation of quantum mechanics. We do not focus on the trajectories of individual particles, but we accept that the square of the wavefunction gives a probability distribution. We therefore require less specific information than (for example) an astronomer.

Going back to the 'independent electron' model

$$\Psi_0(\mathbf{r}_1, \mathbf{r}_2, \ldots, \mathbf{r}_n) = \chi_n(\mathbf{r}_1)\chi_2(\mathbf{r}_2)\ldots\chi_n(\mathbf{r}_n)$$

this implies that the probability of finding electron 1 at point $\mathbf{r}_1$, electron 2 at point $\mathbf{r}_2$ ... and electron n at point $\mathbf{r}_n$ is given by the product of the individual probabilities $P_1(\mathbf{r}_1)$, $P_2(\mathbf{r}_2), \ldots, P_n(\mathbf{r}_n)$

$$P(\mathbf{r}_1, \mathbf{r}_2, \ldots, \mathbf{r}_n) = P_1(\mathbf{r}_1)P_2(\mathbf{r}_2)\ldots P_n(\mathbf{r}_n) \tag{15.17}$$

At first sight, this implies that the electron repulsions must be ignored, but what we do is to solve the Schrödinger equation in such a way that the interactions between the electrons are *averaged*. The interactions are still present, but we allow for them, by considering each electron to come under the influence of an *average* potential due to the remaining electrons (and of course the nucleus).

This is the essence of the Hartree Self-Consistent Field (SCF) model. Each electron moves in the influence of an average field due to the nucleus and the remaining electrons. The contribution due to the remaining electrons is determined by the form of their orbitals, which we do not know until we have solved the Hartree SCF equations!

I can give you a feel for the method as follows. We start with a one-electron atom, of nuclear charge Ze. I will call the electron 'number 1', and denote its coordinates in space by the vector symbol $\mathbf{r}_1$. The Schrödinger equation is

$$\left(\frac{-h^2}{8\pi^2 m}\nabla_1^2 - \frac{Ze^2}{4\pi\varepsilon_0 r_1}\right)\chi_1(\mathbf{r}_1) = \varepsilon\chi_1(\mathbf{r}_1)$$

or to use the more familiar symbol U for the mutual potential energy

$$\left(\frac{-h^2}{8\pi^2 m}\nabla_1^2 + U\right)\chi_1(\mathbf{r}_1) = \varepsilon\chi_1(\mathbf{r}_1)$$

I have been very careful to indicate that we are dealing with electron number 1 up to this point. Consider now the case where this electron moves under the influence of the nucleus and a second electron whose coordinates are $\mathbf{r}_2$ and which occupies orbital $\chi_2(\mathbf{r}_2)$. Remember what I have told you about the Born interpretation of quantum mechanics; this second electron can be regarded as a charge distribution $-e[\chi_2(\mathbf{r}_2)]^2$.

We now modify U in order to take the second electron into account by dividing up the space of electron 2 into infinitesimal elements $\mathrm{d}\tau_2$, and summing their contributions to the mutual potential energy of electrons 1 and 2. This is

$$\frac{e^2}{4\pi\varepsilon_0}\int \frac{\chi_2^2(\mathbf{r}_2)}{r_{12}}\,\mathrm{d}\tau_2$$

where r_{12} is the distance between the two electrons. We then add this to the nuclear attraction term to get a one-electron Schrödinger equation for electron 1.

$$\left[\frac{-h^2}{8\pi^2 m} \nabla_1^2 - \frac{Ze^2}{4\pi\varepsilon_0 r_1} + \frac{e^2}{4\pi\varepsilon_0} \int \frac{\chi_2^2(\mathbf{r}_2)\,\mathrm{d}\tau_2}{r_{12}} \right] \chi_1(\mathbf{r}_1) = \varepsilon_1 \chi_1(\mathbf{r}_1)$$

In order to solve the equation, we make a guess at the atomic orbitals, using our chemical knowledge. We then feed in the guesses to these Hartree equations, and solve the SCF equations to get a new set of atomic orbitals.

We then feed this new set of atomic orbitals back into the SCF equations and iterate until the difference between our input orbitals and the output orbitals is negligible. The process turns out to be straightforward and Hartree was able to give numerical solutions to his equations for very many ground state and excited state atoms.

15.6   THE PAULI PRINCIPLE

The fact of the matter is that quantum particles are quite different to classical particles. For one thing, 'identical' particles are exactly that and they cannot be distinguished one from the other. Also, there are important differences between the wavefunctions of fermions and bosons. Fermion wavefunctions have to change sign when we interchange the names of two identical fermions. Boson wavefunctions are unchanged by this interchange of names.

Consider a lithium atom (three electrons).

A simple statement of the *Pauli exclusion principle*, that will be fine for the present discussion, is

No two electrons can have the same set of four quantum numbers n, l, m_l and m_s

We would guess using chemical intuition that the electronic configuration of Li should be $1s^2 2s^1$. I have to confess that the statement of the Pauli principle I have just given you is not entirely true; one-electron atoms do indeed have the four quantum numbers $n\ l$, m_l and m_s but many-electron atoms, molecules and materials certainly do not. We therefore have to modify the Hartree method to take indistinguishability into account. This was done by Fock, and we now speak about the Hartree–Fock model.

15.7   THE HARTREE–FOCK MODEL

I referred above to 'electron 1' and to 'electron 2' as if we could somehow distinguish between them. Electrons are certainly identical (in the sense that all electrons are equal), just as two mass-produced identically colored automobiles are identical when they roll off the production line, but Heisenberg's uncertainty principle puts a limit on the extent to which we can follow the trajectories of individual particles, and all the evidence is that electrons cannot be labelled; they are *indistinguishable*.

To take a specific example, consider a Li ground state calculation where the orbital configuration is $1s^2 2s^1$. This implies that three electrons have to be allocated to two atomic orbitals; according to our rules as written above, the 1s orbital can hold a maximum of two electrons, one of either spin α or β. Another way of looking at this is to say that there are two quantum states with labels $1s\alpha$ and $1s\beta$, each of which can only hold one electron.

The third electron can occupy one of two quantum states, $2s\alpha$ or $2s\beta$ and each of these two quantum states has the same energy. I am going to focus attention on just one of them, the $2s\alpha$ quantum state.

Our problem now is to allocate the three (indistinguishable) electrons to the three quantum states $1s\alpha$, $1s\beta$ and $2s\alpha$. I will label the electrons 1, 2 and 3, and put $1s\alpha(1)$ to mean that electron 1 occupies quantum state $1s\alpha$. If you prefer, this notation means that electron 1 is in a 1s orbital with α spin, think of things whichever way round you prefer.

So, if we insist that the third electron has to occupy quantum state $2s\alpha$, that gives us the following six possibilities

$$1s\alpha(1)1s\beta(2)2s\alpha(3)$$

$$1s\alpha(1)1s\beta(3)2s\alpha(2)$$

$$1s\alpha(2)1s\beta(1)2s\alpha(3)$$

$$1s\alpha(2)1s\beta(2)2s\alpha(1)$$

$$1s\alpha(3)1s\beta(1)2s\alpha(2)$$

$$1s\alpha(3)1s\beta(2)2s\alpha(1)$$

So to remind you, a configuration $1s\alpha(3)1s\beta(2)2s\alpha(1)$ means that electron 3 is in the quantum state $1s\alpha$, electron 2 is in quantum state $1s\beta$ and electron 1 is in quantum state $2s\alpha$.

These six possible configurations only differ from each other in that we have permuted the names of the electrons between them. They all have the same energy. The second configuration $1s\alpha(1)1s\beta(3)2s\alpha(2)$ was obtained from the first by permuting the names of electrons 2 and 3, and so on.

There is a more general statement of the Pauli principle (essentially a consequence of electrons being fermions), which says that

Any electron wavefunction must be antisymmetric to the exchange of pairs of electron names

so we cannot write (for example) a satisfactory electronic wavefunction as

$$1s\alpha(1)1s\beta(2)2s\alpha(3)$$

because it keeps the same sign when we interchange the names of any of the electrons.

The combination

$$1s\alpha(1)1s\beta(2)2s\alpha(3) + 1s\alpha(1)1s\beta(3)2s\alpha(2)$$

turns into itself when we exchange the names of electrons 1 and 2 and so is symmetric to the exchange of the names of electrons 1 and 2.

The combination

$$1s\alpha(1)1s\beta(2)2s\alpha(3) - 1s\alpha(1)1s\beta(3)2s\alpha(2)$$

turns into the negative of itself when we exchange the names of electrons 1 and 2 and so is antisymmetric to the exchange of the names of electrons 1 and 2.

What we have to do now is to come up with a suitable combination of the six possible quantum states that satisfies the generalized Pauli principle.

There are only two possibilities, and these only differ by an overall + or − sign. They are

$$\pm \, (1s\alpha(1)1s\beta(2)2s\alpha(3)$$
$$- \, 1s\alpha(1)1s\beta(3)2s\alpha(2)$$
$$- \, 1s\alpha(2)1s\beta(1)2s\alpha(3)$$
$$+ \, 1s\alpha(2)1s\beta(3)2s\alpha(1)$$
$$+ \, 1s\alpha(3)1s\beta(1)2s\alpha(2)$$
$$- \, 1s\alpha(3)1s\beta(2)2s\alpha(1))$$

which has the correct behavior; if we interchange names of any pair of electrons, it changes sign; if we leave the electron names alone, it keeps the same sign and is we interchange names of three of the electrons, then it again keeps the same sign.

The overall + or − sign is irrelevant, because the quantity of interest is the square of the wavefunction (which is related to a probability).

You may like to know that we can write these allowed combinations in a more elegant shorthand as a *determinant*

$$\begin{vmatrix} 1s\alpha(1) & 1s\beta(1) & 2a\alpha(1) \\ 1s\alpha(2) & 1s\beta(2) & 2s\alpha(2) \\ 1s\alpha(3) & 1s\beta(3) & 2s\beta(3) \end{vmatrix}$$

which is usually called a Slater Determinant, in honour of J. C. Slater.

In the old days, this phenomenon of indistinguishability was not properly understood, and people spoke about the 'exchange' of electrons as if it were some new force between the fundamental particles.

Back to the SCF method then, in order to finish off the story. In a nutshell, Fock realized that Hartree had missed the antisymmetry of electron wavefunctions from his theory. Fock was able to give a simple correction to Hartree's theory, and so we are left with Hartree–Fock Self-Consistent Field (HF-SCF) theory. The HF-SCF theory is widely used today to model atomic and molecular structure, and its energy expression only differs from that derived by Hartree by the addition of a single term, which takes account of the indistinguishability of electrons. The HF-SCF model forms the core of our modern-day *orbital model*, and it enjoys that status because it is sufficiently accurate for most chemical purposes.

A typical Hartree–Fock numerical atom SCF calculation would give as output the orbitals, their energies and usually the radial distribution curve(s), including the total one. Some typical radial distribution curves are given for helium and neon in Figures 15.12 and 15.13. They are normalized in the sense that the area under the curve is equal to the number of electrons. You should remember that these radial distribution functions are plots of $4\pi r^2 \psi^2$ against r. The 'shells' can be clearly distinguished, and the 'radii' of these shells decrease as the nuclear charge Z increases. The radius of the inner shell is essentially a_0/Z.

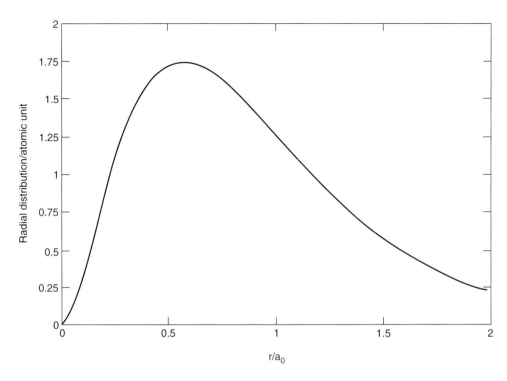

Figure 15.12   Helium atom radial distribution

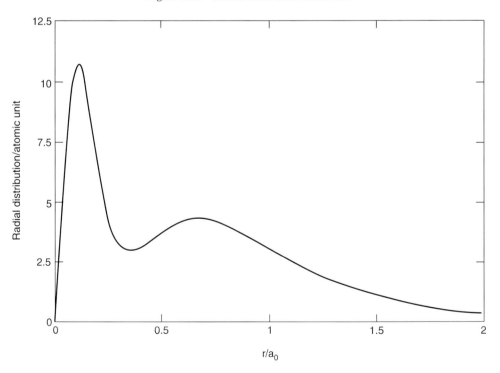

Figure 15.13   Neon radial distribution curve

15.8   THE PERIODIC TABLE

As I mentioned above, the electronic ground state of any atom can be obtained by adding electrons one at a time to the lowest possible orbitals, taking account of the Pauli exclusion principle. As we progress to higher and higher atomic numbers, the ordering of the atomic orbitals is found to change. For example, the chemistry of the first transition row is intimately connected with the relative energies of the 3d and the 4s orbitals.

To a rough first approximation, the atomic orbitals are found to lie in the order

$$1s, \ 2s, \ 2p, \ 3s, \ 3p, \ 3d, \ 4p, \ 5s, \ 4d, \ 5p, \ 6s, \ 4f, \ 5d, \ 6p, \ 7s, \ 6d.$$

In a one-electron atom, the energy only depends on the principal quantum number n so that (for example) the 3s, 3p and 3d atomic orbitals all have the same energy. In a many-electron atom, the outer electrons move in an average field which is the resultant of the field due to the nucleus and the field of the inner electrons. The inner electrons are said to 'screen' the outer electrons from the nucleus. The binding energy for an outer electron will depend on how effectively this electron is screened from the nucleus. For example, if a 3d electron is more completely screened than a 3p, and the 3p electron more completely screened than the 3s, then the order of the orbitals will be 3s, 3p and then 3d and this is exactly the case in many-electron atoms.

We finally have to consider the case where a given atom has partially filled shells. An obvious example is carbon, whose electronic configuration is $1s^2 2s^2 2p^2$.

It is often found experimentally that *Hund's rule of maximum multiplicity* is followed. I can summarise this quite simply by saying that, other things being equal, the lowest atomic energy configuration of electrons is the one corresponding to the largest number of unpaired electrons. So for carbon the '2p-electrons' would have their spins parallel in the lowest energy electronic state. Hund's rule is just that, it is a rule and therefore has many exceptions.

16   Diatomics

In the last chapter, we tackled the quantum mechanical treatment of atomic structure by making use of the Hartree–Fock (HF) method. Just to remind you, the physical basis of the Hartree model is that individual electrons move in a potential that is a sum of that due to the nucleus and an *averaged* one due to the remaining electrons. The HF model incorporates the fermion behavior of electrons. Solution of the HF equations gives the HF orbitals and the orbital energies.

Atoms are special species because of their very high symmetry, and Hartree was able to integrate his differential equations numerically and so arrive at the most accurate answer possible within the framework of the orbital model. We refer to such calculations as being at the *Hartree–Fock limit*; they are the best that can be done within the orbital model.

How to progress to molecules? The simplest molecules are diatomics, and of the diatomics, the simplest is the hydrogen molecule ion H_2^+ with just one electron. This molecule can be produced by passing an electric discharge through hydrogen gas. It is stable but highly reactive. It dissociates according to the chemical equation

$$H_2^+ \quad \rightarrow \quad H + H^+$$

and the dissociation energy (D_e rather than D_0) is known to be $269.6 \, \text{kJ mol}^{-1}$.

It only has one well-established electronic state, the ground state and this has been extensively studied by spectroscopists, who have deduced that it has an equilibrium bond length of 106 pm.

The first simplification is that we can treat separately the overall translational motion of the molecule, and this can be done rigorously. The translational motion of the center of mass need not concern us.

That leaves us to just focus attention on the relative motions of the two nuclei (of mass $m_p = 1.673 \times 10^{-27}$ kg) and the electron (of mass $m_e = 9.109 \times 10^{-31}$ kg) about the center of mass, i.e. three interacting particles. The nuclei repel each other and attract the electron due to their electrostatic potentials. In principle, all we have to do in order to obtain information about this molecule is to write down the time-independent Schrödinger equation and then solve it.

Figure 16.1 shows H_2^+ with position vectors of the two nuclei ($\mathbf{R}_A$ and $\mathbf{R}_B$) and the electron ($\mathbf{r}$) all marked. The wavefunction for the molecule depends on the coordinates of the two nuclei and of the electron and I will write this as $\Psi(\mathbf{R}_A, \mathbf{R}_B, \mathbf{r})$. Born and Oppenheimer pointed out an approximation, based on the relative mass of the electron compared to the nuclei, that permitted the time-independent Schrödinger equation to be solved in steps. According to Born and Oppenheimer, the electron moves much more quickly than the nuclei and so the total wavefunction $\Psi(\mathbf{R}_A, \mathbf{R}_B, \mathbf{r})$ which depends on the position vectors of nuclei A, B and the electron can be written as the product of a nuclear wavefunction Ψ_{nuc} that depends only on the coordinates of the two nuclei and an electronic wavefunction ψ_e that depends on the coordinates of the single electron (assuming that the nuclei are glued in position).

$$\Psi(\mathbf{R}_A, \mathbf{R}_B, \mathbf{r}) = \Psi_{nuc}(\mathbf{R}_A, \mathbf{R}_B)\psi_e(\mathbf{R}_A, \mathbf{R}_B, \mathbf{r}) \tag{16.1}$$

Here the electronic wavefunction is given as the solution of a certain 'electronic' Schrödinger equation.

$$-\frac{h^2}{8\pi^2 m_e}\nabla^2\psi_{el} + U\psi_{el} = \varepsilon_{el}\psi_{el} \tag{16.2}$$

where the potential energy is given by

$$U = \frac{e^2}{4\pi\varepsilon_0}\left(\frac{1}{|\mathbf{r} - \mathbf{R}_A|} + \frac{1}{|\mathbf{r} - \mathbf{R}_B|}\right) \tag{16.3}$$

The physical interpretation is that the term involving the operator ∇^2 represents the kinetic energy of the electron, and U represents the mutual potential energy of the electron and the two nuclei.

We solve this electronic Schrödinger equation for a fixed value of the internuclear separation, R_{AB}. The total energy is then

$$\varepsilon_{el} + \frac{e^2}{4\pi\varepsilon_0 R_{AB}}$$

and this is often referred to colloquially as 'the potential energy'. We then repeat the calculation for a range of values of the internuclear separation to give 'the potential energy curve'.

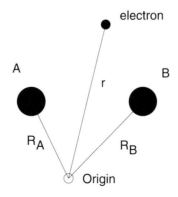

Figure 16.1   Hydrogen molecule ion

If we are interested in the nuclear motions, then the nuclei move under the influence of 'the potential energy', and so we would have to solve a vibrational (nuclear) Schrödinger equation with this potential.

The Born–Oppenheimer approximation is good to almost (but not exactly) the ratio of the masses of the particles, which is 1 to 1836.

The hydrogen molecule ion is unusual in that we can solve the electronic Schrödinger equation exactly, using numerical methods. When we have to consider many-electron systems, then we can no longer solve the equation exactly and so we have to seek approximations.

Perhaps the most obvious approach is to consider the behavior of the molecule for large values of the internuclear distance, where we have a proton and a H atom. Atoms are the building blocks of molecules, and so it can be argued that atomic orbitals should be the building blocks of molecular orbitals. Notice that I am only dealing with a single electron, so that the electronic wavefunction is by definition an orbital.

Imagine a process where we bring a proton (H^+) from infinity up to hydrogen atom A, leaving the proton at a distance greater than the chemical bond length of H_2. The molecular electronic wavefunction would then be (roughly)

$$\psi_{el} = 1s_A$$

I am now going to drop the subscript 'el' from the electronic wavefunction, for the sake of clarity. From now on in this chapter, ψ refers to a molecular orbital.

We now repeat the process of building up an H_2^+ molecule by starting with a fixed hydrogen atom B and bring up the proton from infinity to a position some distance from nucleus B. The molecular orbital would now be (roughly)

$$\psi = 1s_B$$

In the case of a chemically interesting internuclear separation, we might therefore suppose that low energy molecular orbitals for H_2^+ would be of the form

$$\psi = c_A 1s_A + c_B 1s_B$$

where the coefficients c_A and c_B have to be determined. In this particular case, the values of c_A and c_B can be deduced from the symmetry of the problem as follows. According to the Born interpretation of quantum mechanics, $\psi^2 d\tau$ gives the chance that the electron can be found in the volume of space $d\tau$.

$$\psi^2 d\tau = (c_A 1s_A + c_B 1s_B)^2 d\tau = (c_A^2 1s_A^2 + c_B^2 1s_B^2 + 2c_A c_B 1s_A 1s_B) d\tau$$

Electron densities around the two hydrogen atoms A and B are totally equivalent to each other by symmetry, and so the labels A and B can be used interchangeably which means that $c_A^2 = c_B^2$ and so $c_A = \pm c_B$.

What we have done then is to identify two possible molecular orbitals based on the idea of the *linear combination of atomic orbitals* (LCAO). Since we argued in terms of the hydrogen atom $1s$ orbitals (the lowest energy atomic orbitals), we might reasonably expect to find that the two possible LCAO molecular orbitals (LCAO MOs) would describe the ground state and a low excited state.

16.1 SOME USEFUL JARGON

The cs are called *LCAO coefficients*. The atomic orbitals used in making up an LCAO orbital are often referred to as *basis functions* and the set of basis functions is called the *basis set*. For the minute, I will label the LCAO MOs ψ_+ and ψ_- so that

$$\psi_+ = c_+(1s_A + 1s_B)$$
$$\psi_- = c_-(1s_A - 1s_B)$$

(16.4)

There are more formal ways of labeling molecular orbitals; if the molecule in question has symmetry, then the molecular orbitals are labeled according to the irreducible representations of the molecular point group. Of all known molecules, the vast majority are organic and have no symmetry to speak of. I do not want to make an issue of this topic, since a study of individual MOs is not particularly rewarding unless you happen to be interested in photoelectron spectroscopy, where it is possible to identify individual ionizations as if they were from molecular orbitals. In a later section, I will show you how to make sense of the electron density as a whole rather than the individual contributions to the electron density from the occupied MOs.

I have written the LCAO coefficients as c_+ and c_- rather than c_A and c_B because they are obviously related by symmetry. We can determine them as follows.

Once again, I want to remind you of the Born interpretation of quantum mechanics. The probability of finding the electron in a volume element $d\tau$ is given by $\psi_+^2 \, d\tau$ for the ψ_+ LCAO combination. Probabilities have to add to 1, and so

$$\int \psi_+^2 \, d\tau = \int c_+^2(1s_A + 1s_B)^2 \, d\tau$$

must come to 1.

Expanding the terms in parentheses and integrating gives

$$1 = c_+^2 \left(\int 1s_A^2 \, d\tau + \int 1s_B^2 \, d\tau + 2 \int 1s_A 1s_B \, d\tau \right)$$

Each of the individual atomic orbitals is normalized, and so we have

$$1 = c_+^2 \left(1 + 1 + 2 \int 1s_A 1s_B \, d\tau \right)$$

Integrals such as $\int 1s_A 1s_B \, d\tau$ are called *overlap integrals*, and they are given the general symbol S. They give a numerical value for the space that two atomic orbitals ($1s_A$ and $1s_B$ in this case) occupy together. Overlap integrals have values from -1 to $+1$.

After a little rearrangement, we find

$$\psi_+ = \frac{1}{\sqrt{2(1 + S)}} (1s_A + 1s_B)$$
$$\psi_- = \frac{1}{\sqrt{2(1 - S)}} (1s_A - 1s_B)$$

(16.5)

Sometimes it can be shown by symmetry arguments that the overlap between two atomic orbitals is zero, but in general the overlap integrals have to be calculated. In the case of two hydrogen 1s atomic orbitals, the overlap integral is

$$S = \exp\left(-\frac{R}{a_0}\right)\left(1 + \frac{R}{a_0} + \frac{R^2}{3a_0^2}\right) \tag{16.6}$$

where R is the internuclear distance and a_0 the first Bohr radius (52.9 pm).

16.2 POTENTIAL ENERGY CURVES

These two LCAO MO solutions to the H_2^+ molecule ion are obviously not completely correct because they are not exact solutions of the electronic Schrödinger equation, but they seem to be very reasonable on the basis of physical intuition.

How do we test whether our approximations are reasonable or not? What criteria do we have? The answer is given by the *variation method* as discussed in Chapter 5; if we have a real approximation φ to the ground state energy of H_2^+ and we want to find out how good an approximation it is, we calculate the variational integral

$$\frac{\int \varphi\left(-\frac{h^2}{8\pi^2 m_e}\nabla^2 + U\right)\varphi\,d\tau}{\int \varphi^2\,d\tau}$$

which is always greater than or equal to the correct energy, and compare it with experiment if we can.

I mentioned in Chapter 5 that the variation principle applied only to the lowest energy state of a given symmetry. The plus and minus combinations are symmetric and antisymmetric to reflection in any plane passing through the bond mid point and perpendicular to the molecular axis, and so they have different symmetries and so the variation principle applies for both these LCAO MOs separately.

We can calculate the variational integral for a range of internuclear separation, and remember to then add on the internuclear repulsion. This gives a molecular potential energy curve that can be compared (in principle) with the results of experiment

From the ϕ_+ curve we can read off the predicted dissociation energy and the predicted ground state equilibrium bond distance. Spectroscopic studies suggest a dissociation energy D_e of 2.795 eV and an equilibrium separation of 105 pm. The simple LCAO treatment gives 1.76 eV and 132 pm.

The ψ_- curve corresponds to an excited state which is unstable for every value of the internuclear separation.

This LCAO method is widely used in the study of molecular electronic structure. The HF method can be used to find LCAO coefficients, and professional computer packages are available, as I will show you later.

16.3 VISUALIZATION OF THE LCAO MOs

How do we think about LCAO MOs? The classic way is to make a plot of the values of the wavefunction along the internuclear axes. Here they are for ψ_+ and then for ψ_- for the hydrogen molecule ion, calculated at the equilibrium bond distance. I have used 'atomic units' throughout in my calculations of the graphs, in keeping with normal practice. The atomic unit of length, for example, is just the first Bohr radius a_0. Do not get these plots wrong; they are not three-dimensional objects. They show the magnitude of the wavefunction as we move along the internuclear axis, nothing more.

Figure 16.2 is a plot of the lowest energy LCAO MO, ψ_+, which describes the ground state of H_2^+, along the internuclear axis (taken by convention as the z axis). The nuclei can be seen by the positions of the cusps in the wavefunction (at the nuclear positions). Figure 16.3 is a plot of the higher energy LCAO-MO, which describes the first excited state of H_2^+. The horizontal axis corresponds to the internuclear axis, as above. In view of the Born interpretation of quantum mechanics, you might like to see the plots of the squares of the wavefunctions.

Figure 16.4 plots ψ_+^2. The point to notice is that there is a substantial probability of finding the electron in between the two nuclei. In the case of ψ_- the probability of finding the electron at the mid point is zero (the wavefunction has a node there), and it looks intuitively as if there is a reduced chance of finding the electron in the general region between the nuclei compared with the plot given above in Figure 16.4. Figure 16.5 is the electron density plot for antibonding LCAO MO. We can also plot the electron densities as two-dimensional contours. The corresponding three dimensional contour diagram can be obtained by rotating the two-dimensional section about the internuclear axis (Figure 16.6), which is a contour plot of ψ_+^2 the lower energy LCAO MO, or as three-dimensional figures as in the atomic case.

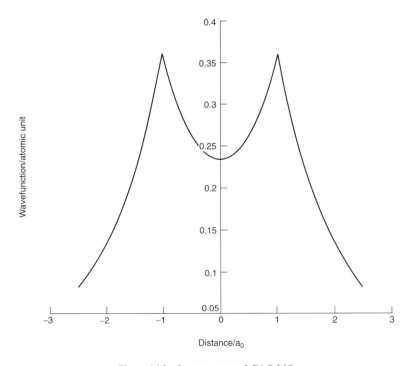

Figure 16.2 Lowest energy LCAO MO

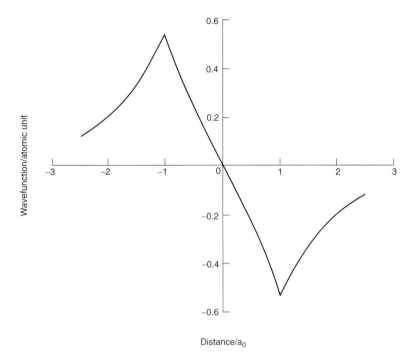

Figure 16.3 Antibonding LCAO MO

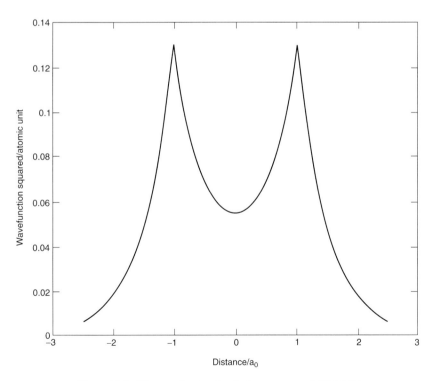

Figure 16.4 Electron density plot for lowest energy LCAO MO

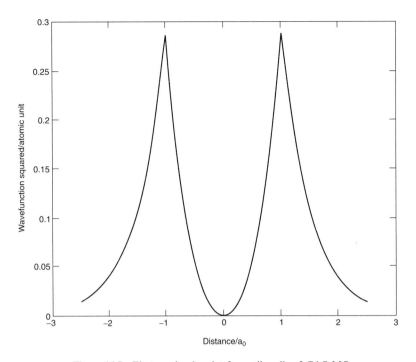

Figure 16.5 Electron density plot for antibonding LCAO MO

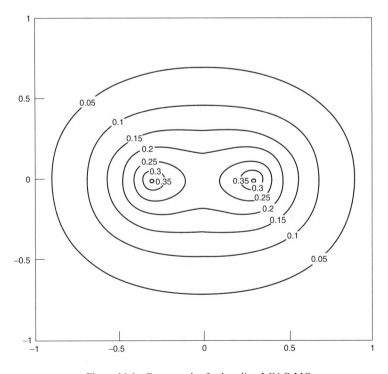

Figure 16.6 Contour plot for bonding LCAO MO

16.4 DENSITY DIFFERENCES

Over the years, many people have tried to give a simple explanation as to why H_2^+ (or indeed any other molecule) is stable. One particular explanation for H_2^+ is to compare the electron density in H_2^+ with the electron density in two mythical hydrogen ions each of which contains a half electron. I can illustrate this most easily for a plot of the magnitude of the electron density for ψ_+ along the internuclear axis minus the electron density expected from such two pseudo hydrogen atoms each carrying half a 1s electron (Figure 16.7).

Negative values on the vertical axis in Figure 16.7 show where electron density has been removed from the two constituent atoms, and positive regions show where electron density has been gained on molecule formation. You can see quite easily that the stability of the molecule can be ascribed to electron density being shifted from the regions 'behind' the nuclei into the bonding region.

If we now consider the ψ_- combination (which corresponds to an unstable electronic state), you can see that electron density has been removed from between the nuclei and placed 'behind' them. For this reason, we mentally classify orbitals as 'bonding' and 'antibonding'. There is a third category, 'non-bonding', that we will meet later (Figure 16.8).

16.5 NOTATION

Bearing in mind all I have told you above, I am going to give you an easy notation for diatomic molecular orbitals. A MO that is composed mostly of 1s atomic orbitals and is a bonding

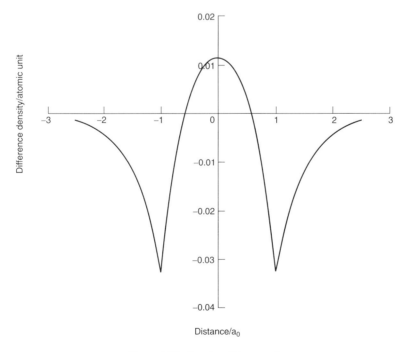

Figure 16.7 Density difference plot

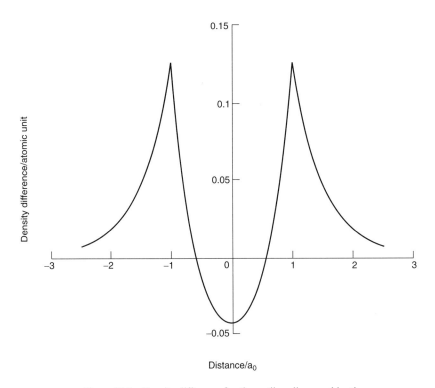

Figure 16.8   Density difference for the antibonding combination

combination is referred to as a $1s\sigma$ MO. The corresponding antibonding combination is called a $1s\sigma^*$ MO. A bonding MO composed mostly of 2s atomic orbitals is a $2s\sigma$ MO, the antibonding combination being $2s\sigma^*$.

16.6   THE NEXT STEPS, H_2, H_2^- AND He_2

The hydrogen molecule-ion is not a molecule of great interest to chemists! Its significance is that it is the simplest possible molecule and that it is stable. To build up a picture of the electronic structure of the next few homonuclear diatomics I want you to recall the Aufbau principle from Chapter 15. We feed electrons into the available orbitals, with spins paired.

H_2 dihydrogen had two electrons, and so these fit nicely into the MO ψ_+ (now referred to as $1s\sigma$) with spins paired. A variational calculation shows that this wavefunction does indeed give a bound molecule, although the agreement between the experimental and predicted dissociation energies leaves a lot to be desired.

H_2^- (or He_2^+) with three electrons would have electronic configuration $\psi_+^2\psi_-^1$, (or in our new notation $(1s\sigma)^2(1s\sigma^*)^1$, and so would have two electrons in the bonding MO and one electron in the antibonding MO. He_2^+ is stable, and has been studied by spectroscopists. It has a dissociation energy of $2.5\,eV$, but is less stable than H_2.

He$_2$ with four electrons has electron configuration $(1s\sigma)^2$ $(1s\sigma*)^2$ and so we might expect that at best it would be unstable; we have no way of knowing from such hand-waving arguments how much a bonding orbital cancels out an antibonding one.

More advanced calculations at this level of theory predict that He$_2$ is unstable, and at first sight this seems to agree with experiment and chemical intuition.

Hartree–Fock theory does not give the full picture, and the fact that helium can be liquefied under extreme circumstances shows that there must be a shallow minimum in the potential energy curve. This is a dispersion effect, which depends on instantaneous changes in the electron distribution. Since Hartree–Fock theory relies on an averaged interaction between electrons, it cannot give an account of such dispersion interactions.

16.7 THE HF–LCAO MODEL

In the case of H$_2^+$, H$_2$, He$_2^+$ and He$_2$ it was easy to derive the LCAO coefficients in the expression

$$\psi = c_A 1s_A + c_B 1s_B$$

because they were determined by symmetry. But how do we proceed to find LCAO MO coefficients for molecules in general? It turns out that there is an LCAO variant of the Hartree–Fock method that allows us to calculate these LCAO MO coefficients. The calculations are not easy, but can be done given a computer and the relevant software is widely available. These ideas form the basis of very many molecular structure calculations done today, and I can illustrate their use by considering diatomic molecules composed of first row atoms.

For the sake of argument, let us look at the homonuclear first row diatomics of formula A_2, where each atom A can be described in terms of atomic orbitals 1s, 2s and 2p. So for example a chemist would write the electron configuration of the F atom $1s^2 2s^2 2p^5$ and so on.

In the original atomic Hartree–Fock model, the relevant Schrödinger equation was solved numerically. This procedure was modified many years ago to cope with the LCAO case. Details are given in many advanced texts, we need not concern ourselves with them at this point. All that matters is the basic physical idea that each and every electron can be described as if it experienced only the potential due to the nuclei and an average of that due to the other electrons. We talk about the SCF-LCAO (or sometimes HF–LCAO) method.

There are several computer packages that will even run on PCs to perform SCF-LCAO calculations. Without more ado, let me show you the results of an SCF-LCAO calculation on F$_2$ at its equilibrium internuclear separation of 143.5 pm. Each F atom has a so-called minimal basis set of 1s, 2s and 2p atomic orbitals, and Table 16.1 below shows the LCAO expansion coefficients and orbital energies in atomic units of energy; 1 au $= e^2/4\pi\epsilon_0 a_0$. The two fluorine atoms are labelled 1F and 2F, and the columns in Table 16.1 show the LCAO coefficients for each molecular orbital.

Thus, the nine columns refer to the nine occupied molecular orbitals, and the orbital energies are -26.0343 au through -0.47200 au. The lowest energy molecular orbital LCAO coefficients are given in the first numerical column in Table 16.1, and so on. The

Table 16.1 LCAO MO coefficients and orbital energies for difluorine

MO energies/au	-26.0343	-26.0334	-1.6095	-1.3524	-0.6013	-0.6013	-0.5410	-0.47200	-0.47200
1 F 1S	0.70339	0.70383	-0.17717	-0.19112	0.00000	0.00000	-0.03906	0.00000	0.00000
2S	0.01545	0.01301	0.67262	0.75834	0.00000	0.00000	0.17795	0.00000	0.00000
2PX	0.00000	0.00000	0.00000	0.00000	0.00000	0.69022	0.00000	0.00000	0.72530
2PY	0.00000	0.00000	0.00000	0.00000	0.69022	0.00000	0.00000	0.72530	0.00000
2PZ	-0.00208	0.00050	-0.08249	0.06855	0.00000	0.00000	0.65099	0.00000	0.00000
2 F 1S	-0.70339	0.70383	-0.17717	0.19112	0.00000	0.00000	-0.03906	0.00000	0.00000
2S	-0.01545	0.01301	0.67262	-0.75834	0.00000	0.00000	0.17795	0.00000	0.00000
2PX	0.00000	0.00000	0.00000	0.00000	0.00000	0.69022	0.00000	0.00000	-0.72530
2PY	0.00000	0.00000	0.00000	0.00000	0.69022	0.00000	0.00000	-0.72530	0.00000
2PZ	-0.00208	-0.00050	0.08249	0.06855	0.00000	0.00000	-0.65099	0.00000	0.00000
MO name	1sσ	1sσ*	2sσ	2sσ*	2pπ	2pπ	2pσ	2pπ*	2pπ*

SCF procedure produces a total of 10 molecular orbitals from 10 atomic orbitals; I have shown only the occupied ones in Table 16.1. We often refer to the unoccupied orbitals as *virtual orbitals*. The occupied molecular orbital with the highest energy is known as the HOMO (think about the acronym), and the lowest energy virtual molecular orbital is the LUMO.

16.7.1 Contour Plots for F_2 at 141.2 pm

The lowest energy F_2 MO and the second lowest MO are composed almost entirely of fluorine 1s atomic orbitals. Note that the MOs are not localized on either atom, they are equally divided between the pair. This is to do with molecular symmetry.

The lowest energy MO has LCAO coefficients given in Table 16.1 as

$$\psi_1 = 0.70399 \ 1s(F_1) + 0.01545 \ 2s(F_1) - 0.00208 \ 2p_z(F_1) - 0.70399 \ 1s(F_2)$$
$$- 0.01545 \ 2s(F_2) + 0.00208 \ 2p_z(F_2)$$

where F_1 and F_2 refer to the two fluorine atoms. In my simplified notation, this is the $1s\sigma$ molecular orbital (Figure 16.9).

I have not shown the second MO, which is just the – combination of the F 1s atomic orbitals. Again, in my simplified notation, this is the $1s\sigma^*$ MO.

The third MO is

$$\psi_3 = -0.17717 1s(F_1) + 0.67262 2s(F_1) - 0.08249 \ 2p_z(F_1) - 0.17117 1s(F_2)$$
$$+ 0.67262 \ 2s(F_2) - 0.08249 \ 2p_z(F_2)$$

and so is mostly a combination of fluorine 2s atomic orbitals (Figure 16.10). This is therefore the $2s\sigma$ MO. A close examination of the electron density shows that it is a bonding orbital (and so it has no * to its name).

Now for the fourth occupied MO, which is mostly the negative combination (apart from numerical factors) of the fluorine 2s atomic orbitals (Figure 16.11). This is an antibonding combination and so is the $2s\sigma^*$ MO.

The fifth and sixth occupied orbitals are combinations of the fluorine $2p_x$ and $2p_y$ orbitals. They have equal energy (they are degenerate), and can be transformed into each other by a rotation of $90°$ about the internuclear z axis. I can best illustrate one of them by rotating the molecule a little, to give a view more or less along the bond axis (Figure 16.12). They are quite different from the σ MOs in that they have a nodal plane in the molecular plane.

They are made up entirely from $2p_x$ and $2p_y$ atomic orbitals; the 0.00000 entries in Table 16.1 are exact zeros. To give a hand-waving argument in explanation, the atomic s and $2p_z$ atomic orbitals behave differently under symmetry operations from the $2p_x$ and $2p_y$ atomic orbitals, which themselves behave equivalently.

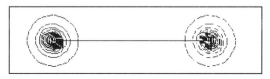

Figure 16.9   Lowest energy LCAO MO for difluorine

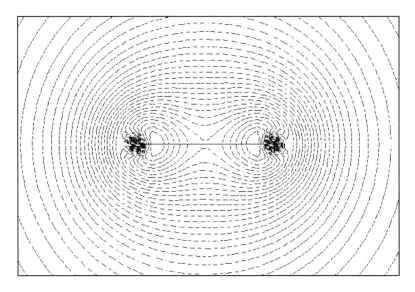

Figure 16.10   Third LCAO MO for difluorine

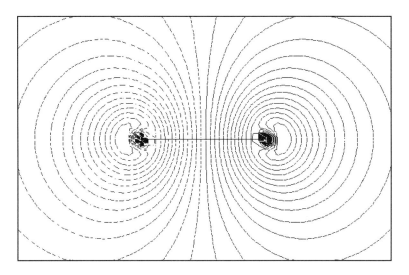

Figure 16.11   Fourth LCAO MO for difluorine

These two MOs are referred to collectively as the 2pπ MOs. Because there are two of them, we can allocate four electrons to the 2pπ molecular orbitals.

Molecular orbital number seven is essentially a combination of fluorine 2p$_z$ atomic orbitals, with some 2s also mixed in. Because the 2p$_z$ contribution dominates, the MO is called the 2pσ MO (Figure 16.13).

The eighth and ninth molecular orbitals are antibonding combinations of the 2p$_z$ atomic orbitals. I have not shown these here, but you will realize that their name is 2pπ*. Because

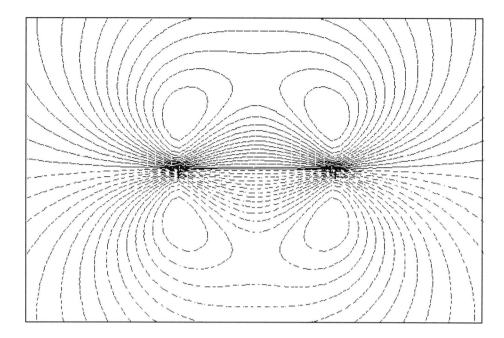

Figure 16.12   One of the degenerate pair

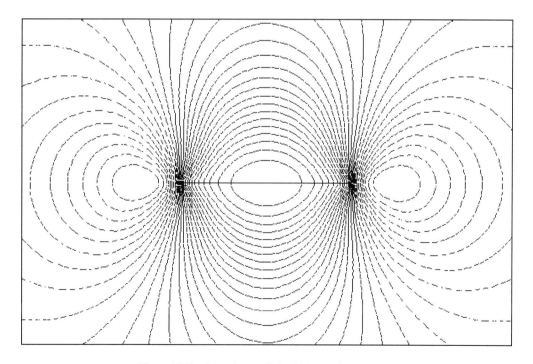

Figure 16.13   Seventh occupied LCAO MO for difluorine

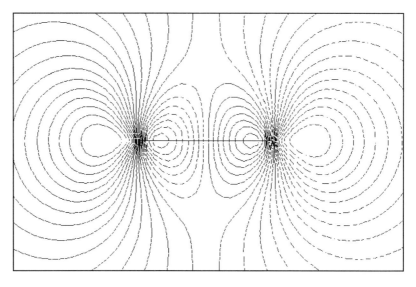

Figure 16.14 LUMO orbital

there are two of them, we can allocate a maximum of four electrons to the $2p\pi^*$ MOs. These two MOs are the HOMOs.

The tenth MO, the LUMO, is the negative combination of MO number 7, and is the $2p\sigma^*$ MO. It is not occupied by electrons in the ground state of the difluorine molecule (Figure 16.14).

Unfortunately, the ordering of the valence orbitals switches about from molecule to molecule. For example, if we repeat the calculation above for dinitrogen at its equilibrium bond length of 109.8 pm we find a different energy level diagram (Table 16.2).

16.8 BOND ORDERS AND POPULATION ANALYSIS

Elementary textbooks give a rough and way of calculating a quantity called the bond order, a quantity that gives a measure of bond strength; the higher the bond order, the stronger a bond. What they do is combine the contribution to the bond order of +1 for each electron in a bonding orbital and −1 for each electron in an antibonding orbital. So for example we calculate +1 for H_2^+, +2 for H_2, $+2 - 1 = +1$ for He_2^+ and $+2 - 2 = 0$ for He_2. This gives a bond strength ordering of $H_2 > H_2^+ > He_2$ in accord with chemical intuition.

The bond order is actually a well defined quantity in the field of molecular electronic structure theory.

Let me return to the simple LCAO treatment of H_2^+, which gave us the two LCAO MOs

$$\psi_+ = \frac{1}{\sqrt{2(1+S)}} \left(1s_A + 1s_B\right)$$

$$\psi_- = \frac{1}{\sqrt{2(1-S)}} \left(1s_A - 1s_B\right)$$

Table 16.2 LCAO MO coefficients and orbital energies for dinitrogen

MO energies/au	−15.5194	−15.517	−1.4462	−0.7220	−0.5756	−0.5756	−0.5404
1 N 1S	0.70306	0.70285	−0.17604	−0.17092	0.00000	0.00000	−0.06894
2S	0.01284	0.02703	0.48772	0.74154	0.00000	0.00000	0.40894
2PX	0.00000	0.00000	0.00000	0.00000	0.00000	0.62441	0.00000
2PY	0.00000	0.00000	0.00000	0.00000	0.62441	0.00000	0.00000
2PZ	−0.00213	−0.00992	−0.24113	0.26536	0.00000	0.00000	0.60378
2 N 1S	0.70306	−0.70285	−0.17604	0.17092	0.00000	0.00000	−0.06894
2S	0.01284	−0.02703	0.48772	−0.74154	0.00000	0.00000	0.40894
2PX	0.00000	0.00000	0.00000	0.00000	0.00000	0.62441	0.00000
2PY	0.00000	0.00000	0.00000	0.00000	0.62441	0.00000	0.00000
2PZ	0.00213	−0.00992	0.24113	0.26536	0.00000	0.00000	−0.60378
MO name	$1s\sigma$	$1s\sigma^*$	$2s\sigma$	$2s\sigma^*$	$2p\pi$	$2p\pi$	$2p\sigma^*$

If we focus attention on the electronic ground state, where the electron occupies the bonding molecular orbital ψ_+, this corresponds to a charge distribution of $-e\psi_+^2$. To emphasize that this depends on the position vector $\mathbf{r}$ of the electron I will write the electron density $\rho(\mathbf{r})$ as

$$\rho(\mathbf{r}) = -e\psi_+^2(\mathbf{r}) \tag{16.7}$$

where $-e$ is the charge on the electron.

Substituting the expression for ψ_+ gives

$$\rho(\mathbf{r}) = -\frac{e}{2(1+S)}\{[1s_A(\mathbf{r})]^2 + [1s_B(\mathbf{r})]^2 + 2 \times 1s_A(\mathbf{r}) \times 1s_B(\mathbf{r})\} \tag{16.8}$$

and this forms the basis for the electron density plots shown above.

From the early days of quantum mechanical molecular modeling, people have tried to divide up this electron density into regions associated with the atoms and regions associated with the bond regions, and to put numbers on these regions.

If we integrate the above expression over the spatial coordinates of the electron, we get such a way of dividing up the electron density.

$$\int \rho(\mathbf{r})\,\mathrm{d}\tau = -\int \frac{e}{2(1+S)}\{[1s_A(\mathbf{r})]^2 + [1s_B(\mathbf{r})]^2 + 2 \times 1s_A(\mathbf{r}) \times 1s_B(\mathbf{r})\}\,\mathrm{d}\tau \tag{16.9}$$

The quantity on the left-hand side is the sum of the electron density in space, and so must come to $-e$. Each atomic orbital $1s_A$ and $1s_B$ is separately normalized, and so the first two terms involving $\mathbf{r}$ under the right-hand integral are equal to 1. The third term is just the overlap integral and so we can write

$$\int \rho(\mathbf{r})\,\mathrm{d}\tau = -\frac{e}{2(1+S)}(1 + 1 + 2S) \tag{16.10}$$

showing that the electron density does indeed add up correctly to $-e$.

We can interpret the integral as a division of the total electron density into a density associated with hydrogen atom 1,

$$-\frac{e}{2(1+S)}[1s_A(\mathbf{r})]^2$$

a density associated with hydrogen atom 2

$$-\frac{e}{2(1+S)}[1s_B(\mathbf{r})]^2$$

and an electron density associated with the two atoms taken together in the overlap region

$$-\frac{2e}{2(1+S)}1s_A(\mathbf{r})1s_B(\mathbf{r})$$

Integration of these densities yields values of $-e/[2(1+S)]$, $-e/[2(1+S)]$ and $-2eS/[2(1+S)]$. People working in this field are usually chemists, and they tend to identify a unit charge as the charge on the electron (which is of course called 'negative'). For this reason, chemists talk about (*net*) *atom and bond populations* of $1/[2(1+S)]$, $1/[2(1+S)]$ and $2S/[2(1+S)]$, and they collect these quantities into a matrix $\mathbf{P}$ called the charge density matrix, which in this case is

$$
\mathbf{P} = \begin{pmatrix} \dfrac{1}{2(1+S)} & \dfrac{S}{2(1+S)} \\[2ex] \dfrac{S}{2(1+S)} & \dfrac{1}{2(1+S)} \end{pmatrix}
\tag{16.11}
$$

Formally the electron density $\rho(\mathbf{r})$ and $\mathbf{P}$ are related in this case by

$$
\rho(\mathbf{r}) = -e(1s_A(\mathbf{r}) \quad 1s_B(\mathbf{r})) \begin{pmatrix} \dfrac{1}{2(1+S)} & \dfrac{S}{2(1+S)} \\[2ex] \dfrac{S}{2(1+S)} & \dfrac{1}{2(1+S)} \end{pmatrix} \begin{pmatrix} 1s_A(\mathbf{r}) \\ 1s_B(\mathbf{r}) \end{pmatrix}
\tag{16.12}
$$

Robert S. Mulliken had the idea to divide up the bond populations into contributions due to each contributing atom, and to add these quantities to the atomic values. He decided arbitrarily to break each bond population into equal contributions from each atom in the bond. In the case of H_2^+ we divide the $2S/[2(1+S)]$ into two, and add each contribution to the atom contributions of $1/[2(1+S)]$. This gives *Mulliken gross populations* of $(1+S)/[2(1+S)] = 1/2$ for each atom, showing that the electron is equally shared between them.

We talk about *Mulliken population analysis*, and Mulliken population analysis indices are routinely given at the end of sophisticated quantum mechanical molecular modeling packages.

In the general case where we have a set of atomic orbitals $\chi_1(\mathbf{r}), \chi_2(\mathbf{r}), \dots, \chi_n(\mathbf{r})$, then the electron density can always be written

$$
\rho(\mathbf{r}) = -e(\chi_1(\mathbf{r}) \quad \chi_2(\mathbf{r}), \dots, \chi_n(\mathbf{r})) \begin{pmatrix} P_{11} & P_{12} & \cdots & P_{1n} \\ P_{21} & P_{22} & \cdots & P_{2n} \\ \vdots & \vdots & \cdots & \vdots \\ P_{n1} & P_{n2} & \cdots & P_{nn} \end{pmatrix} \begin{pmatrix} \chi(\mathbf{r}) \\ \chi_2(\mathbf{r}) \\ \vdots \\ \chi_{n1}(\mathbf{r}) \end{pmatrix}
\tag{16.13}
$$

and in the case of the LCAO expansion, where we have wavefunctions

$$
\psi_R = \sum_{j=1}^{n} c_{Rj} \chi_j(\mathbf{r})
$$

which describe each single electron, then they each make an additive contribution to the matrix elements $\mathbf{P}_{ij}$ of $c_{Ri} c_{Rj}$.

16.9 SLATER ORBITALS

I have been a little vague in this chapter in discussing how to choose the set of atomic orbitals (the *basis set*) for a given atom (for example, fluorine). At first sight, we might simply choose the hydrogenic atomic orbitals given in an earlier chapter, but modified to take account of the nuclear charge of $Z = 9e$. The 1s orbital of fluorine would then be written

$$\chi_{1s}(r) = \frac{1}{\sqrt{\pi}} \left(\frac{Z}{a_0}\right)^{3/2} \exp\left[-\left(\frac{Zr}{a_0}\right)\right] \qquad (16.14)$$

where a_0 is the first Bohr radius, and $Z = 9$.

A polyatomic atom is quite different to a one-electron atom because of the phenomenon called electron shielding, introduced in the previous chapter. The outer electrons in F will 'see' an effective nuclear charge very much reduced from 9, because of the screening due to the inner shell 1s electrons, and also partially the 2s electrons. J. S. Slater gave a set of rules to enable calculation of the effective nuclear charge $Z - S$ as follows.

(1) The effective nuclear charge is $Z - S$, where S represents the screening of an electron by the other electrons in an atom
(2) The orbitals are divided into sets; {1s}, {2s, 2p}, {3s, 3p}, {3d}, and so on.
(3) S is calculated as a sum, as follows
   (i)   Nothing from any electron in a group higher than the one considered.
   (ii) 0.35 for every other electron in the group, except for a 1s electron where 0.30 is taken.
   (iii) For an sp group, an amount 0.85 from each electron in the next lower group and 1.00 for all other electrons nearer the nucleus. For a d group, an amount of 1.00 per inner electron is used.

The atomic orbitals used then have Z replaced by $Z–S$ in Table 15.2 given earlier. Not only that, the radial parts of such orbitals differ from hydrogenic ones in that they have no nodes. So for example, 1s and 2s atomic orbitals are not orthogonal.

Slater's rules are still widely quoted, and they often form the basis of calculations on atomic and molecular systems. Table 16.3 gives values of $(Z–S)/n$ for some atoms, where n is the principal quantum number.

When computers first came on the scene, people decided to use the variation principle and SCF theory in order to obtain the best possible set of atomic orbitals. The idea is to choose a set of $(Z–S)/n$ values, which are usually referred to as orbital exponents, and then repeat an atomic SCF calculation many times until we find the best possible orbital exponents. These are of course the ones corresponding to the lowest energy. Clementi and Raimondi were the first

Table 16.3 Slater's rule for first-row atomic orbital exponents

Atom	Z	$(Z - S)/n$ for $n = 1$	$(Z - S)/n$ for $n = 2$
H	1	1	
He	2	1.70	
Li	3	2.70	0.650
Be	4	3.70	0.975
B	5	4.70	1.300
C	6	5.70	1.625
N	7	6.70	1.950
O	8	7.70	2.275
F	9	8.70	2.600
Ne	10	9.70	2.925

Table 16.4 Clementi and Raimondi's first-row atomic orbital exponents

Atom	Z	$(ZS)/n$ for 1s orbital	$(ZS)/n$ for 2s orbital	$(ZS)/n$ for 2p orbital
H	1	1		
He	2	1.6875		
Li	3	2.6906	0.6396	
Be	4	3.6848	0.9560	
B	5	4.6795	1.2881	1.2107
C	6	5.6727	1.6083	1.5679
N	7	6.6651	1.9237	1.9170
O	8	7.6579	2.2458	2.2266
F	9	8.6501	2.5638	2.5500
Ne	10	9.6241	2.8792	2.8792

to make an accurate study, and here are their results for the first row atoms are given in Table 16.4. Notice that the 2s and the 2p atomic orbitals are not constrained to have the same orbital exponent.

If there is a conclusion to be drawn from this, it is that Slater orbitals are remarkably accurate for atomic calculations.

17 Quantum Modeling of Larger Systems

In Chapter 16, I discussed the electronic structure of diatomic molecules in terms of the Hartree–Fock Linear Combination of Atomic Orbitals (HF–LCAO) procedure. The HF method treats each electron as if it were independent, but acting under the influence of a potential due to the atomic nuclei and an average due to the other electrons. Each electron can therefore be described by an orbital, and according to the Pauli principle each orbital can accommodate a maximum of two electrons, one of either spin. This is the same as saying that there are two quantum states per orbital, one of either spin ($m_s = \pm \frac{1}{2}$).

The LCAO version of HF theory expresses each of these HF orbitals in terms of a linear combination of the atomic orbital basis set. For the diatomic molecules discussed in a previous chapter, the atomic orbital basis set consisted of 1s, 2s and 2p type atomic orbitals on either atomic center, and the HF–LCAO molecular orbitals were given as linear combinations of these.

I did not explain in any detail how the HF–LCAO coefficients had been derived, and I now need to do so in order to make progress. I also want to draw attention to the complexity and computational difficulty of these calculations. So let me work through the HF–LCAO treatment of the molecule adrenaline, $C_9H_{13}O_3N$. Adrenaline (Figure 17.1) is a non-steroid hormone, and was the first such hormone to be isolated in crystalline form in 1901. It raises the blood pressure and is used locally to stop haemorrhage. It is optically active.

First of all, we make use of the Born–Oppenheimer approximation. This tells us that the nuclei can be regarded as being fixed in space when we consider the electronic wavefunction. In other words, we can treat the nuclei as point positive charges fixed at certain positions in space. These positions can be determined by X-ray diffraction, we can calculate them theoretically just as we did for the MM method, or we can take an idealized structure. There is an interesting problem here, as X-ray data generally refers to solids in which there are very strong intermolecular forces and so there is no obvious correlation with quantum chemical calculations, which generally refer to isolated molecules in the gas phase, for which there are no intermolecular forces of any kind.

Experimentalists such as biologists tend to be interested in interactions occurring in solution of a certain pH, usually pH 7. I will return to this point later.

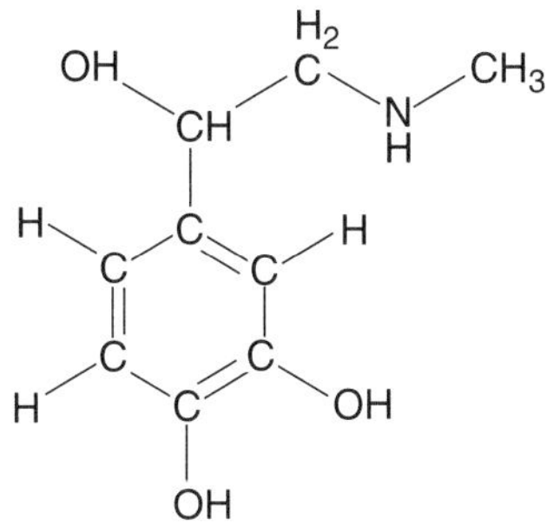

Figure 17.1 Adrenaline

To focus on the problem in hand, I drew the molecule using a two-dimensional Chemical Drawing package and did a MM calculation to get a three-dimensional geometry. As I explained in Chapter 10, this is just one of a very large number of possible geometrical conformations that correspond to local minima on the molecular potential energy curve.

17.1 IMPLEMENTATION OF MOLECULAR HF–LCAO THEORY

Suppose we have a collection of N nuclei, considered as point charges of magnitudes eZ_1, $eZ_2, \ldots, eZ_N$ at position vectors $\mathbf{R}_1$, $\mathbf{R}_2, \ldots, \mathbf{R}_N$.

The mutual electrostatic potential energy U of these point positive charges is

$$U = \frac{e^2}{4\pi\epsilon_0} \sum_{\alpha=1}^{N-1} \sum_{\beta=\alpha+1}^{N} \frac{Z_\alpha Z_\beta}{R_{\alpha\beta}} \qquad (17.1)$$

where $R_{\alpha\beta}$ is the scalar distance between the two nuclei α and β. The summations run over distinct pairs of nuclei, and we do not count each interacting pair of nuclei twice. The summation can equally be written

$$U = \frac{1}{2} \frac{e^2}{4\pi\varepsilon_0} \sum_{\alpha=1}^{N} \sum_{\beta=1,\beta\neq\alpha}^{N} \frac{Z_\alpha Z_\beta}{R_{\alpha\beta}}$$

where the factor of $1/2$ allows for double counting but we have to remember not to include a term where $\alpha = \beta$. In the case of adrenaline there are just $N = 25$ atoms, and calculation of the nuclear repulsion energy is straightforward.

Let me assume also that we have decided on a set of n atomic orbitals for the atoms comprising adrenaline. This set of atomic orbitals could be 1s, 2s and 2p Slater orbitals for the first row atoms, and a 1s orbital on each hydrogen atom. Adrenaline has formula $C_9H_{13}NO_3$, and so we might want have to use an atomic orbital basis set comprising $n = (9 + 1 + 3) \times 5 + 13 = 78$ atomic orbitals. For the sake of generality, I will call the number of atomic orbitals n.

There are $(9 \times 6 + 13 \times 1 + 1 \times 7 + 3 \times 8) = 98$ electrons in this particular molecule, and the electronic ground state of the molecule almost certainly corresponds to a HF–LCAO description with $m = 49$ doubly occupied molecular orbitals. We would write this electronic configuration as

$$\psi_1^2 \psi_2^2 \cdots \psi_{49}^2$$

If we have reason to believe that the ground state of the molecule can indeed be described by such a wavefunction then we would choose the so-called Restricted Hartree–Fock (RHF) variant of HF theory in order to calculate our SCF wavefunction. If for some spectroscopic reason we believed that our molecules had an imbalance between the number of spin α and spin β electrons then we might take a ground state configuration

$$\psi_1^2 \psi_2^2 \cdots \psi_{49}^1 \psi_{50}^1$$

which is called a Restricted Open Shell Hartree–Fock (ROHF) wavefunction, or we might chose to relax the requirement that all electron pairs had the same spatial symmetry, and write an Unrestricted Hartree–Fock (UHF) configuration

$$(\psi_1 \alpha)^1 (\psi_1' \beta)^1 (\psi_2 \alpha)^1 (\psi_2' \beta)^1 \cdots (\psi_{49} \alpha)^1 (\psi_{50} \alpha)^1$$

The UHF notation means that the first electron goes into quantum state $\psi_1 \alpha$, the second electron goes into quantum state $\psi_1' \beta$ and so on until we reach the 97th and 98th electrons. The spatial parts ψ_1 and ψ_1' are not required to be the same. They are allowed to vary independently during the SCF calculation. I have assumed that the spectroscopic evidence suggests a spin triplet state, and so I have allocated them each into α spin quantum states. I have illustrated these ideas in the Figure 17.2. I will concentrate now on the RHF case, which corresponds to the electronic ground state. Notice that the molecule has no particular

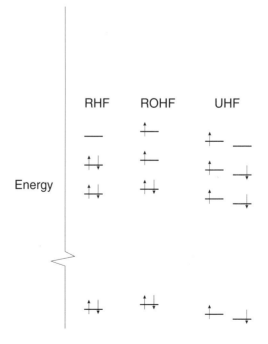

Figure 17.2   RHF, ROHF and UHF cases

symmetry, and so we do not need to worry about giving symmetry labels to the individual molecular orbitals. That is why I have labelled them numerically in order of increasing orbital energy.

Our problem is now to find the LCAO coefficients for each molecular orbital, the orbital energies and the total energy. Once we have determined the molecular wavefunction, we can then interest ourselves in calculating any relevant molecular properties such as the electric multipoles and so on. We will be most interested in the lowest energy 49 molecular orbitals corresponding to the 49 electron pairs, but I will tell you that the HF–LCAO procedure gives 78 solutions, the same as the number of atomic orbitals. The lowest energy 49 correspond to occupied molecular orbitals and $(78 - 49) = 29$ to unoccupied ('virtual') ones. I will use the symbol m for the number of electron pairs in the following discussion.

The HF–LCAO procedure seeks to express each MO ψ_A as a linear combination of the 78 χ_i for example

$$\psi_A = c_{A,1}\chi_1 + c_{A,2}\chi_2 + \cdots + c_{A,78}\chi_{78} \tag{17.2}$$

where the coefficient $c_{A,1}$ is the LCAO coefficient of atomic orbital χ_1 in the LCAO orbital ψ_A and so on. I would normally store the LCAO coefficients in an $n \times n$ matrix $\mathbf{U}$ so that the jth coefficient of the Ath MO is stored at $U_{j,A}$.

$$\mathbf{U} = \begin{pmatrix} c_{A,1} & c_{B,1} & \cdots & c_{n,1} \\ c_{A,2} & c_{B,2} & \cdots & c_{n,2} \\ \vdots & \vdots & \cdots & \vdots \\ c_{A,n} & c_{B,n} & \cdots & c_{n,n} \end{pmatrix} \tag{17.3}$$

The first column of $\mathbf{U}$ collects the LCAO coefficients for the Ath LCAO MO and so on. I can summarize input to the LCAO MO problem in Table 17.1.

We now have to write a mathematical expression for the total molecular wavefunction, and in order to do this we must take account of electron spin, the Pauli principle and electron indistinguishability.

I explained in an earlier chapter how to do this in terms of a Slater Determinant. I do not want to write out such a complicated expression, but I will give the wavefunction the symbol $\Psi(1, 2, 3, \ldots, 2m)$ to show that it depends on the space and spin coordinates of the $2m$ electrons (m electron pairs in this treatment).

The next steps are concerned with calculating the mutual potential energy of the electrons and nuclei. Since the nuclei are fixed in space (according to the Born–Oppenheimer approximation), we just calculate their contribution to the total mutual potential energy as mentioned above. The next substep is to write down the molecular electronic Hamiltonian operator H_{el}.

Table 17.1   Data for an LCAO–MO calculation

N **nuclei fixed at positions** $\mathbf{R}_1, \mathbf{R}_2, \ldots, \mathbf{R}_N$
n **basis functions** $\chi_1, \chi_2, \ldots, \chi_n$ **centred on the atoms**
RHF problem for m **electron pairs**
Find the LCAO MO coefficients to minimize the Hartree–Fock energy

---

This contains the following terms

- the kinetic energy of each of the $2m$ electrons
- the mutual potential energy of each of the $2m$ electrons with each of the N nuclei, and
- the mutual potential energy of each pair of electrons. In symbols, this is

$$\hat{H}_{el} = -\frac{h^2}{8\pi^2 m}\sum_{i=1}^{2m}\nabla_i^2 - \frac{e^2}{4\pi\epsilon_0}\sum_{i=1}^{2m}\sum_{\alpha=1}^{N}\frac{Z_\alpha}{R_{\alpha i}} + \frac{e^2}{4\pi\epsilon_0}\sum_{i=1}^{2m-1}\sum_{j=2}^{2m}\frac{1}{r_{ij}} \qquad (17.4)$$

where $R_{\alpha i}$ is the scalar distance between electron i and nucleus α and r_{ij} is the scalar distance between electrons i and j.

This notation is cumbersome and it is usual to simplify the Hamiltonian into a certain sum of so-called one-electron operators $h(\mathbf{r}_i)$ that depend only on the spatial coordinates of the electrons taken one at a time

$$\hat{h}(\mathbf{r}_i) = -\frac{h^2}{8\pi^2 m}\nabla_i^2 - \frac{e^2}{4\pi\epsilon_0}\sum_{\alpha=1}^{N}\frac{Z_\alpha}{R_{\alpha i}} \qquad (17.5)$$

and two-electron operators that depend on the spatial coordinates of pairs of electrons

$$\hat{g}(\mathbf{r}_i, \mathbf{r}_j) = \frac{e^2}{4\pi\epsilon_0 r_{ij}} \qquad (17.6)$$

With this notation, the Hamiltonian becomes

$$\hat{H}_{el} = \sum_{i=1}^{2m}\hat{h}(\mathbf{r}_i) + \sum_{i=1}^{2m-1}\sum_{j=2}^{2m}\hat{g}(\mathbf{r}_i, \mathbf{r}_j)$$

The next step is to work out a variational electronic energy expression

$$\varepsilon_{el} = \frac{\displaystyle\int \Psi\hat{H}\Psi\, d\tau}{\displaystyle\int \Psi^2\, d\tau} \qquad (17.7)$$

for the molecular wavefunction $\Psi(1, 2, 3, \ldots, 2m)$. Remember that each molecular orbital ψ_1 through ψ_m is doubly occupied, and that I have used a Slater Determinant in order to take account of indistinguishability. Without going into details, I will just quote the result.

$$\varepsilon_{el} = 2\sum_{R=1}^{m}\int \psi_R(\mathbf{r}_1)\hat{h}(\mathbf{r}_1)\psi_R(\mathbf{r}_1)\, d\tau_1$$

$$+ \sum_{R=1}^{m}\sum_{S=1}^{m}2\iint \psi_R(\mathbf{r}_1)\psi_R(\mathbf{r}_1)\hat{g}(\mathbf{r}_1, \mathbf{r}_2)\psi_S(\mathbf{r}_2)\psi_S(\mathbf{r}_2)\, d\tau_1\, d\tau_2$$

$$- \sum_{R=1}^{m}\sum_{S=1}^{m}\iint \psi_R(\mathbf{r}_1)\psi_S(\mathbf{r}_1)\hat{g}(\mathbf{r}_1, \mathbf{r}_2)\psi_R(\mathbf{r}_2)\psi_S(\mathbf{r}_2)\, d\tau_1\, d\tau_2 \qquad (17.8)$$

The first term represents the average value of the kinetic and nuclear attraction energy for each electron. The first of the two-electron terms has a classical interpretation; if I write it as

$$-\frac{e^2}{4\pi\epsilon_0}\iint \psi_R^2(\mathbf{r}_1)\frac{1}{r_{12}}\psi_S^2(\mathbf{r}_2)\,\mathrm{d}\tau_1\,\mathrm{d}\tau_2$$

you should recognize this as the mutual potential energy of a pair of electron distributions $-e\psi_R^2$ and $-e\psi_S^2$. The second term arises because of the fermion nature of electrons. It is usually called the *exchange* term, and I will have more to say about it in Chapter 18.

These integrals all refer to the HF–LCAO orbitals, and in order to finally evaluate the energy we normally expand the LCAO orbitals into the constituent basis set of atomic orbitals. If we look at a typical one-electron term

$$\int \psi_A(\mathbf{r}_1)\hat{h}(\mathbf{r}_1)\psi_A(\mathbf{r}_1)\,\mathrm{d}\tau_1$$

then if we substitute the LCAO expression $\psi_A = c_{A,1}\chi_1 + c_{A,2}\chi_2 + \cdots + c_{A,n}\chi_n$ we find a contribution to the mutual potential energy of

$$\sum_{i=1}^{n}\sum_{j=1}^{n}\left(c_{A,i}c_{A,j}\int \chi_i(\mathbf{r}_1)\hat{h}(\mathbf{r}_1)\chi_j(\mathbf{r}_1)\,\mathrm{d}\tau_1\right)$$

We have to do this sum for all of the occupied orbitals in order to calculate the energy and so the total one-electron contribution to the electronic energy is

$$2\sum_{R=1}^{m}\sum_{i=1}^{n}\sum_{j=1}^{n}\left(c_{R,i}c_{R,j}\int \chi_i(\mathbf{r}_1)\hat{h}(\mathbf{r}_1)\chi_j(\mathbf{r}_1)\,\mathrm{d}\tau_1\right)$$

we can switch the orders of the summation signs to give

$$2\sum_{i=1}^{n}\sum_{j=1}^{n}\sum_{R=1}^{m}(c_{R,i}c_{R,j})\left(\int \chi_i(\mathbf{r}_1)\hat{h}(\mathbf{r}_1)\chi_j(\mathbf{r}_1)\,\mathrm{d}\tau_1\right)$$

Recalling the electron density matrix $\mathbf{P}$ from Chapter 16, it turns out that we can write this one electron contribution as

$$\sum_{i=1}^{n}\sum_{j=1}^{n}\left(P_{ij}\int \chi_i(\mathbf{r}_1)\hat{h}(\mathbf{r}_1)\chi_j(\mathbf{r}_1)\,\mathrm{d}\tau_1\right) \tag{17.9}$$

After applying similar arguments to the two-electron terms, the total energy expression becomes

$$\varepsilon_{el} = \sum_{i=j}^{n}\sum_{j=1}^{n}P_{ij}\int \chi_i(\mathbf{r}_1)\hat{h}(\mathbf{r}_1)\chi_j(\mathbf{r}_1)\,\mathrm{d}\tau_1$$

$$+\frac{1}{2}\sum_{i=1}^{n}\sum_{j=1}^{n}P_{ij}\sum_{k=1}^{n}\sum_{l=1}^{n}P_{kl}\left(\iint \chi_i(\mathbf{r}_1)\chi_j(\mathbf{r}_1)\hat{g}(\mathbf{r}_1,\mathbf{r}_2)\chi_k(\mathbf{r}_2)\chi_l(\mathbf{r}_2)\,\mathrm{d}\tau_1\,\mathrm{d}\tau_2\right.$$

$$\left.-\frac{1}{2}\iint \chi_l(\mathbf{r}_1)\chi_k(\mathbf{r}_1)\hat{g}(\mathbf{r}_1,\mathbf{r}_2)\chi_j(\mathbf{r}_2)\chi_l(\mathbf{2})\,\mathrm{d}\tau_1\,\mathrm{d}\tau_2\right) \tag{17.10}$$

This equation should ring alarm bells with you; there are an awful lot of contributions to the electronic energy, and also the individual contributions seem to be very complicated integrals. Even the one-electron integrals are three-dimensional ones, and the two-electron integrals involve the coordinates of two electrons and so are six-dimensional objects.

It is probably easier if I introduce a certain matrix notation at this point. I will collect the one-electron integrals over the basis functions χ_i into an $n \times n$ matrix $\mathbf{h}$ whose ijth element is

$$\int \chi_i^*(\mathbf{r}_1)\hat{h}(\mathbf{r}_1)\chi_j(\mathbf{r}_1)\,\mathrm{d}\tau_1$$

I have added the complex conjugate sign * at this point to remind you that we should strictly cater for complex quantities in quantum mechanical calculations. In fact, very few people need to work with complex atomic orbitals and I will ignore this possibility from now on.

We also group together the two-electron contributions into an $n \times n$ matrix $\mathbf{G}$ whose elements depend on the elements of the electron density matrix $\mathbf{P}$ in a complicated way. The ijth element of this electron repulsion matrix $\mathbf{G}$ is

$$\sum_{k=1}^{n}\sum_{l=1}^{n} P_{kl}\left(\iint \chi_i(\mathbf{r}_1)\chi_j(\mathbf{r}_1)\hat{g}(\mathbf{r}_1,\mathbf{r}_2)\chi_k(\mathbf{r}_2)\chi_l(\mathbf{r}_2)\,\mathrm{d}\tau_1\,\mathrm{d}\tau_2\right.$$

$$\left. -\frac{1}{2}\iint \chi_i(\mathbf{r}_1)\chi_k(\mathbf{r}_1)\hat{g}(\mathbf{r}_1,\mathbf{r}_2)\chi_j(\mathbf{r}_2)\chi_l(\mathbf{r}_2)x_l(\mathbf{r}_2)\,\mathrm{d}\tau_1\,\mathrm{d}\tau_2\right) \tag{17.11}$$

and the energy expression can be compactly written as

$$\varepsilon_{\mathrm{el}} = \sum_{i=1}^{n}\sum_{j=1}^{n} P_{ij}h_{ij} + \frac{1}{2}\sum_{i=1}^{n}\sum_{j=1}^{n} P_{ij}G_{ij} \tag{17.12}$$

The final step is to seek an energy minimum. We use the variation principle to find the best LCAO coefficients in the sense that they give an energy that is a minimum.

To do this, we let each LCAO coefficient vary more or less independently. The only constraint is that the LCAO MOs are kept normalized and orthogonal to each other. The atomic orbitals are fixed throughout the calculation, and these are rarely orthogonal. For many purposes, we need to consider the $n \times n$ matrix of atomic orbital overlap integrals

$$\mathbf{S} = \begin{pmatrix} S_{11} & S_{12} & \cdots & S_{1n} \\ S_{21} & S_{22} & \cdots & S_{2n} \\ \vdots & \vdots & & \vdots \\ S_{n1} & S_{n2} & \cdots & S_{nn} \end{pmatrix}$$

where a typical term is

$$S_{ij} = \int \chi_i(\mathbf{r})\chi_j(\mathbf{r})\,\mathrm{d}\tau$$

Again, I will cut very many corners and just tell you the result.

To find the best LCAO coefficients and associated orbital energies, we have to find the eigenvectors and eigenvalues of a matrix called the HF matrix

$$\mathbf{h}^{\mathrm{F}} = \mathbf{h} + \mathbf{G}$$

The HF matrix is $n \times n$, and what we do is to solve the matrix equation

$$\mathbf{h}^F \mathbf{c} = \varepsilon \mathbf{S} \mathbf{c}$$

or, written out more explicitly

$$\begin{pmatrix} h_{11}^F & h_{12}^F & \cdots & h_{1n}^F \\ h_{21}^F & h_{22}^F & \cdots & h_{2n}^F \\ \vdots & \vdots & \cdots & \vdots \\ h_{n1}^F & h_{n2}^F & \cdots & h_{nn}^F \end{pmatrix} \begin{pmatrix} c_1 \\ c_2 \\ \vdots \\ c_n \end{pmatrix} = \varepsilon \begin{pmatrix} S_{11} & S_{12} & \cdots & S_{1n} \\ S_{21} & S_{22} & \cdots & S_{2n} \\ \vdots & \vdots & \cdots & \vdots \\ S_{n1} & S_{n2} & \cdots & S_{nn} \end{pmatrix} \begin{pmatrix} c_1 \\ c_2 \\ \vdots \\ c_n \end{pmatrix} \qquad (17.13)$$

Mathematicians refer to such equations as 'generalized' matrix eigenvalue equations; if the atomic orbitals had been orthogonal then we would have had to solve

$$\mathbf{h}^F \mathbf{c} = \varepsilon \mathbf{c}$$

which is a standard matrix eigenvalue equation.

In any case, we get exactly n solutions of the matrix eigenvalue problem. The m lowest energy solutions correspond to the occupied orbitals for the ground electronic state (which has $2m$ electrons, arranged into m electron pairs).

For the record, the elements of the HF–LCAO Hamiltonian $\mathbf{h}^F$ are given by

$$h_{ij}^F = \int \chi_i(\mathbf{r}_1) \left(-\frac{h^2}{8\pi^2 m} \nabla^2 - \frac{e^2}{4\pi\epsilon_0} \sum_{\alpha=1}^{NUC} \frac{Z_\alpha}{r_\alpha} \right) \chi_j(\mathbf{r}_1) \, d\tau_1$$

$$+ \sum_{k=1}^{n} \sum_{l=1}^{n} P_{kl} \left[\iint \chi_i(\mathbf{r}_1) \chi_j(\mathbf{r}_1) \left(\frac{e^2}{4\pi\epsilon_0 r_{12}} \right) \chi_k(\mathbf{r}_2) \chi_l(\mathbf{r}_2) \, d\tau_1 \, d\tau_2 \right.$$

$$\left. - \frac{1}{2} \iint \chi_i(\mathbf{r}_1) \chi_k(\mathbf{r}_1) \left(\frac{e^2}{4\pi\epsilon_0 r_{12}} \right) \chi_j(\mathbf{r}_2) \chi_l(\mathbf{r}_2) \, d\tau_1 \, d\tau_2 \right) \qquad (17.14)$$

It should be clear from the discussion above that the elements of the Hartree–Fock matrix depend themselves on the HF–LCAO coefficients (because the matrix $\mathbf{G}$ itself depends on the elements of $\mathbf{P}$). What we do is to have a guess at the HF–LCAO coefficients, and then iterate until we reach self consistency. This can involve starting from the LCAO coefficients, finding the electron density matrix, calculating the HF matrix, solving the generalized eigenvalue problem to give a new and (hopefully improved) set of LCAO coefficients with which to start the next cycle in the iterative procedure.

17.1.1   Gaussian Orbitals

A major problem in the early days of molecular electronic structure theory was the two-electron integrals described above, typified by

$$\iint \chi_i(\mathbf{r}_1) \chi_j(\mathbf{r}_1) \frac{1}{4\pi\varepsilon_0 r_{12}} \chi_k(\mathbf{r}_2) \chi_l(\mathbf{r}_2) \, d\tau_1 \, d\tau_2$$

In a HF–LCAO calculation with n atomic orbitals $\chi_1, \chi_2, \ldots, \chi_n$ there are at first sight n^4 of these integrals for evaluation, and in the case of our typical molecule adrenaline this gives $78^4 = 37\,015\,056$. There is an immediate simplification; some of the integrals are equal just because of the indistinguishability of electrons. So that

$$\iint \chi_i(\mathbf{r}_1)\chi_j(\mathbf{r}_1) \frac{1}{4\pi\epsilon_0 r_{12}} \chi_k(\mathbf{r}_2)\chi_l(\mathbf{r}_2)\,d\tau_1\,d\tau_2$$

$$= \iint \chi_k(\mathbf{r}_1)\chi_l(\mathbf{r}_1) \frac{1}{4\pi\epsilon_0 r_{12}} \chi_i(\mathbf{r}_2)\chi_j(\mathbf{r}_2)\,d\tau_1\,d\tau_2$$

and so on, which reduces the number of integrals to roughly $n^4/8$ which is still about 4×10^6. This is still a formidable number, and not only that the integrals are difficult to evaluate because

- they are six-dimensional integrals
- they have a singularity; as the distance r_{12} between electrons 1 and 2 gets smaller the integrand gets bigger, and it becomes infinite as the coordinates of the two electrons approach each other

There are also technical problems with the integrals. If the four atomic orbitals are centred on four different atoms, then there is no apparent choice of co-ordinate origin and no obvious choice for a 'change of variables'. In fact, these integrals turn out to be impossibly difficult for hydrogenic and Slater orbitals.

The breakthrough came with a classic paper due to S. F. Boys in 1950. Boys proposed the use of Gaussian orbitals for molecular calculations rather than Slater orbitals. Gaussian orbitals have an exponential form of $\exp(-\alpha r^2/a_0^2)$ rather than $\exp(-\xi r/a_0)$ as in Slater orbitals.

The factor α is called the Gaussian exponent, and workers in the field refer to 'Gaussian-type orbitals, GTO's'. In full, a normalised s-type GTO has the functional form

$$G(\alpha) = \left(\frac{2\alpha}{\pi a_0^2}\right)^{3/4} \exp\left(-\alpha\,\frac{r^2}{a_0^2}\right) \tag{17.15}$$

and a graph of a $1s$ Gaussian G ($\alpha = 0.283$) *versus* r/a_0 is shown in Figure 17.3, compared to the 'corresponding' Slater type 1s orbital with $\xi = 1$. (The Gaussian exponent of 0.283 gives the best fit in a least squares sense to the Slater exponent of 1.)

Gaussian orbitals have the 'wrong' behavior at $r = 0$; they are flat whereas they should show a cusp. They give rather poor energies. They also fall off much too quickly with large r. Nevertheless, GTOs lead to very simple two-electron integrals, and so a great deal of time and effort has been devoted to making use of them in molecular structure calculations.

A favourite way of implementing Gaussian orbitals in such calculations is to fit the STOs to a fixed linear combination of GTOs. Thus for example, a 1s STO of exponent ξ can be fitted to a linear combination of three GTOs of exponents α_1, α_2 and α_3 by least squares means as

$$\text{STO}(\xi = 1) = 0.444635\,\text{GTO}(\alpha = 0.109818) + 0.535328\,\text{GTO}(\alpha = 0.405771)$$

$$+ 0.154329\,\text{GTO}(\alpha = 2.22766). \tag{17.16}$$

The expansion coefficients and exponents are then kept constant for molecular applications.

Each of these so-called primitive Gaussians are plotted in Figure 17.4.

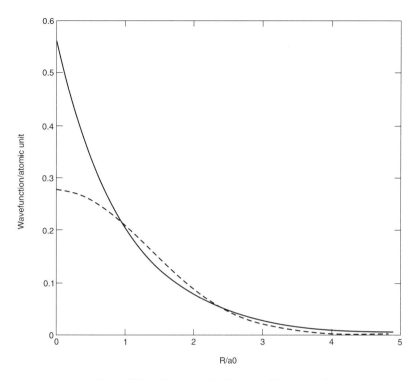

Figure 17.3 Gaussian orbital *versus* a Slater orbital

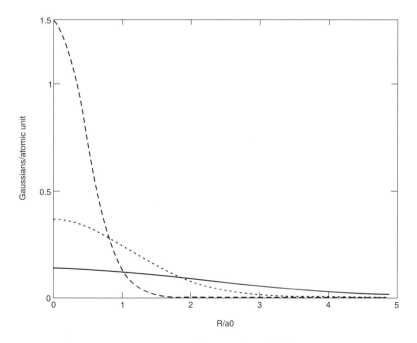

Figure 17.4 Three Gaussian orbitals

You can see roughly that each of the primitive Gaussians gives a good representation of part of the STO. Plot the three of them together as a linear combination and you will find a very good agreement, apart from the region near to the nucleus where the STO has a cusp and the GTOs have a point of inflexion.

The same comments apply to p, d and f Gaussians.

There are many different basis sets in the literature; the one described above, where each atomic orbital is represented as a fixed linear combination of three Gaussians, is the STO-3G basis set (Figure 17.5). A basis set such as STO/6-31G for carbon would represent the 1s orbital as a fixed linear combination of six Gaussians, and each of the four valence atomic orbitals ($2s$, $2p_x$, $2p_y$ and $2p_z$) is represented as two sets of Gaussians; a set of three and a set of one.

Atomic orbitals that are not occupied in the atomic electronic ground state are found to make small but significant contributions to molecular wavefunctions. So for example, it is usual to include d orbitals on first row atoms in a molecular calculation. This is denoted by *, so the two basis sets given above would be written STO-3G* and STO/6-31G*.

17.1.2 A Modern Implementation

There are several immensely powerful professional packages on the market for Gaussian orbital HF–LCAO calculations. Some of these packages run on top-end PCs.

To finish off our adrenaline calculation, let me take you through the output from a package called GAUSSIAN94. This particular package is poor on graphical input and output, but it

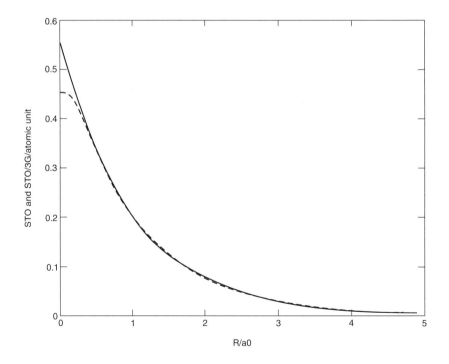

Figure 17.5 The STO-3G idea

does just about every possible type of calculation that anyone could reasonably want to do. Third party products for visualization are readily available.

Input to the program is very similar to that given earlier; the Cartesian coordinates of each atom. We do not input the connectivity data, this is information that should come out of the calculation! If two atoms are close together and there is a substantial amount of electron density between them, then they are bonded.

The first few lines in Table 17.2 remind us that the package is copyright, and was written by a large team. The line #HF/STO-3G defines the route through the package. We are running a HF–LCAO calculation at the STO-3G level of theory as discussed above.

The statement SCF = Direct means that all atomic integrals are calculated afresh each and every cycle of the HF procedure. An alternative procedure is to calculate them once only (as they do not vary through the calculation) and save them on disk. Strange as it may seem, it is usually better to calculate them every cycle because of the finite time taken for disk read-writes. Such considerations are part of the folk-lore of molecular structure calculations.

The next part of the calculation (Table 17.3) is concerned with reorienting the molecule to best effect in order to identify any elements of symmetry, and finally performing the HF–LCAO calculation.

You might now be expecting a molecular orbital by orbital description of the molecule, and indeed you can print out the LCAO coefficients if you wish. Not only that, many such packages and third-party products allow for the visualization of these molecular orbitals. The problem is that there are 49 occupied ones. They can indeed be printed out and molecular orbital by molecular orbital descriptions given of the stability and properties of the molecule under study. Apart from the orbital energies, most users of such a package would look at the electron distribution as given by the Mulliken population analysis indices, discussed in the previous chapter (Table 17.4).

Table 17.2   Start of an LCAO MO calculation

Cite this work as:
Gaussian 94, Revision B.2,
M. J. Frisch, G. W. Trucks, H. B. Schlegel, P. M. W. Gill,
B. G. Johnson, M. A. Robb, J. R. Cheeseman, T. Keith,
G. A. Petersson, J. A. Montgomery, K. Raghavachari,
M. A. Al-Laham, V. G. Zakrzewski, J. V. Ortiz, J. B. Foresman,
C. Y. Peng, P. Y. Ayala, W. Chen, M. W. Wong, J. L. Andres,
E. S. Replogle, R. Gomperts, R. L. Martin, D. J. Fox,
J. S. Binkley, D. J. Defrees, J. Baker, J. P. Stewart,
M. Head-Gordon, C. Gonzalez, and J. A. Pople,
Gaussian, Inc., Pittsburgh PA, 1995.

****
Gaussian 94: 486-Windows-G94RevB.2 3-May-1995
            18-Sep-1997
****
Default route: MaxDisk = 700MB SCF = Direct
- - - - - - - - - -
HF/STO-3G
- - - - - - - - - -

Table 17.3 Start of an LCAO MO calculation

Stoichiometry C9H13NO3
Framework group C1[X(C9H13NO3)]
Deg. of freedom   72
Full point group   C1   NOp   1
Largest Abelian subgroup   C1   NOp   1
Largest concise Abelian subgroup C1   NOp   1
              Standard orientation:
- -
Center   Atomic   Coordinates (Angstroms)
Number Number   X   Y   Z
- -
   1   6   0.532411   -1.366890   -0.016928
   2   6   1.896451   -1.570126   -0.280190

... and so on until ...

   24   1   4.686088   -0.021001   -0.466870
   25   1   -1.569814   1.190195   1.014796
   26   1   -1.713215   -0.582381   2.376170
- -
Rotational constants (GHZ):   1.9133302   0.3406653   0.3169506
Isotopes: C-12,C-12,C-12,H-1,C-12,C-12,C-12,H-1,C-12,O-16,O-16,C-12,H-1,H-1,O-16,
N-14,H-1,H-1,H-1,H-1,C-12,H-1,H-1,H-1,H-1,H-1
Standard basis: STO-3G (5D, 7F)
There are   78 symmetry adapted basis functions of   A symmetry.
Crude estimate of integral set expansion from redundant integrals = 1.000.
Integral buffers will be   262144 words long.
Raffenetti 1 integral format.
Two-electron integral symmetry is turned on.
  78 basis functions   234 primitive gaussians
  49 alpha electrons   49 beta electrons
    nuclear repulsion energy   782.2585721412 Hartrees.
One-electron integrals computed using PRISM.
The smallest eigenvalue of the overlap matrix is   1.829D-01
Projected INDO Guess.
Warning! Cutoffs for single-point calculations used.
Requested convergence on RMS density matrix = 1.00D-04 within 64 cycles.
Requested convergence on MAX density matrix = 1.00D-02.
Requested convergence on   energy = 5.00D-05.
SCF Done: E(RHF) = −619.422661768   A.U. after 5 cycles
      Convg = 0.8664D-04   -V/T = 2.0093
      S**2 = 0.0000

Finally, the package calculates the molecular dipole moment and every other electric moment one might reasonably want to know.

All such calculations incur costs, both human and machine. This particular calculation took 8 min on a GATEWAY P5-166 with 64 Mbyte RAM and 1 GB of available disk space.

Table 17.4   Next step in an LCAO MO calculation

Sum of Mulliken charges =   0.00000
Atomic charges with hydrogens summed into heavy atoms:

            1
  1 C   -0.005739
  2 C   -0.003942
  3 C   -0.018210
  4 H   0.000000

... and so on until ...

 24 H   0.000000
 25 H   0.000000
 26 H   0.000000
Sum of Mulliken charges $= 0.00000$
Electronic spatial extent (au): $< R**2 >= 3476.1509$
Charge $= 0.0000$ electrons
Dipole moment (Debye):
  X = -0.1493   Y =   2.4575   Z =   -0.1889 Tot =   2.4693
Quadrupole moment (Debye-Ang):
   XX =   -74.1564   YY =   -67.7689   ZZ =   -67.8586
   XY =   4.6846   XZ =   5.1802   YZ =   0.4376
Octapole moment (Debye-Ang**2):
XXX =   7.8879   YYY =   19.5955   ZZZ =   14.4449   XYY =   11.9250
XXY =   20.4425   XXZ =   -12.6314   XZZ = -4.3885   YZZ =   4.7936
YYZ =   0.9566   XYZ =   2.0767
Hexadecapole moment (Debye-Ang**3):
XXXX = -3177.3275 YYYY =   -433.0652 ZZZZ = -200.5357 XXXY =   78.6664
XXXZ =   43.5348 YYYX =   49.6119 YYYZ =   6.7902 ZZZX =   -30.3411
ZZZY = -12.7655 XXYY = -650.9110 XXZZ =   -576.3283 YYZZ =   -121.2626
XXYZ = -21.1266 YYXZ =   12.7761 ZZXY =   3.0668
N-N = 7.822585721412D + 02 E-N = -3.018808244614D + 03 KE = 6.136984198373D + 02

Job cpu time: 0 days 0 hours 7 minutes 57.0 seconds.
File lengths (MBytes): RWF =   10 Int =   0 D2E =   0 Chk =   1 Scr =   1
Normal termination of Gaussian 94

17.2   KOOPMANS' THEOREM

The orbital energies do not sum to give the electronic energy, but they do have a particular relevance. According to a theorem due to Koopmans, the negative of the orbital energy corresponds to the ionization energy

$$M \quad \rightarrow \quad M^+ + e^-$$

provided there is no change in the molecular orbitals on ionization. So for ionization from molecular orbital ψ_R with orbital energy ε_R, Koopmans' theorem gives an estimate of $-\varepsilon_R$ for the ionization energy.

Electron relaxation very often does occur, but Koopmans' theorem calculations have been widely used to study such ionization processes.

17.3 GEOMETRY OPTIMIZATION

17.3.1 The Floating Spherical Gaussian Orbital (FSGO) Model

I can collect together many of the concepts discussed above by describing a very simple model that was used in the 1960s, the Floating Spherical Gaussian Orbital (FSGO) model.

The idea of an electron pair, and the idea of an electron pair bond dominate descriptive chemistry; to a chemist, ammonia (NH_3) consists of a nitrogen $1s^2$ inner shell and three equivalent bonds comprising an electron pair shared between the N atom and one of the H atoms. The three NH bonds are thought of as equivalent, since they are equally distributed in space. The extra electron pair is said to be a lone pair, and is visualized as having the same directional properties as a NH bond. The four valence electron pairs are visualized as making a (rough) tetrahedral shape.

At first sight, these pictures do not appear from HF–LCAO theory. Table 17.5 gives the LCAO coefficients (at the HF/STO-3G level of theory) for NH_3.

The lowest MO is certainly the N inner shell, but examination of the remaining LCAO coefficients shows that the LCAO MOs are completely delocalized. For example, the third MO has the density plot shown in Figure 17.6.

In fact the only invariant in such a calculation is the total electron density, and this can be broken down into the normal delocalized LCAO MOs or into a totally equivalent set of localized ones, which have to be calculated from the normal ones by a so-called localization procedure.

The FSGO method in its simplest form applies only to molecules with an even number of electrons, and in their lowest energy singlet spin states. Each pair of electrons occupies a single spherical (i.e. s-type) Gaussian orbital centred at an arbitrary position in the molecule. There is no HF problem, and no LCAO coefficients to determine. For methane, we would take five Gaussian orbitals $\chi_1, \chi_2, \ldots, \chi_5$ and write a suitable antisymmetric wavefunction $\Psi(1, 2, \ldots, 10)$ based on the orbital configuration $\chi_1^2, \chi_2^2, \ldots, \chi_5^2$ (i.e. a Slater

Table 17.5   Final output from an LCAO MO calculation

Orbital energy (eV)			-15.313	-1.08662	-0.56155	-0.56147	-0.35846
S	N	1	-0.9935	0.2189	0.0000	0.0000	0.0959
S	N	1	-0.0313	-0.7398	0.0000	-0.0000	-0.4817
PX	N	1	0.0002	0.0064	-0.5923	0.0216	-0.0396
PY	N	1	0.0032	0.0879	0.0337	0.4631	-0.5423
PZ	N	1	-0.0041	-0.1103	-0.0076	0.3704	0.6802
S	H	2	0.0065	-0.1591	0.4496	-0.2211	0.1382
S	H	3	0.0065	-0.1591	-0.4163	-0.2788	0.1382
S	H	4	0.0065	-0.1591	-0.0334	0.5000	0.1382

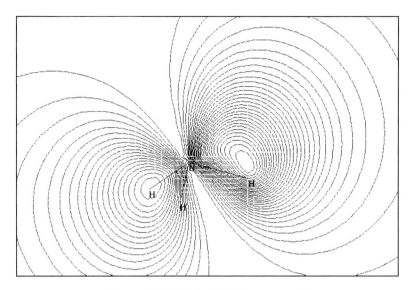

Figure 17.6   LCAO calculation on ammonia

Determinant). We also choose a molecular geometry. We then calculate the variational electronic energy

$$\varepsilon_{\text{elec}} = \frac{\displaystyle\int \Psi \hat{H} \Psi \, d\tau}{\displaystyle\int \Psi^* \Psi \, d\tau}$$

and finally add on the mutual potential energy of the nuclei.

The energy calculation is straightforward. If we write the matrix of overlap integrals $\mathbf{S}$ and denote its inverse by $\mathbf{T}$ then the denominator is

$$\int \Psi^* \Psi \, d\tau = \frac{(2n)!}{[\det(\mathbf{S})]^2} \tag{17.18}$$

where $\det(\mathbf{S})$ is the determinant of $\mathbf{S}$ and the numerator is

$$\sum_{i=1}^{n} \sum_{j=1}^{n} T_{ij} \int \chi_i(\mathbf{r}_1) \hat{h}(\mathbf{r}_1) \chi_j(\mathbf{r}_1) \, d\tau_1$$

$$+ \sum_{i=1}^{n} \sum_{j=1}^{n} \sum_{k=1}^{n} \sum_{l=1}^{n} (2T_{ij} T_{kl} - T_{ik} T_{jl}) \iint \chi_i(\mathbf{r}_1) \chi_j(\mathbf{r}_1) \hat{g}(\mathbf{r}_1, \mathbf{r}_2) \chi_k(\mathbf{r}_2) \chi_l(\mathbf{r}_2) \, d\tau_1 \, d\tau_2 \tag{17.19}$$

A normalized 1s GTO at position $\mathbf{R}$ is defined as

$$\chi(\mathbf{r}) = \left(\frac{2\alpha}{\pi}\right)^{3/4} \exp(-\alpha |\mathbf{R} - \mathbf{r}|^2) \tag{17.20}$$

where α is the orbital exponent. The integrals in the variational formula are straightforward.

Table 17.6   FSGO calculation on NH_3

Nucleus	Charge	$X(\text{Å})$	$Y(\text{Å})$	$Z(\text{Å})$
N	7	0.0000	0.0000	0.0000
H1	1	0.0000	1.4995	1.0013
H2	1	1.2986	−0.7500	1.0013
H3	1	−1.2986	−0.7500	1.0013
Orbital	Radius	$x(\text{Å})$	$y(\text{Å})$	$z(\text{Å})$
N 1s	0.2769	0.0000	0.0000	0.0008
N lone pair	1.5493	0.0000	0.0000	0.0080
Bond1	1.5007	0.0000	0.5712	0.4000
Bond2	1.5007	0.4946	−0.2856	0.4000
Bond3	1.5007	−0.4946	−0.2856	0.4000

Notice that $\alpha^{-1/2}$ has the dimensions of length and so it is often referred to as the orbital radius. Through integration it is found that 74% of the electron density in a given FSGO is within a sphere of radius $1/\sqrt{\alpha}$, and so it is a useful measure of the size of the FSGO.

It is usual in FSGO calculations to regard the orbital radii, the orbital positions and the nuclear positions as variational parameters, and to optimize each of them. For example, a calculation on NH_3 gave the results shown in Table 17.6.

The nitrogen 1s orbital is almost exactly localized on the N nucleus, the nitrogen lone pair is very close to the N nucleus but oriented along the z axis and each of the bond orbitals is placed along the corresponding N–H bonds.

The total energy is particularly poor compared to a conventional HF-LCAO calculation, but trends across series are correctly reproduced.

17.3.2   General Comments on Geometry Optimization

I said a great deal about the topic of geometry optimization in Chapter 10. The key formula for gradient searches is

$$\mathbf{x}_2 = \mathbf{x}_1 - \mathbf{H}^{-1}\mathbf{g} \qquad (17.21)$$

where $\mathbf{x}_1$ is the 'old' geometry vector, $\mathbf{x}_2$ the new estimate, $\mathbf{H}$ the hessian and $\mathbf{g}$ the gradient. In the early days of molecular structure such quantities were evaluated numerically. Over the years, a great deal of effort has gone into finding algebraic expressions for the gradient and the hessian for wavefunctions of ever-increasing complexity. Most molecular structure packages offer gradient searching techniques for geometry optimization.

17.4   LOWER LEVELS OF CALCULATION

Only in the last 30 years have computers become available in order to perform such resource intensive calculations. Workers have striven since the 1930s to tackle large molecules by quantum mechanical means, and they developed a variety of less stringent methods based on the accurate theories. The underlying idea was to 'calibrate' the rigorous theories by appeal to experiment, in order to avoid complicated integrals.

I want to now spend time discussing such landmark theories. Different authors classify these theories in different ways. I am going to give you a simple classification based on the HF model.

An *Ab Initio* calculation implies that we have chosen a set of atomic orbitals, a molecular geometry, calculated all the integrals correctly and done all the other calculations to the accuracy required. The HF calculation above on adrenaline is an *Ab Initio* calculation.

The term *Ab Initio* does not imply that the calculation will automatically give perfect agreement with experiment, or that it is 'correct'. What it means is that you choose a set of atomic orbitals, a molecular geometry, evaluate all integrals correctly and perform all the remaining calculations such as a HF–LCAO energy, a geometry search and so on with full mathematical rigour.

The HF–LCAO model I gave you above is not rigorously correct; first of all there is an error introduced by the choice of atomic orbital basis set. The electronic energy can be improved by a careful and systematic choice of this basis set. In the case of atoms, the HF equations can be integrated numerically without recourse to a basis set, to give what is called the HF limit.

Even when the HF limit is reached, there is still an error because the many-electron wavefunction is based on the *orbital* model. Hartree–Fock orbitals are calculated by making use of the assumption that each electron moves in an average field due to the other electrons (and of course the nuclei). Thus HF theory cannot cater for the non-zero interaction of two helium atoms, which is due to instantaneous interactions. This is illustrated in Figure 17.7.

The top calculation shows a typical HF–LCAO calculation. A careful choice of basis set will improve the energy (i.e. make it lower), and eventually the HF limit will be reached. This is the best energy that can be attained for an orbital wavefunction. For atomic calculations, the HF limit is routinely reached by suitable numerical integration of the HF equations. For

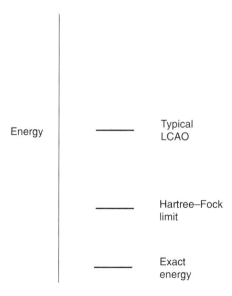

Figure 17.7   Electron correlation

molecules, the HF equations cannot be integrated numerically and so the HF limit is usually unknown.

Even the HF limit energy still lies above the true energy. We call the difference between the HF limit and the true energy the 'correlation energy', and we refer to the instantaneous interactions between electrons as electron correlation.

A treatment of electron correlation is generally necessary when studying processes where bonds are formed and broken, or where dispersion processes are thought to dominate. The subject of electron correlation is a mature one, and general methods for treating electron correlation are described in other texts. I have selected just two topical methods (density functional theory and MPn) for a chapter by themselves.

Techniques that are formally equivalent to *Ab Initio* techniques in that they are thought to be based on some correct theory (such as the orbital model), but where complicated integrals are replaced by experimental data are called *semi-empirical*.

Semi-empirical calculations on a range of related molecules may well give vastly better agreement with experiment than *Ab Initio* calculations, simply because of the calibration.

I can conveniently discuss semi-empirical theories as if they were all variants on *Ab Initio* HF theory.

17.5   THE HÜCKEL π-ELECTRON MODEL

From the earliest days of quantum chemistry, a vast amount of theoretical work has been performed on unsaturated and conjugated hydrocarbon molecules. Aromatic molecules have been subject to countless studies, and much success has accrued in rationalizing and correlating aromatic properties.

Chemists argue that the interesting chemistry of species such as ethene, benzene, naphthalene is due in large measure to the multiple bonds and so the simplest theories were developed by treating the so-called π or unsaturated electrons alone. In ethene we have an underlying planar structure

$$\left(\begin{array}{cc} H & H \\ \diagdown & \diagup \\ C\!\!-\!\!-\!\!-\!\!C \\ \diagup & \diagdown \\ H & H \end{array} \right)^{2+}$$

which can be neatly described using sp^2 hybrids on the carbons and s orbitals on the hydrogens. This leaves for the last bond between the two carbons, two electrons and one p_z orbital on each carbon (taking the z axis to be perpendicular to the molecular plane). This is the π bond. In benzene we have 6π electrons, 10 in naphthalene and so on.

In the simplest theories, the π electrons are treated apart from the rest, and it is assumed that the effect of the other σ electrons can be somehow lumped in together with the effect of the nuclei to give an effective potential in which the π electrons move. Each atom contributes a number of electrons (1 for C, 1 for N in pyridine, 2 for N in pyrrole, and so on) to the π system leaving an effective atomic nucleus that carries a formal positive charge (+1 for C atoms, +1 for the N in pyridine, +2 for the N in pyrrole, and so on).

Hückel π-electron theory treats butadiene, which has 4π electrons, as a HF–LCAO problem with atomic orbitals that resemble the carbon $2p_z$ orbitals but are strictly undefined. They are taken to be orthonormal, i.e. they or mutually orthogonal and normalized to unity.

The HF–LCAO eigenvalue equations for four atomic orbitals are

$$\mathbf{h}^F \mathbf{c} = \varepsilon \mathbf{S} \mathbf{c}$$

or, written out in full (with $n = 4$)

$$
\begin{pmatrix} h_{11}^F & H_{12}^F & H_{13}^F & h_{14}^F \\ h_{21}^F & h_{22}^{\Gamma} & h_{23}^{\Gamma} & h_{24}^{\Gamma} \\ h_{31}^F & h_{32}^F & h_{33}^F & h_{34}^F \\ h_{41}^F & h_{42}^F & h_{43}^F & h_{44}^F \end{pmatrix}
\begin{pmatrix} c_1 \\ c_2 \\ c_3 \\ c_4 \end{pmatrix}
= \varepsilon
\begin{pmatrix} S_{11} & S_{12} & S_{13} & S_{14} \\ S_{21}' & S_{22} & S_{23} & S_{24} \\ S_{31} & S_{32} & S_{33} & S_{34} \\ S_{41} & S_{42} & S_{43} & S_{44} \end{pmatrix}
\begin{pmatrix} c_1 \\ c_2 \\ c_3 \\ c_4 \end{pmatrix}
\qquad (17.22)
$$

Since the atomic orbitals are orthonormal

$$
\begin{pmatrix} h_{11}^F & h_{12}^F & h_{13}^F & h_{14}^F \\ h_{21}^F & h_{22}^F & h_{23}^F & h_{24}^F \\ h_{31}^F & h_{32}^F & h_{33}^F & h_{34}^F \\ h_{41}^F & h_{42}^F & h_{43}^F & h_{44}^F \end{pmatrix}
\begin{pmatrix} c_1 \\ c_2 \\ c_3 \\ c_4 \end{pmatrix}
= \varepsilon
\begin{pmatrix} 1 & 0 & 0 & 0 \\ 0 & 1 & 0 & 0 \\ 0 & 0 & 1 & 0 \\ 0 & 0 & 0 & 1 \end{pmatrix}
\begin{pmatrix} c_1 \\ c_2 \\ c_3 \\ c_4 \end{pmatrix}
$$

Solution of this matrix eigenvalue equation would give four possible values of ε and for each ε there would be one vector of the $\mathbf{c}$s.

In Hückel π-electron theory we make a number of assumptions.

(i) The orbitals are not strictly defined, but they can be thought of as being closely related to ordinary atomic $2p_z$ orbitals. They have the additional feature that they are orthonormal.

(ii) Elements of the $\mathbf{h}^F$ matrix that refer to non-bonded atoms are taken as zero.

so in this case the 1,3, 3,1, 1,4, 4,1, 2,4, and 4,2 elements are all zero.

(iii) Elements of the $\mathbf{h}^F$ matrix that refer to bonded atoms are taken as constants for each type of atom. The constant is written β_{CC} for a pair of bonded carbon atoms, and can be determined if necessary by comparison of a suitable molecular property with experiment. β_{CC} turns out to be a negative quantity.

(iv) Elements of the $\mathbf{h}^F$ matrix that refer to atoms are taken as constants for each type of atom. The constant is written α_C for a carbon atom, and it can be determined in principle by comparison of a suitable molecular property with experiment. So for butadiene, the HF–LCAO equations are

$$
\begin{pmatrix} \alpha_C & \beta_{CC} & 0 & 0 \\ \beta_{CC} & \alpha_C & \beta_{CC} & 0 \\ 0 & \beta_{CC} & \alpha_C & \beta_{CC} \\ 0 & 0 & \beta_{CC} & \alpha \end{pmatrix}
\begin{pmatrix} c_1 \\ c_2 \\ c_3 \\ c_4 \end{pmatrix}
= \varepsilon
\begin{pmatrix} 1 & 0 & 0 & 0 \\ 0 & 1 & 0 & 0 \\ 0 & 0 & 1 & 0 \\ 0 & 0 & 0 & 1 \end{pmatrix}
\begin{pmatrix} c_1 \\ c_2 \\ c_3 \\ c_4 \end{pmatrix}
$$

This set of equations is quite easily solved to give the LCAO MOs in Table 17.7.

I have arranged them in the order lowest energy... highest energy, and so in its electronic ground state butadiene has the first two MOs occupied each by a pair of electrons (Figure 17.8).

Rather than deal with each individual MO, it is profitable to focus on the electron density matrix $\mathbf{P}$ discussed earlier. This is defined for a system with m electron pairs as

$$P_{ij} = 2 \sum_{R=1}^{m} c_{Ri} c_{Rj}$$

and so for butadiene

$$\mathbf{P} = \begin{pmatrix} 1.0000 & 0.8944 & 0.0000 & -0.4472 \\ 0.8944 & 1.0000 & 0.4472 & 0.0000 \\ 0.0000 & 0.4472 & 1.0000 & 0.8944 \\ -0.4472 & 0.0000 & 0.8944 & 1.0000 \end{pmatrix} \tag{17.23}$$

The charge on each C atom is therefore zero (+1 electron from the formal C^+ and -1.0000 from the π electron density).

Table 17.7 Hückel calculation buta-1,3-diene

i	ε	c_1	c_2	c_3	c_4
1	$\alpha_C - 1.618\beta_{CC}$	0.3717	0.6015	0.6015	0.3717
2	$\alpha_C - 0.618\beta_{CC}$	−0.6015	−0.3717	0.3717	0.6015
3	$\alpha_C + 0.618\beta_{CC}$	−0.6015	0.3717	0.3717	−0.6015
4	$\alpha_C + 1.618\beta_{CC}$	−0.3717	0.6015	−0.6015	0.3717

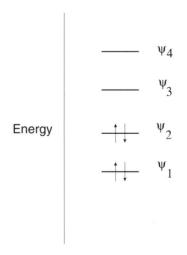

Figure 17.8 Hückel treatment of butadiene

17.5.1 Applications

Over the years, a great deal of work was published having at its heart the Hückel π-electron theory. Such theories are best used to correlate properties across a range of related molecules. Two examples follow:

(i) A widely reported property for such molecules is the 'delocalization energy'. This is the extra stability that results from delocalization of electrons originally confined to isolated double bonds. In Hückel π-electron theory the energy of butadiene is $2 \times (\alpha - 1.618\beta) + 2 \times (\alpha - 0.618\beta) = 4\alpha - 4.472\beta$ and the corresponding energy of two ethene molecules works out as $4\alpha - 4\beta$. The delocalization energy of butadiene is therefore 0.472β. Benzene has a delocalization energy of 2β, and so on. These can then be correlated against experimental estimates in order to make new predictions and incidentally to give an estimate for β. The aromatic sextet has long been a part of organic chemistry, and most aromatic compounds can be divided into rings that share six electrons per ring. Hückel found a satisfying explanation for this in terms of his simple MO theory. The rule can be stated as 'those monocyclic coplanar systems of trigonally hybridized atoms which contain $4n + 2$ π electrons will possess relative electronic stability'.

(ii) The off-diagonal elements of the charge density matrices corresponding to bonded atom pairs are found to correlate quite well with bond lengths. Coulson derived a famous equation

$$R/10^{-10}m = s - \frac{s - d}{1 + k\left(\dfrac{1 - P}{P}\right)} \tag{17.24}$$

where s is the 'natural' C–C single bond distance, d the 'natural' double bond distance and k a constant.

17.5.2 Heteroatoms

Hückel theory caters for heteroatoms by varying the fundamental parameters α and β. We write the variation in terms of the parameters for C, as follows.

$$\alpha_X = \alpha_C + h_X\beta_{CC}$$
$$\beta_{XY} = k_{XY}\beta_{CC} \tag{17.25}$$

Tables of suitable h_X and k_{XY} are available in the literature. For example, we might use $h_N = 1$ and $k_{CN} = 1$ for pyridine. This gives a charge density matrix as follows (atom 1 is the nitrogen atom)

$$\mathbf{P} = \begin{pmatrix} 1.3697 & 0.6188 & -0.0414 & -0.3069 & -0.0414 & 0.6188 \\ & 0.8548 & 0.6766 & 0.1121 & -0.3234 & -0.1452 \\ & & 1.0082 & 0.6613 & 0.0000 & -0.3234 \\ & & & 0.9044 & 0.6601 & 0.1121 \\ & & & & 1.0082 & 0.6766 \\ & & & & & 0.8548 \end{pmatrix}$$

which we can compare to benzene

$$
\mathbf{P} = \begin{pmatrix}
1.0000 & 0.6667 & 0.0000 & -0.3333 & 0.0000 & 0.6667 \\
 & 1.0000 & 0.6667 & 0.0000 & -0.3333 & 0.0000 \\
 & & 1.0000 & 0.6667 & 0.0000 & -0.3333 \\
 & & & 1.0000 & 0.6667 & 0.0000 \\
 & & & & 1.0000 & 0.6766 \\
 & & & & & 1.0000
\end{pmatrix}
$$

The formal charge on each C atom in benzene is identically zero, but pyridine displays a negative nitrogen with alternate $+$ and $-$ charges on the surrounding ring carbons.

17.6 ZERO DIFFERENTIAL OVERLAP (ZDO) π-ELECTRON THEORIES

An immense failing of Hückel theory is its cavalier treatment of electron repulsion (which it effectively ignores). This means for example that corresponding singlet and triplet excited states in conjugated molecules are predicted to have the same energy whilst we know from experiment that the triplet state will normally have lower energy than the corresponding singlet state.

Chronologically, the next type of theories were those that still applied to π-electron molecules, and still treated just the π electrons, but took better account of the details of the HF model and in particular the electron repulsion (two-electron) integrals. Such theories are associated with the names of Pariser, Parr and Pople, and have been called PPP models. For reference, I reproduce the elements of the HF matrix

$$
\hat{h}_{ij}^{F} = \int \chi_i(\mathbf{r}_1) \left(-\frac{h^2}{8\pi^2 m} \nabla^2 - \frac{e^2}{4\pi\epsilon_0} \sum_{\alpha=1}^{NUC} \frac{Z_\alpha}{r_\alpha} \right) \chi_j(\mathbf{r}_1) \, d\tau_1
$$

$$
+ \sum_{k=1}^{n} \sum_{l=1}^{n} P_{kl} \left(\iint \chi_i(\mathbf{r}_1)\chi_j(\mathbf{r}_1) \left(\frac{e^2}{4\pi\epsilon_0 r_{12}} \right) \chi_k(\mathbf{r}_2)\chi_l(\mathbf{r}_2) \, d\tau_1 \, d\tau_2 \right.
$$

$$
\left. - \frac{1}{2} \iint \chi_i(\mathbf{r}_1)\chi_k(\mathbf{r}_1) \left(\frac{e^2}{4\pi\epsilon_0 r_{12}} \right) \chi_j(\mathbf{r}_2)\chi_l(\mathbf{r}_2) \, d\tau_1 \, d\tau_2 \right)
$$

In PPP models we make the following simplifications and appeals to experiment.

(i) The atomic orbitals χ_i are not strictly defined, but they can be thought of as being closely related to ordinary atomic $2p_z$ orbitals. They have the additional feature that they are orthonormal (and so the overlap matrix $\mathbf{S}$ is the unit matrix).

(ii) Two-electron integrals involving atomic orbitals are treated explicitly as follows

$$
\iint \chi_i(\mathbf{r}_1)\chi_j(\mathbf{r}_1) \frac{1}{4\pi\epsilon_0 r_{12}} \chi_k(\mathbf{r}_2)\chi_l(\mathbf{r}_2) \, d\tau_1 \, d\tau_2
$$

is taken to be zero unless $i = j$ and $k = l$. The integral is then written

$$
\gamma_{ik} = \iint \chi_i(\mathbf{r}_1)\chi_i(\mathbf{r}_1) \frac{1}{4\pi\epsilon_0 r_{12}} \chi_k(\mathbf{r}_2)\chi_k(\mathbf{r}_2) \, d\tau_1 \, d\tau_2 \tag{17.26}
$$

and it is treated as a parameter which depends on the types of atoms that the atomic orbitals χ_i and χ_j are centred on.

(iii) The one-electron integrals are broken up into constituent kinetic energy and nuclear attraction components, and combinations of these integrals are treated as parameters to be determined from experiment. So for a typical one-electron integral involving two different atomic orbitals

$$\int \chi_i(\mathbf{r}) \left(-\frac{h^2}{8\pi^2 m} \nabla_i^2 - \frac{e^2}{4\pi\epsilon_0} \sum_{\alpha=1}^{N} \frac{Z_\alpha}{R_i} \right) \chi_j(\mathbf{r}) \, \mathrm{d}\tau$$

we treat it as a constant that depends only on the type and length of the bond joining those atoms on which the atomic orbitals χ_i and χ_j are centred. The constant is invariably written β_{ij}, although it is not the same constant as in Hückel theory. The diagonal elements

$$\int \chi_i(\mathbf{r}) \left(-\frac{h^2}{8\pi^2 m} \nabla_i^2 - \frac{e^2}{4\pi\epsilon_0} \sum_{\alpha=1}^{N} \frac{Z_\alpha}{R_{\alpha i}} \right) \chi_i(\mathbf{r}) \, \mathrm{d}\tau$$

are decomposed to separate out the contribution from the nucleus on which the atomic orbital is centred and the remaining nuclei to give

$$\int \chi_i(\mathbf{r}) \left(-\frac{h^2}{8\pi^2 m} \nabla_i^2 - \frac{Z_\alpha}{R_{\alpha i}} \right) \chi_i(\mathbf{r}) \, \mathrm{d}\tau$$

$$-\frac{e^2}{4\pi\epsilon_0} \int \chi_i(\mathbf{r}) \left(\sum_{\beta \neq A}^{N} \frac{Z_\beta}{R_{\beta i}} \right) \chi_i(\mathbf{r}) \, \mathrm{d}\tau$$

The first term is called the valence state ionization energy, written ω_i and treated as a parameter. Each remaining term is given a value $-Z_j \gamma_{ij}$ where Z_j is essentially the number of electrons donated to the conjugated system by atom j.

With these approximations, the HF matrix has elements

$$h_{ii}^{\mathrm{F}} = \omega_i + \frac{1}{2} P_{ii} \gamma_{ii} + \sum_{j \neq i} (P_{ij} - Z_j) \gamma_{ij}$$

$$h_{ij}^{\mathrm{F}} = \beta_{ij} - \frac{1}{2} P_{ij} \gamma_{ij} \tag{17.27}$$

Various sets of parameters are available in the literature. They are all slightly different, since they have been obtained by fitting different experimental quantities. The electron densities calculated by these more advanced theories are comfortably similar to those obtained from Hückel theory.

17.7   ALL-VALENCE ELECTRON ZDO THEORIES

Treating the π-electrons of conjugated molecules in isolation is all good fun, but eventually one has to bite the bullet of the remaining electrons! The next significant advances were made in the 1960s with the introduction of all-valence electron ZDO theories.

In order to progress from π electron treatments to all-valence electron treatments, it proved necessary to tackle the problem of *invariance*; the energy calculated for a molecular structure study must be the same whatever the orientation of the axis system used to study the molecule, and it must be the same if we choose to work with hybrid orbitals and/or symmetry combinations of atomic orbitals rather than with ordinary atom centred atomic orbitals. Many of the early attempts to parameterize HF theory to cover the case of all-valence electrons did not satisfy these criteria, with catastrophic results.

The place to start our study is with the Complete Neglect of Differential Overlap models (CNDO/1 and CNDO/2), which applied originally to molecules containing no more than first row atoms. Each atom therefore can be assumed to contribute a 1s orbital (H) or 2s, $2p_x$, $2p_y$ and $2p_z$ orbitals (first row atom) to the molecular wavefunction, and the requisite number of valence electrons.

The ZDO approximation was applied to two-electron integrals;

$$\iint \chi_i(\mathbf{r}_1)\chi_j(\mathbf{r}_1) \frac{1}{4\pi\epsilon_0 r_{12}} \chi_k(\mathbf{r}_2)\chi_l(\mathbf{r}_2)\, d\tau_1\, d\tau_2$$

this means that all repulsion integrals are taken to be zero unless $i = j$ and $k = l$, as in π-electron theories, which leaves a set of non-zero repulsion integrals

$$\gamma_{ik} = \iint \chi_i(\mathbf{r}_1)\chi_i(\mathbf{r}_1) \frac{1}{4\pi\epsilon_0 r_{12}} \chi_k(\mathbf{r}_2)\chi_k(\mathbf{r}_2)\, d\tau_1\, d\tau_2$$

In this case, some of the atomic orbitals will be on the same atomic centre, and some will be on different centers. So for example in a hydrocarbon we would find non-zero repulsion integrals involving the orbitals on a given carbon atom, such as $\gamma_{2s,2s}$, $\gamma_{2p_x,2p_x}$, $\gamma_{2p_y,2p_y}$ and $\gamma_{2p_z,2p_z}$. Exact calculation shows that the last three integrals are indeed equal, but that they are different from the first one. Nevertheless, the essence of CNDO theory is that all four integrals are taken to be equal and in general we write

$$\gamma_{AB} = \iint \chi_i(\mathbf{r}_1)\chi_i(\mathbf{r}_1) \frac{1}{4\pi\epsilon_0 r_{12}} \chi_k(\mathbf{r}_2)\chi_k(\mathbf{r}_2)\, d\tau_1\, d\tau_2$$

for all 2s and 2p orbitals on center A and for all 2s and 2p orbitals χ_k on atom B.

The repulsion integrals are calculated exactly assuming that the atomic orbitals are Slater type orbitals, but overlap is then ignored in solving the HF equations. I will just tell you that the elements of the HF matrix are written in a very similar way to those for the π-electron case

$$h_{ii}^F = U_{ii} + \left(P_{AA} - \frac{1}{2}P_{ii}\right)\gamma_{AA} + \sum_{B \neq A}(P_{BB}\gamma_{AB} - V_{AB})$$

$$h_{ij}^F = \beta_{AB}^0 S_{ij} - \frac{1}{2}P_{ij}\gamma_{AB} \tag{17.28}$$

U is analogous to ω, it is an atomic quantity and can be obtained from atomic spectroscopic data. The term V_{AB} represents the interaction of an electron on atom A with 'nucleus' B, and is called a penetration term. Terms like P_{AA} mean that we collect together all the diagonal elements of the **P** matrix that involve atom A. The β term of π-electron theory is replaced by a more general term that depends on the nature of the two atoms A and B, and the overlap between the particular Slater type orbitals on that center.

In the first version of CNDO (i.e. CNDO/1), the V_{AB} terms were calculated exactly. It turned out that CNDO/1 gave very poor predictions for molecular geometries, and this failing was analyzed and found to be due to the treatment of the U and V_{AB} terms. CNDO/2 attempted to correct these failings by making more realistic approximations.

17.8   PSEUDOPOTENTIAL MODELS

The ground state silver atom has electronic configuration

$$\text{Ag } (1s)^2(2s)^2(2p)^6(3s)^2(3p)^6(3d)^{10}(4s)^2(4p)^6(4d)^{10}(5s)^1$$

The chemistry of silver is mostly determined by the outer 5s electron, and the remaining electrons seem to form an inert core. This suggests that we should try to model the silver atom as a single electron constrained by the influence of a nucleus of charge $+47$ and an electron cloud of 46 electrons. An early potential due to Hellman (1935) models the potential due to the inner shell electrons in an atom as

$$U_{\text{core}} = -\frac{n_v}{4\pi\epsilon_0 r} + \frac{A\,\exp(-2kr)}{r} \tag{17.29}$$

and so we just treat silver as a hydrogen atom with a modified core. Here, n_v is the number of valence electrons, and A and k are constants that have to be determined by fitting an atomic property. Use of such a potential means that we have to calculate many fewer integrals than would otherwise be the case in a standard HF-LCAO calculation.

Adrenaline has formula $C_9H_{13}O_3N$, and so has 98 electrons. A full *Ab Initio* study would treat all 98 of those electrons; an all-valence electron study at the CNDO/1 level of theory would treat 72 valence electrons only (98–13 electron pairs), and a Hückel π-electron calculation would treat only the 10π electrons (six from the benzene ring and two from each of the ring OHs). The idea that electrons can be partitioned into chemically distinct groups such as inner shells, σ-electron cores and so on is an old one and has proved very useful in chemistry.

From the computational point of view, any calculation that reduces the number of basis functions that are explicitly taken into account is very attractive, and the idea of a *pseudopotential* that would represent the effect of an inner shall was first introduced in 1935. The idea is superficially simple; for the case of adrenaline we might partition the electrons into a core set of 26 (13 electron pairs, identified with the inner shells on the first row atoms) and a valence set of the remaining 72. If I write the coordinates of each electron as $\mathbf{x}_i$ (space and spin) then we are interested in a total wavefunction

$$\Psi(\mathbf{x}_1, \mathbf{x}_2, \ldots, \mathbf{x}_{98}) = \Psi_{\text{core}}(\mathbf{x}_1, \mathbf{x}_2, \ldots, \mathbf{x}_{26})\Psi_{\text{valence}}(\mathbf{x}_{27}, \mathbf{x}_{28}, \ldots, \mathbf{x}_{98})$$

and the aim is to treat explicitly the valence electrons only. These valence electrons are assumed to come under the influence of the nuclei and the inner shell electrons. The inner shells are represented by a so-called *effective core potential*.

First of all, we recognize that the total wavefunction has to satisfy the Pauli principle and so must be antisymmetric to exchange of the names of any pair of electrons. Even if the core and valence wavefunctions individually satisfy this condition, we have to allow for the interchange of (say) the electrons labelled 26 and 27. The second requirement is more subtle, and I can best explain it by appealing to the HF model of silver discussed above. The core consists of atomic orbitals 1s, 2s, 2p and so on up to 4d, whilst the valence shell consists of a single 5s atomic orbital. In order to develop a pseudopotential model, we have to ensure that the valence orbital is always orthogonal to all of the core orbitals; since we do not treat the core orbitals explicitly in pseudopotential theory, this simple requirement turns out to be a real problem. Many molecular structure packages now give the option of Pseudopotential calculations.

17.9 THERMODYNAMIC PROPERTIES

In Chapter 14 I gave you a beginner's guide to the complicated and difficult subject of statistical thermodynamics. I focused on the molecular partition function q, and showed you how to evaluate q given a knowledge of molecular properties such as the geometry (needed for the rotational constants), the force constants (needed for the vibrational frequencies) and the electronic energy levels. For an ideal gas of N particles, the canonical partition function Q is related to q as

$$Q = q^N$$

or

$$Q = \frac{q^N}{N!}$$

depending on whether the N particles making up the ideal gas are taken to be distinguishable or indistinguishable. Q is an important quantity, because it relates to thermodynamically important things like equilibrium constants.

Many molecular structure packages give as part of their output, quantities that are useful in such calculations of q.

The first step is to decide on the level of theory, and from that point on it is wise to be consistent with that decision; do not do parts of a study at one level of theory, and other parts at another level because you will almost certainly find that your conclusions are invalidated by the 'errors' inherent in trying to combine together such results. So for the sake of argument, I will take the HF/6-31G* level of theory for all molecules mentioned in this section. The 6-31G* notation means that 1s atomic orbitals are represented as fixed linear combinations of six primitive Gaussians, and the valence atomic orbitals are each represented by fixed linear combinations of three primitives (for the inner part of the atomic orbital) and a further one primitive for the outer part. In addition, d orbitals are added to each first row atom.

It may well be that electron correlation plays an important part; if so, you will have to redo my calculations at a higher level of theory.

The next step is a geometry optimization. This is mandatory if your study includes force constants, because they are defined as second derivatives of the energy at the potential minimum under study. As I mentioned earlier, a large molecule will have very many minima, one of which is the global minimum. Each of the minima will have a set of force constants, but they have to be found by investigating the second derivatives of each minimum point in the potential energy surface. You certainly cannot assume that the force constants are the same for each minimum point.

In the case of a diatomic molecule like HCl, the global minimum for the ground electronic state is easily found. The coordinates from a HF/6-31G* level calculation and related material, taken from the popular GAUSSIAN package are given in Table 17.8.

The next step is to calculate the force constants, defined as the second derivatives of the potential energy curve (Table 17.9).

Amongst other self-explanatory items, the package gives the harmonic wavenumber, the integrated infra-red and Raman intensities and the normal mode of vibration. In this case the first and only normal mode is given by $z_{Cl} - 0.03z_H$.

Thermochemical data are produced next. There are several items of interest. First of all the principal moments of inertia; for a collection of masses m_1 at position $\mathbf{R}_1$, m_2 at $\mathbf{R}_2$ and so on

Table 17.8   Stationary point found

<table>
<tr><td colspan="4">Optimization completed.
Stationary point found.
- - - - - - - - - - - - - - - - - -
Optimized Parameters
(Angstroms and Degrees)</td></tr>
<tr><td>Name Definition</td><td>Value</td><td colspan="2">Derivative Info.</td></tr>
<tr><td>R1   R(2,1)</td><td>1.2662</td><td>-DE/DX =</td><td>0.</td></tr>
</table>

Table 17.9   Harmonic frequencies at the stationary point

Harmonic frequencies (cm**-1), IR intensities (KM/Mole),
Raman scattering activities (A**4/AMU), Raman depolarization ratios,
reduced masses (AMU), force constants (mDyne/A) and normal coordinates:

		1		
		SG		
Frequencies	–	3186.1358		
Red. masses	–	1.0360		
Frc consts	–	6.1965		
IR Inten	–	24.3352		
Raman Activ	–	126.2809		
Depolar	–	0.3966		
Atom AN	X	Y	Z	
1 1	0.00	0.00	1.00	
2 17	0.00	0.00	-0.03	

the moment of inertia tensor $\mathbf{I}$ about x, y and z axes passing through the coordinate origin is often written

$$\mathbf{I} = \left\{ \begin{array}{ccc} \displaystyle\sum_{i=1}^{N} m_i(y_i^2 + z_i^2) & \displaystyle\sum_{i=1}^{N} m_i x_i y_i & \displaystyle\sum_{i=1}^{N} m_i x_i z_i \\[2em] \displaystyle\sum_{i=1}^{N} m_i y_i x_i & \displaystyle\sum_{i=1}^{N} m_i(x_i^2 + z_i^2) & \displaystyle\sum_{i=1}^{N} m_i y_i z_i \\[2em] \displaystyle\sum_{i=1}^{N} m_i z_i x_i & \displaystyle\sum_{i=1}^{N} m_i z_i y_i & \displaystyle\sum_{i=1}^{N} m_i(x_i^2 + y_i^2) \end{array} \right) \tag{17.30}$$

For a linear molecule lying along the z axis, the $\mathbf{I}$ matrix is diagonal and we speak of the principal moments of inertia. We normally discuss rotational problems with respect to rotation about the center of mass, rather than an arbitrary coordinate origin. This explains the items relating to the moment of inertia.

The so-called rotational temperature θ_r is often used in texts discussing statistical thermodynamics; the idea is to write the corresponding molecular rotational partition function for a diatomic molecule

$$q_{\text{rot}} = \frac{8\pi^2 \mu_{\text{AB}} R_{\text{AB}}^2 k_{\text{B}} T}{\sigma h^2} \tag{17.31}$$

from an earlier Chapter in terms of the temperature, the symmetry factor and a collection of constants called θ_r (which turn out to have dimension of temperature) as

$$q_{\text{rot}} = \frac{T}{\sigma \theta_r} \tag{17.32}$$

The so-called vibrational temperature is defined in a similar way for a diatomic molecule; from an earlier chapter we have

$$q_{\text{vib}} = \frac{1}{\left[1 - \exp\left(-\dfrac{hc_0 \omega_e}{k_{\text{B}} T} \right) \right]} \tag{17.33}$$

which we write using θ_v as

$$q_{\text{vib}} = \frac{1}{\left[1 - \exp\left(-\dfrac{\theta_v}{T} \right) \right]} \tag{17.34}$$

Obviously, each molecule has its own rotational and vibrational temperatures because they are defined in terms of molecular properties. They are certainly not universal constant, just convenient collections of molecular constants.

The block of results shown in Table 17.10 makes use of the calculated vibrational frequencies to calculate the vibrational molecular partition function.

For a larger molecule such as adrenaline, we have to go through the same series of calculations, but there is now no obvious molecular arrangement corresponding to the electronic ground state, and for many molecules there will be no obvious electronic ground state either. I have chosen to use the geometric configuration discussed earlier, not because I know for sure

Table 17.10   Thermodynamic properties of HCl

```
- - - - - - - - - - - - - - - - - -
-Thermochemistry of HCl-
- - - - - - - - - - - - - - - - - -
```

Temperature 298.150 K   Pressure   1.00000 Atm.
Atom 1 has atomic number   1 and mass   1.00783
Atom 2 has atomic number 17 and mass 34.96885
Molecular mass: 35.97668 amu.
Principal axes and moments of inertia in atomic units:

	1	2	3
EIGENVALUES - -	0.00000	5.60863	5.60863
X	0.00000	0.00000	1.00000
Y	0.00000	1.00000	0.00000
Z	1.00000	0.00000	0.00000

THIS MOLECULE IS A PROLATE SYMMETRIC TOP.
ROTATIONAL SYMMETRY NUMBER 1.
ROTATIONAL TEMPERATURE/K   15.44289
ROTATIONAL CONSTANT/GHz   321.779407
Zero-point vibrational energy   19057.3 (J/mol)
                  4.55481 (kcal/mol)
VIBRATIONAL TEMPERATURES: 4584.11 K

Zero-point correction =	0.007259 (Hartree/Particle)
Thermal correction to Energy =	0.009619
Thermal correction to Enthalpy =	0.010563
Thermal correction to Gibbs Energy =	-0.010600
Sum of electronic and zero-point Energies =	-460.052718
Sum of- electronic and thermal Energies =	-460.050357
Sum of electronic and thermal Enthalpies =	-460.049413
Sum of electronic and thermal Gibbs Energies =	-460.070576

	U kcal/mol	C_v cal/mol-K	S cal/mol-K
TOTAL	6.036	4.968	44.541
ELECTRONIC	0.000	0.000	0.000
TRANSLATIONAL	0.889	2.981	36.671
ROTATIONAL	0.592	1.987	7.870
VIBRATIONAL	4.555	0.000	0.000

	q
TOTAL BOT	0.750790D + 05
TOTAL V = 0	0.163756D + 09
VIB (BOT)	0.458480D-03
VIB (V = 0)	0.100000D + 01
ELECTRONIC	0.100000D + 01
TRANSLATIONAL	0.848188D + 07
ROTATIONAL	0.193066D + 02

Table 17.11   Start of thermodynamics calculations

```
- - - - - - - - - - - - - - - - - - - - - - -
-Thermochemistry of adrenaline -
- - - - - - - - - - - - - - - - - - - - - - -
Temperature    298.150 Kelvin. Pressure     1.00000 Atm.
Atom 1 has atomic number 6 and mass     12.00000
. . . until we get to . . .
Atom 25 has atomic number 1 and mass    1.00783
Atom 26 has atomic number 1 and mass    1.00783
Molecular mass:    183.08954 amu.
Principal axes and moments of inertia in atomic units:
                  1        2        3
   EIGENVALUES - -   884.587805019.683615478.60968
      X        .99998    .00114    .00591
      Y       -.00119    .99998    .00695
      Z       -.00591   -.00695    .99996
THIS MOLECULE IS AN ASYMMETRIC TOP.
ROTATIONAL SYMMETRY NUMBER 1.
ROTATIONAL TEMPERATURES/K    .09791    .01725    .01581
ROTATIONAL CONSTANTS/GHz    2.04021    .35953    .32942
```

that it corresponds to the global electronic ground state minimum, because I certainly do not know that. Just for illustration then I will take the geometric configuration discussed earlier.

I will not give the optimized geometric parameters, because they only relate to one of very many local minima on the potential energy curve.

The first part of the output relates to the nuclei (Table 17.11).

There are $3N - 6$ vibrations, and each makes a contribution to the thermodynamic properties. The next part of the output is shown in Table 17.12.

17.10   INCLUDING THE SOLVENT (THE SELF-CONSISTENT REACTION FIELD)

Most electronic structure calculations in the literature relate to isolated molecules in the gas phase. Chemistry generally takes place in solution, and in order to allow for solute–solvent interactions it is necessary to take the solvent into account. This can be done in principle by including solvent molecules in a periodic box, as discussed earlier. The problem is that the size of the calculation rapidly becomes prohibitive for *Ab Initio* studies.

The most commonly used approximation is to model the solvent as a continuum of dielectric material which polarizes the solute and is itself polarized by the solute. The solute molecule occupies a cavity within the dielectric (Figure 17.9). Such models are referred to as *continuum models* and the simplest case is where we consider a dipole embedded in a spherical cavity. The problem can be thought of in electrostatic terms; we have a molecular charge density $\rho(\mathbf{r})$ that is completely enclosed within the cavity, where the relative permittivity is 1. Outside the cavity the relative permittivity is taken as a value representative of a solvent. Relative permittivity data can be found in the literature, and a selection of values are given at 25°C in Table 17.13.

Table 17.12   Thermodynamic properties of adrenaline

Zero-point vibrational energy   609875.1 (Joules/Mol)
                                  145.76365 (Kcal/Mol)

VIBRATIONAL TEMPERATURES: 60.13   72.61   114.91   181.36   263.36
    (KELVIN)   277.89   307.29   323.66   377.95   468.22
              476.55   495.60   554.36   564.12   610.36
              653.94   698.12   737.00   777.78   916.86
. . . and so on until . . .
              4796.08   4867.59   4892.19   5429.87   5887.01
              5899.61   5935.30

Zero-point correction =	.232289 (Hartree/Particle)
Thermal correction to Energy =	.245011
Thermal correction to Enthalpy =	.245955
Thermal correction to Gibbs Free Energy =	.192676
Sum of electronic and zero-point Energies =	-627.149076
Sum of electronic and thermal Energies =	-627.136355
Sum of electronic and thermal Enthalpies =	-627.135410
Sum of electronic and thermal Free Energies =	-627.188689

	E (Thermal) KCAL/MOL	C CAL/MOL-KELVIN	S CAL/MOL-KELVIN
TOTAL	153.747	47.221	112.134
ELECTRONIC	.000	.000	.000
TRANSLATIONAL	.889	2.981	41.521
ROTATIONAL	.889	2.981	31.565
VIBRATIONAL	151.969	41.259	39.048
VIBRATION 1	.594	1.980	5.172
VIBRATION 2	.595	1.977	4.799
. . . and so on until . . .			
VIBRATION 16	.813	1.351	.782
VIBRATION 17	.841	1.283	.696
VIBRATION 18	.867	1.223	.628
VIBRATION 19	.896	1.160	.564

	q
TOTAL BOT	.237471D-88
TOTAL V = 0	.166143D + 19
VIB (BOT)	.138122-102
VIB (BOT) 1	.494980D + 01
VIB (BOT) 2	.409599D + 01
. . . and so on until . . .	
VIB (V = 0) 19	.107948D + 01
ELECTRONIC	.100000D + 01
TRANSLATIONAL	.973769D + 08
ROTATIONAL	.176560D + 07

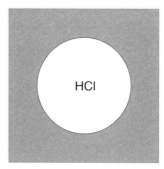

Figure 17.9   Molecule in solvent dielectric cavity

Table 17.13   Relative permittivity

Substance	Relative permittivity at 25°C
Free Space	1
Cyclohexane	2.015
Carbon Tetrachloride	2.228
Chlorobenzene	5.621
Methanol	32.63
Nitrobenzene	34.82
Water	78.54

First of all, let me mention the classical electromagnetic solution to this problem. One naturally reaches for the Poisson equation, which relates the electrostatic potential $\phi(\mathbf{r})$ at points in space $\mathbf{r}$ to the charge density $\rho(\mathbf{r})$ at those points.

$$\nabla^2 \phi(\mathbf{r}) = -\frac{\rho(\mathbf{r})}{\epsilon_0} \qquad (17.35)$$

The spherical symmetry of the problem suggests that we should write the potential as a linear combination of Legendre polynomials, and the familiar boundary conditions (ϕ must be continuous across the boundary between free space and the dielectric, and the component of the electric displacement perpendicular to the boundary must be continuous) help us solve the problem.

It turns out that the electrostatic problem can be solved, and the potential for the dipolar molecule (HCl in this case) differs from the free-space potential by a quantity called the *reaction field*. This is the Onsager model, and it is particularly easy to incorporate in a molecular structure calculation, once the size of the cavity is known.

Many molecular structure packages use a Monte Carlo numerical integration in order to find a suitable molecular volume. For HCl at the HF/4-31G* level of theory I found a molecular volume of 26.43 Å^3. This suggests a radius of 1.85 Å, and we normally add an extra 0.5 Å in such calculations to ensure that all the charge is enclosed by the spherical cavity. At the SCF level of theory, an extra term is added to the HF Hamiltonian to account for the reaction field and the calculation runs as usual. GAUSSIAN94 runs on HCl with three different solvents, each with a cavity radius of 2.35 Å, are summarized in Table 17.14.

Table 17.14   Reaction field calculations

Solvent	HF energy (au)	Solvation energy (kJ mol^{-1})
Free space	−460.059 976	0
Cyclohexane $\epsilon_r = 2.02$	−460.060 834	2.39
Water $\epsilon_r = 78.54$	−460.062 258	6.99

The Onsager model has obvious limitations; systems having no dipole moment cannot exhibit solvent effects, and the use of a spherical cavity is questionable for larger molecules of complex shape.

Over the years, more sophisticated models have been developed; the Polarized Continuum Model (PCM) attempts to obtain a more realistic cavity shape by treating the molecule (and hence the cavity) as a series of overlapping atomic spheres. The Isodensity PCM defines the cavity as an isodensity surface of the molecule; this surface is determined iteratively as part of (say) a HF cycle. Finally, the Self-Consistent Isodensity Polarized Continuum Model takes full account of the relationship between the shape of the cavity and the molecular electron density. The cavity shape is developed as the calculation proceeds, and it is thought to be a more realistic model for large molecules.

18 Describing Electron Correlation

In the previous chapter, I referred to electron correlation, and told you that it could be thought of as the instantaneous interaction between electrons in an atomic or molecular system. I defined the correlation energy as the difference between the best possible Hartree–Fock (HF) energy and the exact energy.

At the heart of the HF model is the physical assumption that we can usefully treat the interaction between electrons as some kind of average. Each electron moves in an average potential due to the nuclei and the remaining electrons. This does not pose a problem with the nuclei, because we fix them in space in accord with the Born–Oppenheimer approximation. But there is obviously an instantaneous interaction between electrons which should not be averaged out for exact calculations.

Nevertheless, the HF model is remarkably accurate for a wide range of chemical applications such as the prediction of equilibrium molecular geometries and force constants, electric properties and so on. It usually fails dramatically when we have to consider chemical processes where covalent bonds are broken. The lowest energy bond breaking process in dihydrogen is known to be

$$\mathrm{H_2(g)} \;\rightarrow\; 2\mathrm{H(g)}$$

yet HF calculations give a dramatically incorrect value for the dissociation energy because they refer to different dissociation products

$$\mathrm{H_2(g)} \;\rightarrow\; \tfrac{1}{2}(2\mathrm{H(g)} + \mathrm{H^+(g)} + \mathrm{H^-(g)})$$

and so the predicted dissociation energy refers to a quite different chemical process than the experiment.

The HF model also fails when we have to consider chemical processes that depend explicitly on the instantaneous interaction between electrons. For example, we know that neon does indeed liquefy under suitable conditions of temperature and pressure, and so the short range interatomic forces between two neon atoms must show a minimum. HF calculations do not give a minimum for this interaction because it is due to just those instantaneous interactions that are averaged out in HF theory.

I want to give you three methods for addressing the problem of electron correlation. A traditional method (*Configuration Interaction*, CI), a family of methods based on perturbation theory (MPn where $n = 2, 3, \ldots$) and the most up-to-the minute approach, density functional theory.

18.1   CONFIGURATION INTERACTION

Output from the adrenaline HF–LCAO calculation given in Chapter 17 comprises a set of HF–LCAO coefficients, orbital energies, the total energy and various details of the electron density distribution. I did this illustrative calculation at the HF/STO-3G level of theory, and therefore used 78 basis functions. The electronic ground state of the molecule is described by the electronic configuration

$$\Psi_0 = \psi_1^2 \psi_2^2 \cdots \psi_{49}^2$$

The HF–LCAO molecular orbitals ψ_1 through ψ_{49} are doubly occupied orbitals. Because we used 78 basis functions in the calculation, the HF–LCAO procedure produces a total of 78 molecular orbitals. The remaining $(78 - 49) = 29$ are called the virtual orbitals. In order to describe excited electronic states of the molecule, we might consider singly excited states with electronic configurations such as

$$\psi_1^2 \psi_2^2 \cdots \psi_{49}^1 \psi_{50}^1$$

where I have promoted a single electron from ψ_{49} to ψ_{50}, doubly excited states with electronic configurations such as

$$\psi_1^2 \psi_2^2 \cdots \psi_{48}^1 \psi_{49}^1 \psi_{50}^1 \psi_{51}^1$$

where I have promoted electrons from ψ_{48} and ψ_{49} to ψ_{50} and ψ_{51} and so on. I have been careful not to mention electron spin so far. The electronic ground state is a singlet state. For the singly excited state shown above, there are four possible quantum states once we take spin into account; the electron in orbital ψ_{48} could be α or β spin and the electron in ψ_{50} could be α or β spin. For the doubly excited state there are up to 16 possible quantum states.

Only certain combinations of these quantum states will correspond to singlet spin states. How we go about taking combinations of the spin states in order to get a suitable spin eigenfunction is a very advanced topic, much too complicated for this text. It can be done, but it turns out not to be necessary for very many advanced calculations.

A traditional way to treat electron correlation is the CI method. Configuration interaction gives a reasonable description of excited electronic states together with an improved description of the electronic ground state. Starting once again with adrenaline, I will consider the HF–LCAO approximation to the electronic ground state

$$\Psi_0 = \psi_1^2 \psi_2^2 \cdots \psi_{49}^2$$

a typical singly excited state

$$\Psi_1 = \psi_1^2 \psi_2^2 \cdots \psi_{49}^1 \psi_{50}^1$$

and a typical doubly excited state

$$\Psi_2 = \psi_1^2 \psi_2^2 \cdots \psi_{48}^1 \psi_{49}^1 \psi_{50}^1 \psi_{51}^1$$

According to the variation principle, the energy associated with a combination such as

$$\Psi_{\text{better}} = c_0 \Psi_0 + c_1 \Psi_1 + c_2 \Psi_2$$

might turn out to give an energy for the ground state lower than that associated with Ψ_0. If c_1 and c_2 both come to zero, then the energy will be exactly equal to that associated with Ψ_0. In order to find the coefficients c_0, c_1 and c_2 then we would have to solve the 3×3 matrix eigenvalue problem

$$\begin{pmatrix} H_{00} & H_{01} & H_{02} \\ H_{10} & H_{11} & H_{12} \\ H_{20} & H_{21} & H_{22} \end{pmatrix} \begin{pmatrix} c_0 \\ c_1 \\ c_2 \end{pmatrix} = E \begin{pmatrix} c_0 \\ c_1 \\ c_2 \end{pmatrix} \tag{18.1}$$

where for example

$$H_{00} = \int \Psi_0 \hat{H} \Psi_0 \, d\tau$$

Do not get this matrix equation confused with the HF–LCAO matrix equation. The solutions give energies of three possible electronic states, not orbitals. Elements of this Hamiltonian matrix generally involve integrals over the HF–LCAO orbitals rather than integrals over the basis functions, and this is where the problem begins.

A typical two-electron integral over the HF–LCAO orbitals ψ_A, ψ_B, ψ_C and ψ_D is

$$\int\int \psi_A(\mathbf{r}_1) \psi_B(\mathbf{r}_1) g(\mathbf{r}_1, \mathbf{r}_2) \psi_C(\mathbf{r}_2) \psi_D(\mathbf{r}_2) \, d\tau_1 \, d\tau_2$$

Each HF–LCAO is a linear expansion of the basis functions typically

$$\psi_A = c_{A,1} \chi_1 + c_{A,2} \chi_2 + \cdots + c_{A,78} \chi_{78}$$

and so the two-electron integral over HF–LCAO orbitals is a four-fold sum over the 78 basis functions (in this particular case)

$$\sum_{i=1}^{78} \sum_{j=1}^{78} \sum_{k=1}^{78} \sum_{l=1}^{78} \int\int c_{A,i} \chi_i(\mathbf{r}_1) c_{B,j} \chi_j(\mathbf{r}_1) \hat{g}(\mathbf{r}_1, \mathbf{r}_2) c_{C,k} \chi_k(\mathbf{r}_2) c_{D,l} \chi_l(\mathbf{r}_2) \, d\tau_1 \, d\tau_2 \tag{18.2}$$

This so-called transformation to HF–LCAO orbitals from basis functions is a very time consuming step. Not only that, the virtual orbitals often give a very poor representation of excited states and so very many terms indeed are needed in a typical CI calculation.

Over the years, people have investigated ways of truncating the CI expansion in order to make the calculation tractable. If we consider the ground state plus all singly excited states (but no others) then we talk about the SECI model (sometimes known as the Tamm–Dancoff approximation). Sometimes it is useful to exclude inner shells from consideration. The results of such a calculation for adrenaline are given in Table 18.1.

The NStates = 10 option means that I have chosen to output only the first 10 excited singlet states.

Table 18.1   Start of CIS calculations

```
****************************************************
Gaussian 94: 486-Windows-G94RevB.2    3-May-1995
           26-Jun-1998
****************************************************
%chk = d:\adren.chk
Default route: MaxDisk = 700MB SCF = Direct MP2 = Stingy
- - - - - - - - - - - - - - - - - - - - -
# CIS(Nstates = 10)/STO-3G guess = read
- - - - - - - - - - - - - - - - - - - - -
```

The first step is a straightforward HF–LCAO calculation, as given in an earlier chapter. The next step is transformation of the integrals from basis functions to HF–LCAO orbitals. The inner shell orbitals were not used in this calculation, hence the statement 'Range of MOs used for correlation 14 78' (Table 18.2). The remaining information is to do with disk housekeeping in the integral transformation.

The final step is construction of the Hamiltonian matrix and solution of the eigenvalue problem for the number of states of interest (10 in this case). The sample output is given in Table 18.3.

The notation should be self-explanatory; the energy of each excited state relative to the ground state is given (in eV), and the photon wavelength needed for a transition is also given (in nm). The oscillator strength is f, a measure of integrated transition intensity; the larger f the more intense the transition. The molecule chosen has no symmetry, hence the messages. The entry $48 \rightarrow 50$ 0.18129 gives the expansion coefficient for the singly excited state where an electron has been promoted from ψ_{48} to ψ_{50}.

CI calculations are very expensive in computer resource, and the CI series converges very slowly.

Table 18.2   Next step in CIS calculation

```
SCF Done: E(RHF) = -619.422687134    A.U. after    13 cycles
        Convg =    0.7389D-08         -V/T = 2.0093
        S**2 =    0.0000
Range of M.O.s used for correlation:    14    78
NBasis =    78 NAE =    49 NBE =    49 NFC =    13 NFV =    0
NROrb =    65 NOA =    36 NOB =    36 NVA =    29 NVB =    29
Semi-Direct transformation.
ModeAB =    4 MOrb =    36 LenV =    6370947
 LASXX =    2202201 LTotXX =    2202201 LenRXX =    4720707
LTotAB =    2518506 MaxLAS =    6547320 LenRXY =    0
NonZer =    6922908 LenScr =    16755309 LnRSAI =    6547320
LnScr1 =    16191927 MaxDsk =    91750400 Total =    44215263
SrtSym =    T
```

Table 18.3 Output from CIS calculations

Excitation energies and oscillator strengths:

CIS wavefunction symmetry could not be determined.
Excited State   1:   Singlet-?Sym   6.9970 eV   177.20 nm f = 0.0667
    48 – > 50   0.18129
    48 – > 51   0.34104
    49 – > 50   0.53140
    49 – > 51   -0.23726
This state for optimization and/or second-order correction.
Copying the Cisingles density for this state as the 1-particle RhoCI density.

CIS wavefunction symmetry could not be determined.
Excited State   2:   Singlet-?Sym   7.4604 eV   166.19 nm f = 0.1116
    48 – > 50   -0.34733
    48 – > 51   0.14045
    49 – > 50   0.25472
    49 – > 51   0.51287

CIS wavefunction symmetry could not be determined.
Excited State   3:   Singlet-?Sym   9.5611 eV   129.67 nm f = 1.5619
    48 – > 51   0.57028
    49 – > 50   -0.35643

CIS wavefunction symmetry could not be determined.
Excited State 4:   Singlet-?Sym   9.6477 eV   128.51 nm f = 0.9644
    48 – > 50   0.55098
    49 – > 51   0.38467

18.2 PERTURBATION THEORY

It seems intuitively clear that we should try and model electron correlation as some sort of 'perturbation', to be added to the HF model, and this leads me to my next topic. Perturbation theory is as old as the hills, and certainly does not owe its origins to quantum mechanics. For example, if you were to study the motion of the moon about the earth then I would advise you to begin by ignoring the sun and all the planets (apart from the earth). This procedure has the distinct advantage of reducing an insoluble problem in classical mechanics to an exactly soluble one. In this case, the exactly soluble problem is that of the motion of the moon around the earth in isolation. Once we have solved the simpler problem, then we can go back to the more complicated problem but with ideal solutions for a simpler system to hand. Rather than deal with electron correlation directly, I want to start the discussion with a simple example of the use of perturbation theory. In Chapter 5 I gave a treatment of the quantum mechanical description of a harmonically vibrating diatomic molecule. For a reduced mass of μ and an extension of x from the equilibrium internuclear separation R_e then the Hamiltonian operator is

$$\hat{H}_0 = -\frac{h^2}{8\pi^2\mu}\frac{d^2}{dx^2} + \frac{1}{2}k_s x^2 \qquad (18.3)$$

Here k_s is the Hooke's law spring constant. I have written H_0 rather than H_{vib} as in Chapter 5, for a reason that will become clear. It turns out that this simple problem is soluble, although the algebra involved is far from easy. Energy is quantized and the general formula for the allowed energies is

$$\varepsilon_{vib} = \frac{h}{2\pi} \sqrt{\frac{k_s}{mu}} \left(v + \frac{1}{2} \right) \tag{18.4}$$

where the vibrational quantum number v can take values $0, 1, 2, \ldots$.

The normalized vibrational wavefunctions are given by the general expression

$$\psi_v(\xi) = \left(\frac{\sqrt{\beta/\pi}}{2^v v!} \right)^{1/2} H_v(\xi) \exp(-\xi^2/2) \tag{18.5}$$

where $\beta = 2\pi \sqrt{\mu k_s}/h$ and $\xi = \sqrt{\beta} x$. We call the polynomials H_v the hermite polynomials (see Table 5.1).

The Hooke's law model for a vibrating diatomic is not a particularly accurate one, and you might want to refine the treatment by (for example) adding a cubic and a quartic term to the potential to give

$$\hat{H}_{vib} = -\frac{h^2}{8\pi^2 \mu} \frac{d^2}{dx^2} + \frac{1}{2} k_s x^2 + T x^3 + F x^4 \tag{18.6}$$

A careful analysis shows that we cannot solve this new problem exactly, because of the powers of x beyond the second, yet spectroscopic studies tell us that the effect of these extra terms cannot be large. Physical intuition suggests that we study the eigenvalues and eigenvectors of H_{vib} in terms for those of the simpler problem H_0.

Imagine then a simple one-dimensional system whose Hamiltonian H_0 we know and which system we can solve. That is to say, we can find solutions to the problem

$$\hat{H}_0 \Psi(x, t) = j \frac{h}{2\pi} \frac{\partial \Phi(x, t)}{\partial t}$$

Time-dependent perturbation theory is the correct tool when we have to deal with situations where the potential is time dependent. In Chapter 5, I showed how to write the solution as a product of a time and space part, provided that the potential does not contain any time-dependent terms. If this is the case then we are interested in solution of the time-independent equation

$$\hat{H}_0 \psi(x) = \varepsilon \phi(x)$$

and time-independent perturbation theory is the correct tool in this case. To simplify the treatment, I am going to assume that all the solutions to the time-independent equation are discrete and so can be labelled with the subscript i as follows

$$\hat{H}_0 \psi_i(x) = \varepsilon_i \psi_i(x)$$

and also that there are no degeneracies. For a reason that will shortly become clear, I will add a superscript (0) to the solutions as follows

$$\hat{H}_0 \psi_i^{(0)}(x) = \varepsilon_i^{(0)} \psi_i^{(0)}(x) \tag{18.7}$$

The question is now to investigate the solutions of the more complicated problem

$$(\hat{H}_0 + \lambda \hat{V})\psi_i(x) = \varepsilon_i \psi_i(x) \tag{18.8}$$

in terms of the simpler one, with V a time-independent perturbation. λ is a parameter that I have added in order to keep track of the order of the perturbation, and it goes with the subscript and superscript (0) introduced above. Eventually I shall set λ to 1, or identify it with a physical quantity such as an electric field strength.

We now make Taylor expansions of the ψ_i and the ε_i

$$\varepsilon_i = \varepsilon_i^{(0)} + \lambda \varepsilon_i^{(1)} + \lambda^2 \varepsilon_i^{(2)} + \cdots \tag{18.9}$$

The individual terms are referred to as corrections of various orders so that $\lambda \varepsilon_i^{(1)}$ is the first order correction to the energy and so on. The energy $\varepsilon_i^{(0)} + \lambda \varepsilon_i^{(1)}$ is called the first-order approximation to the energy and so on.

In a similar way we assume a perturbation expansion of the eigenfunctions of H

$$\psi_i = \psi_i^{(0)} + \lambda \psi_i^{(1)} + \lambda^2 \psi_i^{(2)} + \cdots \tag{18.10}$$

In order to develop formal expressions for the corrections, we can substitute the expansions into the eigenvalue equation to give

$$(H_0 + \lambda V)(\psi_i^{(0)} + \lambda \psi_i^{(1)} + \lambda^2 \psi_i^{(2)} + \cdots)$$

$$= (\varepsilon_i^{(0)} + \lambda \varepsilon_i^{(1)} + \lambda^2 \varepsilon_i^{(2)} + \cdots)(\psi_i^{(0)} + \lambda \psi_i^{(1)} + \lambda^2 \psi_i^{(2)} + \cdots)$$

Equating like powers of λ leads to the equations

$$\hat{H}_0 \psi_i^{(0)} = \varepsilon_i^{(0)} \psi_i^{(0)}$$

$$V \psi_i^{(0)} + \hat{H}_0 \psi_i^{(1)} = \varepsilon_i^{(0)} \psi_i^{(1)} + \varepsilon_i^{(1)} \psi_i^{(0)}$$

$$\vdots$$

$$V \psi_i^{(j-1)} + \hat{H}_0 \psi_i^{(j)} = \sum_{k=0}^{j} \varepsilon_i^{(k)} \psi_i^{(j-k)} \tag{18.11}$$

The first order equation is exactly the one we started from. In order to solve the second equation, we expand $\psi_i^{(1)}$ in terms of the zero-order functions

$$\psi_i^{(1)} = \sum_{k \neq i} c_{ik}^{(0)} \psi_k^{(0)}$$

where the coefficients $c_{ik}^{(1)}$ have to be determined. Without loss of generality we can assume that

$$\int \psi_i^{(0)} \psi_i^{(1)} \, d\tau = 0$$

and substitution of the sum into the second perturbation equation followed by multiplication by $\psi_i^{(1)}$ and integration yields the first order correction to the energy

$$\varepsilon_i^{(1)} = \int \psi_i^{(0)} V \psi_i^{(0)} \, d\tau$$

The first order approximation to the energy is therefore

$$\varepsilon_i^{(0)} + \int \psi_i^{(0)} V \psi_i^{(0)} \, d\tau \tag{18.12}$$

This is an interesting formula because it involves only the zero-order energy, the zero-order wavefunction and the perturbation. You can calculate the first order correction to the energy directly from the zero-order wavefunction and the perturbation. Solution for the first order coefficients gives the less memorable formula

$$c_{ik}^{(1)} = \frac{\int \psi_k^{(0)} V \psi_i^{(0)} \, d\tau}{\varepsilon_k^{(0)} - \varepsilon_i^{(0)}} \tag{18.13}$$

The important thing to note is that the first order correction to the wavefunction ψ_i involves a knowledge of all the states ψ_k (where $k \neq i$) yet calculation of the first order energy requires a knowledge only of the state of interest ψ_i.

The process can be continued to second and higher orders.

Let us return to the case of the one-dimensional oscillator where we take $V = Tx^3 + Fx^4$ as the perturbation. For the sake of illustration, I will only consider the effect of V on the vibrational state with $v = 0$, for which

$$\psi_0^{(0)} = \sqrt{\frac{\beta}{\pi}} \exp\left(-\frac{\xi^2}{2}\right)$$

$$\varepsilon_i^{(0)} = \frac{1}{2} \frac{h}{2\pi} \sqrt{\frac{k_s}{\mu}}$$

The first-order correction to the energy is

$$\varepsilon_0^{(1)} = \int \psi_0^{(0)} V \psi_0^{(0)} \, d\tau$$

$$= \int_0^\infty \psi_0^{(0)} (Tx^3 + Fx^4) \psi_0^{(0)} \, dx$$

Simple symmetry considerations tell us that the Tx^3 term gives zero contribution to the integral (thus we can take $T = 0$), and so we only need consider the quartic term Fx^4. The remaining quartic integral is a standard one, and comparison with tables of integrals gives the result

$$\varepsilon_0^{(1)} = \frac{3Fh^2}{16\pi^2 \mu k_s}$$

The perturbed wavefunction is given by

$$\psi_0^{(1)} = \sum_{k \neq 0} c_{0k}^{(0)} \psi_k^{(0)}$$

where the perturbation coefficients are

$$c_{0k}^{(1)} = \frac{\int \psi_k^{(0)} V \psi_0^{(0)} \, d\tau}{\varepsilon_k^{(0)} - \varepsilon_0^{(0)}}$$

These coefficients can be evaluated, tediously, from tables of standard integrals.

18.3   THE MP*n* METHOD FOR TREATING ELECTRON CORRELATION

The HF model treats electron repulsion in an average fashion; each electron moves in an average field due to the nuclei and the remaining electrons. In his original model Hartree treated the electrons as independent particles; there was no correlation whatever between their motions. The HF model gives correlation between electrons of like spin, but no correlation between electrons of unlike spin.

A relatively inexpensive way of addressing electron correlation even in large molecules is afforded by the Møller–Plesset perturbation procedure. The idea is to take the zero-level approximation H_0 as the Hartree–Fock Hamiltonian, and the perturbation V as the difference between the true Hamiltonian and the HF one. V thus contains details of the electron correlation, and you can perform the Møller–Plesset calculation to higher and higher levels of perturbation theory, depending on how much computer resource you have at your command.

If you decide to stop the calculation at the second level of perturbation theory, then we refer to MP2 calculations. If we progress to third order then we are dealing with the MP3 level of theory, and so on.

Table 18.4 is an example of MP2 computer output from GAUSSIAN94 for the sample molecule that I have taken, adrenaline. I have abbreviated the output considerably, since most of it is common with the output files shown in earlier chapters.

The first step in any perturbation calculation normally involves the solution of a zero-level problem which is exactly soluble. In the MP*n* model, this zero-level problem is taken as the HF calculation. In the context of a large molecule, the LCAO version of HF theory is invariably used and so the first part of an MP*n* calculation involves us solving the HF–LCAO equations.

I ran many of the calculations reported in this text on my office PC, and I saved the results of the HF–LCAO calculation into the checkpoint file D:\adren.chk. I restarted calculations using this information, hence the reference to the 'checkpoint file'.

As I mentioned above, the first-order perturbed wavefunction involves a knowledge of all excited states, and in particular quantities such as

$$c_{0k}^{(1)} = \frac{\int \psi_k^{(0)} V \psi_0^{(0)} \, d\tau}{\varepsilon_k^{(0)} - \varepsilon_0^{(0)}} \tag{18.14}$$

In a traditional HF–LCAO perturbation calculation, we would first determine the HF–LCAO coefficients in order to determine the ground electronic state and describe the excited states in terms of singly and doubly excited states formed by promoting one, two, ... electrons from the occupied HF–LCAO orbitals to the virtual ones. Evaluation of the perturbation coefficients would involve calculating all the possible atomic two-electron integrals and then transforming them to the HF–LCAO orbitals.

Table 18.4   Start of MPn calculation

Standard basis: STO-3G (5D, 7F)
There are   78 symmetry adapted basis functions of A   symmetry.
Crude estimate of integral set expansion from redundant integrals = 1.000.
Integral buffers will be   262144 words long.
Raffenetti 1 integral format.
Two-electron integral symmetry is turned on.
  78 basis functions   234 primitive gaussians
  49 alpha electrons   49 beta electrons
     nuclear repulsion energy   782.2585721412 Hartrees.
One-electron integrals computed using PRISM.
The smallest eigenvalue of the overlap matrix is 1.829D-01
Initial guess read from the checkpoint file:
D:\adren.chk
Requested convergence on RMS density matrix = 1.00D-08 within 64 cycles.
Requested convergence on MAX density matrix = 1.00D-06.
Keep R1 integrals in memory in canonical form, NReq =   5172582.
SCF Done: **E(RHF) = -619.422687134**   A.U. after   14 cycles
     Convg =   0.3779D-08   -V/T = 2.0093
     S**2 = 0.0000

It turns out that this process is much simpler in MP2 calculation, but the integrals do have to be manipulated and this process is costly both in terms of computer time and disk space.

One way to cut down on the computer resource is to leave the atomic cores frozen, and this is the meaning of the first comment in the Table 18.5. There are 13 heavy atoms, and their inner $1s^2$ shells have been frozen for the calculation.

Finally, the MP2 perturbation calculation gives the improved energy, the new estimates of the electric multipole moments and the electron density broken down according to the methods of population analysis. I have not included these, but the primary quantity of interest is usually the energy

The HF energy is -619.4226871 hartree, and the MP2 energy is -620.077544 hartree. MPn calculations are expensive in terms of computer time and computer space, but they are many orders of magnitude less costly than the more traditional CI method of dealing with electron correlation.

18.4   DENSITY FUNCTIONAL THEORY

The time-independent Schrödinger equation contains the space and spin coordinates of all the nuclei and electrons that go to make up the system of interest. Application of the Born–Oppenheimer approximation allows us to treat the nuclei as if they were point charges fixed in space (as far as the electrons are concerned). In Chapter 17, I discussed the Hartree and the HF models for molecular electronic structure. Most molecular physical properties such as the molecular electric dipole moment depend only on the coordinates of (any) one electron, and so on the charge density which I have denoted by the matrix **P**. I explained how to calculate the

Table 18.5   Next step in MPn calculation

Range of M.O.s used for correlation: 14   78
NBasis =   78 NAE =   49 NBE =   49 NFC =   13 NFV =   0
NROrb =   65 NOA =   36 NOB =   36 NVA =   29 NVB =   29
Semi-Direct transformation.
ModeAB =   2 MOrb =   36 LenV =   6373745
LASXX =   2626065 LTotXX =   2626065 LenRXX =   2626065
LTotAB =   2972646 MaxLAS =   7209540 LenRXY =   7209540
NonZer =   6547320 LenScr =   16194725 LnRSAI =   0
LnScr1 =   0 MaxDsk =   104857600 Total =   26030330
SrtSym =   F
JobTyp = 0 Pass   1:   I =   1 to 36.
Spin components of T(2) and E(2):
     alpha-alpha T2 =   0.2742588032D-01 E2 =   -0.5877961081D-01
     alpha-beta T2 =   0.2342635499D + 00 E2 =   -0.5372980334D + 00
     beta-beta T2 =   0.2742588032D-01 E2 =   -0.5877961081D-01
ANorm =   0.1135392140D + 01
E2 =   -0.6548572550D + 00 EUMP2 =   **-0.62007754438872D + 03**
Would need an additional   3342042 words for in-memory AO integral storage.
DD1Dir will call FoFDir   2 times, MxPair =   712
NAB =   666 NAA =   0 NBB =   0.
Spin AB IR =   106 I =   4 J =   4 Mu =   61 Nu =   61 IOp = 1 MuP =   61 NuP =   61
FactIJ = 0.0 Fact = 0.0 A = -5.9185949272D-02 AP = -5.9185949272D-02
Integrals replicated using symmetry in FoFDir.
MinBra = 0 MaxBra = 1 MinRaf = 0 MaxRaf = 1.
IRaf =   0 NMat = 712 IRICut =   82 DoRegI = T DoRafI = T ISym2E = 2 JSym2E = 2.
Spin AB IR =   310 I =   36 J =   36 Mu =   21 Nu =   21 IOp = 1 MuP =   21 NuP =   21
FactIJ = 0.0 Fact = 0.0 A = -9.0248708264D-02 AP = -9.0248708264D-02
MP4(D) = -0.37030125D-01
MP4 = −0.41897604D − 02
MP4(R + Q) = 0.11123512D-01
T4Å = -0.49540733D-03
T4(AAB) =   -0.73222960D-02
Time for triples =   12411.00 seconds.
MP4(T) =   -0.15635407D-01
E3 =   -0.10485659D + 00   EUMP3 =   -0.62018240097D + 03
E4(DQ) =   -0.25906613D-01   UMP4(DQ) =   -0.62020830759D + 03
E4(SDQ) = -0.30096374D-01   UMP4(SDQ) =   -0.62021249735D + 03
E4(SDTQ) = -0.45731780D-01   UMP4(SDTQ) = -0.62022813276D + 03
Largest amplitude = 7.42D-02

electronic energy ε_{el} for a system comprising doubly occupied orbitals, and presented the electronic energy equation

$$
\varepsilon_{\text{el}} = \sum_{i=1}^{n} \sum_{j=1}^{n} P_{ij} \int \chi_i(\mathbf{r}_1) h(\mathbf{r}_1) \chi_j(\mathbf{r}_1) \, \mathrm{d}\tau_1
$$

$$
+ \frac{1}{2} \sum_{i=1}^{n} \sum_{j=1}^{n} P_{ij} \sum_{k=1}^{n} \sum_{l=1}^{n} P_{kl} \left(\iint \chi_i(\mathbf{r}_1) \chi_j(\mathbf{x}_1) g(\mathbf{r}_1, \mathbf{r}_2) \chi_k(\mathbf{r}_2)_l(\mathbf{r}_2) \, \mathrm{d}\tau_1 \tau_2 \right.
$$

$$
\left. - \frac{1}{2} \iint \chi_i(\mathbf{r}_1) \chi_k(\mathbf{r}_1) g(\mathbf{r}_1, \mathbf{r}_2) \chi_l(\mathbf{r}_2) \, \mathrm{d}\tau_1 \, \mathrm{d}\tau_2 \right)
$$

There are n atomic orbitals and the $n \times n$ matrix $\mathbf{P}$ is the electron density matrix. This complicated equation is often written in terms of the so-called 'Coulomb' and 'Exchange' matrices $\mathbf{J}$ and $\mathbf{K}$, which both depend on the elements of $\mathbf{P}$ and are defined by

$$J_{ij} = \sum_{k=1}^{n} \sum_{l=1}^{n} P_{kl} \iint \chi_i(\mathbf{r}_1)\chi_j(\mathbf{r}_1)g(\mathbf{r}_1,\mathbf{r}_2)\chi_k(\mathbf{r}_2)\chi_l(\mathbf{r}_2)\,\mathrm{d}\tau_1\,\mathrm{d}\tau_2$$

and

$$K_{ij} = \sum_{k=1}^{n} \sum_{l=1}^{n} P_{kl} \iint \chi_i(\mathbf{r}_1)\chi_k(\mathbf{r}_1)g(\mathbf{r}_1,\mathbf{r}_2)\chi_j(\mathbf{r}_2)\chi_l(\mathbf{2})\,\mathrm{d}\tau_1\,\mathrm{d}\tau_2$$

The energy formula can then be written more compactly as

$$\varepsilon_{el} = \sum_{i=1}^{n} \sum_{j=1}^{n} P_{ij}h_{ij} + \frac{1}{2}\sum_{i=1}^{n} \sum_{j=1}^{n} P_{ij}G_{ij}$$

(in terms of the matrix $\mathbf{G} = \mathbf{J} - \frac{1}{2}\mathbf{K}$ introduced in Chapter 17), or

$$\varepsilon_{el} = \sum_{i=1}^{n} \sum_{j=1}^{n} P_{ij}h_{ij} + \frac{1}{2}\sum_{i=1}^{n} \sum_{j=1}^{n} P_{ij}J_{ij} - \frac{1}{4}\sum_{i=1}^{n} \sum_{j=1}^{n} P_{ij}K_{ij}$$

We have to add the nuclear repulsion energy in order to get the total. The first term in the expression above is the one-electron energy; it gives us the kinetic energy of the electrons and the mutual potential attractions between the various nuclei and the various electrons. The second term represents the classical Coulomb repulsion between the electrons taken a pair at a time. The final term is absent in the Hartree treatment, but appears in HF theory when we allow for the Pauli Principle. For shorthand, I will write the energy equation as

$$\varepsilon_{el} = \varepsilon^{(1)} + \frac{1}{2}\varepsilon_C - \frac{1}{4}\varepsilon_{Ex} \tag{18.15}$$

in a fairly obvious notation. The matrix $\mathbf{P}$ defines the electron density, which varies from place to place within the molecular system. The terms ε_C and ε_{Ex} depend on the integral of this electron density over the molecule, and they are referred to as functionals.

According to a remarkable result known as the Hohenberg–Kohn theorem, the electron density of a ground state determines uniquely the energy of that electronic state. In principle, we should be able to concentrate on the electron density (which depends only on the coordinates of a single electron) rather than the electronic wavefunction (which depends on the coordinates of all electrons present) in order to solve the electronic Schrödinger equation. The problem with the Hohenberg–Kohn theorem is that it does not give us any practical clue as to how to go about calculating the electron density; it is what mathematicians call an 'existence theorem'.

In modern density functional theory, we focus attention on the HF energy equation

$$\varepsilon_{el} = \varepsilon^{(1)} + \frac{1}{2}\varepsilon_C - \frac{1}{4}\varepsilon_{Ex} + \varepsilon_{Corr} \tag{18.16}$$

The exchange term ε_{Ex} is replaced by a more general expression which can involve the electron density and its gradient, and we add a correlation contribution ε_{Corr} that is absent from HF theory, and which again can involve the electron density and its gradient.

Thus for example Slater took an exchange potential proportional to the one-third power of the local electron density, and so the exchange contribution to the energy is proportional to

$$\iiint P(x,y,z)^{4/3}\,\mathrm{d}\tau \tag{18.17}$$

Table 18.6   Start of density functional calculation

```
**************************************************
Gaussian 94: 486-Windows-G94RevB.2 3-May-1995
           19-Jun-1998
**************************************************
%chk = d:\adren.chk
Default route: MaxDisk = 700MB SCF = Direct
- - - - - - - - - - - - - - - - - - - - - - -
 # BLYP/STO-3G Guess = Read
- - - - - - - - - - - - - - - - - - - - - - -
1/38 = 1/1;
2/12 = 2,17 = 6,18 = 5/2;
3/11 = 2,25 = 1,30 = 1/1,2,3;
4/5 = 1/1;
5/5 = 2,32 = 1,42 = 42/2;
6/7 = 2,8 = 2,9 = 2,10 = 2,19 = 1,22 = −2,28 = 1/1;
99/5 = 1,9 = 1/99;
- - - - - - - - -
Adrenaline
- - - - - - - - -
```

Table 18.7   Output from density functional calculations

```
There are 78 symmetry adapted basis functions of A    symmetry.
Crude estimate of integral set expansion from redundant integrals = 1.000.
Integral buffers will be    262144 words long.
Raffenetti 2 integral format.
Two-electron integral symmetry is turned on.
   78 basis functions    234 primitive gaussians
   49 alpha electrons    49 beta electrons
 nuclear repulsion energy 782.2585721412 Hartrees.
One-electron integrals computed using PRISM.
The smallest eigenvalue of the overlap matrix is   1.829D-01
Initial guess read from the checkpoint file:
D:\adren.chk
Warning! Cutoffs for single-point calculations used.
Requested convergence on RMS density matrix = 1.00D-04 within 64 cycles.
Requested convergence on MAX density matrix = 1.00D-02.
Requested convergence on     energy = 5.00D-05.
SCF Done: E(RB-LYP) =    -622.678755748   A.U. after    7 cycles
        Convg =   0.3172D-04        -V/T = 2.0161
        S**2 = 0.0000
```

A practical molecular density functional calculation is in many ways like a traditional HF one, and the final outcome is a set of molecular orbitals. These are referred to as the Kohn–Sham orbitals, and they are often expanded in LCAO form. A typical commercial package such as GUASSIAN94 has a choice of exchange functionals and a choice of correlation functionals. Following a traditional Hartree–Fock calculation, any extra contributions are evaluated analytically.

Table 18.6 gives the major features from a BLYP calculation at the STO-3G level on adrenaline. BLYP is shorthand for the Becke 1988 exchange functional combined with the Lee–Yang–Parr correlation functional.

The calculation is very similar to a traditional HF–LCAO run.

At each cycle in the HF–LCAO procedure, a numerical integration is done in order to take account of the extra exchange and/or correlation terms. A density functional calculation therefore adds an additional resource to the calculation, and these calculations can be memory intensive depending on the number of points used in the numerical integration.

The final output is very similar to that of a HF–LCAO calculation (Table 18.7). The energy reported is the density functional energy.

19 The Band Theory of Solids

The ability of solids to conduct electricity varies over an immense range; metallic copper has an electrical conductivity of about 10^8 S m^{-1} whilst polystyrene has an electrical conductivity of about 10^{-15} S m^{-1}. In this chapter, I want to show you how we can model the electrical properties of materials. By making a suitable choice of material, it is possible to design a solid having any desired electrical conductivity.

19.1 DRUDE'S MODEL OF A METAL

Figure 19.1 shows Drude's model of a metal. The larger circles are the cations, fixed in a rigid array. Each metal atom is assumed to lose z valence electrons represented by the smaller circles, which are the so-called conduction electrons. The forces between the electrons are ignored, and the forces between the electrons and the metallic cations are ignored except during a collision.

The number density n of these conduction electrons is

$$n = zL\frac{\rho}{R_{\mathrm{m}}} \tag{19.1}$$

where L is the Avogadro constant, R_{m} the molar mass and ρ the density of the metal. These free electrons behave as a classical electron gas. That is to say they have a Maxwell–Boltzmann energy distribution and the mean speed of the electrons $\langle v \rangle$ is

$$\langle v \rangle = \sqrt{\frac{8k_{\mathrm{B}}T}{\pi m_{\mathrm{e}}}}$$

According to the Drude model, the conduction electrons travel freely in any direction and so there is no net flow of charge.

If a battery is connected to the metal, each of the conduction electrons experience an

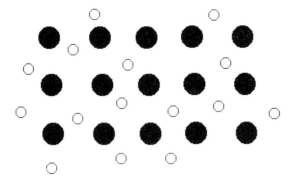

Figure 19.1   Drude's model of a metal

electrostatic force due to a uniform applied electric field **E**, given by

$$\mathbf{F} = -e\mathbf{E}$$

Force is rate of change of momentum and so

$$\frac{d\mathbf{p}}{dt} = -e\mathbf{E}$$

In the time interval Δt, the momentum change is therefore

$$\Delta\mathbf{p} = -e\mathbf{E}\Delta t$$

and so the change in the velocity of the electron is

$$\Delta\mathbf{v} = -\frac{e}{m_e}\Delta t\mathbf{E}$$

Because of the random motion, we need to take an average of all the electrons at a particular instant of time. If I denote this average by $\langle\ldots\rangle$ then

$$\langle\Delta\mathbf{v}\rangle = -\frac{e}{m_e}\langle\Delta t\rangle\mathbf{E} \tag{19.2}$$

The term on the left-hand side is the average change in the velocity vector induced by the electric field. In the absence of such a field the term is zero, and we refer to the average change as the *drift velocity* $\mathbf{v}_d$. The drift velocity is zero in the absence of an external electric field. Thus we have

$$\mathbf{v}_d = -\frac{e}{m_e}\langle\Delta t\rangle\mathbf{E} \tag{19.3}$$

I now need an expression for $\langle\Delta t\rangle$. The mean free path λ of the electrons is by definition the average distance that they travel between collisions and so, if the average speed of the electrons is $\langle v\rangle$ then we have

$$\lambda = \langle\Delta t\rangle\langle v\rangle$$

and so

$$\mathbf{v}_d = -\frac{e\lambda}{m_e\langle v\rangle}\mathbf{E} \tag{19.4}$$

The net effect of an external electric field is therefore to superimpose a drift velocity on the random motion of the electrons. Because electrons are negatively charged, the direction of the drift velocity vector is opposite to the applied electric field. This drift of charged particles constitutes an electric current.

19.2   OHM'S LAW

I want to demonstrate a relationship between the electric current and the applied field. To do this, I work out a quantity called the *current density* $\mathbf{j}$.

Consider a cylinder of length $v_d t$ and cross-sectional area A drawn in the metallic conductor (Figure 19.2).

Electric current is defined as the electric charge flowing through a fixed area per time, so the current crossing the black plane (which has area A) in Figure 19.2 is

$$i = \frac{nev_{\mathrm{d}}tA}{t}$$

where n is the number density of electrons.

The current density $\mathbf{j}$ is a vector with magnitude i/A and with the direction of the current. Since the electrons are negative charges, they drift in the opposite direction to the conventional current flow and I can write

$$\mathbf{j} = -ne\mathbf{v}_{\mathrm{d}}$$

Substitution for the drift velocity gives

$$\mathbf{j} = \frac{ne^2\lambda}{m_{\mathrm{e}}\langle v \rangle}\mathbf{E} \qquad (19.5)$$

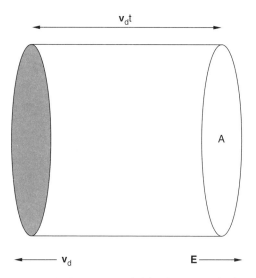

Figure 19.2   Construct needed for the current density $\mathbf{j}$

The constant of proportionality is called the *electrical conductivity*, and it is written σ. Sometimes we discuss the *electrical resistivity* $\rho = 1/\sigma$. The SI unit of σ is S m^{-1} and the SI unit of ρ is therefore Ωm.

I can rewrite the final equation in a more familiar form; if I assume that the direction of **j** and **E** is the same then I have

$$j = \frac{ne^2\lambda}{m_e\langle v\rangle}E$$

$$i = A\frac{ne^2\lambda}{m_e\langle v\rangle}E \tag{19.6}$$

and if the electric field is constant across the length l of the metal, $E = V/l$ where V is the potential difference. This gives

$$V = \left(\frac{\rho l}{A}\right)i \tag{19.7}$$

which is Ohm's law.

This was a major success of the Drude model. I mentioned in Chapter 13 that the Drude model is fundamentally flawed; if we work out the distance that an electron will travel, on average, between collisions, this distance turns out to be much too great as to be reasonable. We can easily calculate the mean free paths, and these can turn out to be many hundreds of bond lengths. It cannot be possible that an electron would not collide with a cation during this time.

A more serious difficulty comes when we try to calculate the heat capacity of a metal. If the conduction electrons are free particles, then each should have an average energy of $3/2\, k_B T$, according to equipartition of energy. Experimentally, metals do not show such behavior and the conciliation between theory and experiment only came with the application of quantum mechanics to the problem.

19.3 PAULI'S FREE ELECTRON MODEL

Drude's model is a classical model, and so it does not cater for energy quantization, the Pauli exclusion principle and so on.

In Chapter 13, I showed you how we can explain many of the properties of a metal in terms of Pauli's free electron theory. According to this theory, the potential inside a solid is uniform. This assumption gives the free electron energy level diagrams of previous chapters. Figure 19.3 shows a one-dimensional potential energy well. Note that the quantum states get further apart as the quantum number increases. The quantum states do not band together. According to the Pauli principle, each state can hold at most two electrons, one of either spin.

19.3.1 Conductivity

Just to remind you of two key results. The energy levels for a three-dimensional potential energy well of length L are given by

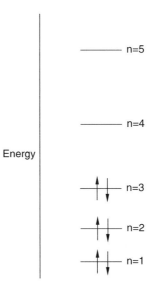

Figure 19.3   Free-electron model

$$\psi_{n,k,l} = \left(\frac{2}{L}\right)^{3/2} \sin\left(\frac{n\pi}{L}x\right) \sin\left(\frac{k\pi}{L}y\right) \sin\left(\frac{l\pi}{L}z\right)$$

$$\varepsilon_{n,k,l} = (n^2 + k^2 + l^2)\frac{h^2}{8mL^2}$$

(19.8)

and each energy level can accommodate two electrons, one of either spin. At 0 K we allocate pairs of electrons into each energy level, starting with the lowest and working up, until all electrons are accounted for. The energy of the highest occupied level is the Fermi energy ε_F.

The density of quantum states $D(\varepsilon)$ is

$$D(\varepsilon) = 2\frac{1}{8}2\pi\left(\frac{8mL^2}{h^2}\right)^{3/2}\varepsilon^{1/2}$$

(19.9)

I have added an extra factor of 2 to the result because each quantum state can hold a maximum of 2 electrons, one of either spin. This gives finally

$$D(\varepsilon) = \frac{1}{2}\pi\left(\frac{8mL^2}{h^2}\right)^{3/2}\varepsilon^{1/2}$$

(19.10)

At a temperature T, the probability that a certain quantum state is occupied is given by the Fermi probability

$$p(\varepsilon) = \frac{1}{\exp[(\varepsilon - \varepsilon_F)/k_B T] + 1}$$

(19.11)

At 0 K all available quantum states are occupied up to the Fermi level, and all the states above the Fermi level are empty. The total number of states at 0 K is

$$\int_0^{\varepsilon_F} D(\varepsilon)\,\mathrm{d}\varepsilon$$

(19.12)

which must equal the total number of electrons N. Many authors focus on the number density n of the electrons, which is N/L^3.

Integrating and solving for ε_F we get

$$\varepsilon_F = \frac{h^2}{2m_e}\left(\frac{3n}{\pi}\right)^{2/3} \tag{19.13}$$

In order to use Pauli's model for the purpose of discussing the electrical conductivity, we still ask how each individual electron will respond to an applied electric field $\mathbf{E}$, and the key equation

$$\frac{d\mathbf{p}}{dt} = -e\mathbf{E}$$

is still valid. But as a consequence of the Pauli principle, it is completely incorrect to assume that an electron can have its velocity reduced (for example) to zero by the effect of a collision; this cannot happen if the energy level associated with zero energy is completely filled.

19.3.2 The Wavevector

A completely free particle with momentum $\mathbf{p}$ (of magnitude p) has a de Broglie wavelength $\lambda_{dB} = h/p$. We define its wavevector $\mathbf{k}_w$ as follows

$$k_w = \frac{2\pi}{\lambda_{dB}} = \frac{2\pi p}{h}$$
$$\mathbf{k}_w = \frac{2\pi \mathbf{p}}{h} \tag{19.14}$$

and the energy of the wave is $p^2/2m$, which works out as

$$\varepsilon_w = k_w^2 \frac{h^2}{8\pi^2 m}$$

The symbol normally used in the literature for the wavevector is $\mathbf{k}$. I have chosen to use $\mathbf{k}_w$ in order to avoid confusion with my use of k for a quantum number in the electron in a box, and also to avoid confusion with the Boltzmann constant (written in this text as k_B) which many authors also write k.

A completely free electron can have any value for its momentum and hence any value for its wavevector. The direction of the wavevector is that of the momentum.

In the case of a constrained electron, the momentum is quantized. If we consider the three dimensional potential well of length L, then the wavevector components must satisfy

$$(\mathbf{k}_w)_x = 2\pi \frac{n_x}{L}$$
$$(\mathbf{k}_w)_y = 2\pi \frac{n_y}{L} \tag{19.15}$$
$$(\mathbf{k}_w)_z = 2\pi \frac{n_z}{L}$$

where n_x, n_y and n_z are integers $0, \pm 1, \pm 2 \dots$.

The energy of the wavevector is still given by

$$\varepsilon_w = k_w^2 \frac{h^2}{8\pi^2 m}$$

which is just the translational energy of an electron constrained to a finite box. If I substitute for the Fermi energy then

$$\varepsilon_F = k_F^2 \frac{h^2}{8\pi^2 m} \tag{19.16}$$

where I have written k_F for the wavevector corresponding to the Fermi energy.

In order to discuss the electrical conductivity in a metal, we have to examine the effect of an applied electric field on the occupancies of the quantum states. It is usual to tackle the problem in terms of the *wavevector*, given the connection between wavevector and momentum.

I can illustrate the connection between **k** and the energy. The components of $\mathbf{k}_w$ are plotted in three dimensions in Figure 19.4.

Let us start at $0\,\mathrm{K}$ where all states up to the Fermi level are occupied. If I plot out the wavevector components on a three-dimensional Cartesian grid, then since

$$k_w^2 = (k_w)_x^2 + (k_w)_y^2 + (k_w)_z^2$$

all wavevector states corresponding to a given value of k_w lie on the surface of the sphere drawn. I have drawn an octant of the sphere for convenience, but quantum numbers can now be negative and positive and so we must consider the entire sphere. All momentum states whose wavevectors have magnitudes less than or equal to the Fermi wavevector are occupied, and so lie within the sphere drawn above.

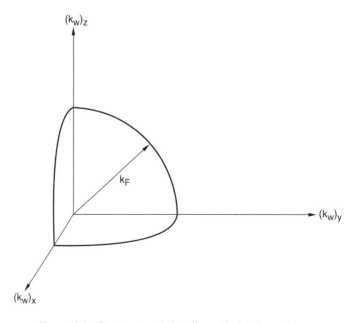

Figure 19.4 Construct needed to discuss the Pauli model

For temperatures greater than 0 K, some of the lattice points outside the sphere will correspond to occupied states, and some states inside the sphere will be unoccupied. We call these unoccupied states *holes*.

At low temperatures, the occupied states outside the sphere will correspond to lattice points close to the sphere. Likewise the holes will correspond to lattice points inside the sphere but close to the surface.

The distribution of the states is symmetrical about the sphere origin, and for every state with a wavevector $\mathbf{k}_w$ there is another state with wavevector $-\mathbf{k}_w$.

The momentum $\mathbf{p}$ is related to the wavevector $\mathbf{k}_w$ by

$$\mathbf{p} = \frac{h}{2\pi}\mathbf{k}_w$$

and so the total momentum of all the electrons is zero.

If we now switch on a constant external electric field $\mathbf{E}$, the electrons experience a force $-e\mathbf{E}$ in the direction of the field. Hence

$$-e\mathbf{E} = \frac{\mathrm{d}}{\mathrm{d}t}\left(\frac{h}{2\pi}\mathbf{k}_w\right)$$

The wavevector of each electron therefore changes at a constant rate, and if I replace the differentials such as dt by small quantities Δt, we would find a change in the wavevector over time interval Δt of

$$\Delta\mathbf{k}_w = -\frac{2\pi e\mathbf{E}\Delta t}{h} \tag{19.17}$$

Essentially, the electrons now occupy a sphere that is offset by Δk from the coordinate origin of Figure 19.4.

This applied field breaks the symmetry of the state distribution. There are more electrons with wavevectors opposite to the direction of the applied field than with the field and so there is a net momentum in that direction.

A motion of electrons against the applied field is equivalent to a conventional current in the direction of the field.

So, the Pauli picture is this. The applied electric field causes a shift in the wavevector of each electron, which is equivalent to a change in the momentum, exactly as in the classical picture.

I will not give the derivation between the current density $\mathbf{j}$ and the applied electric field, the result turns out as

$$\mathbf{j} = \sigma\mathbf{E}$$

exactly as in Drude theory. The exact form of this equation is

$$j = \frac{ne^2\lambda}{m_e\langle v\rangle}E \tag{19.18}$$

which again seems to agree exactly with Drude's theory. However, both the mean free path λ and the mean speed $\langle v\rangle$ are quite different in the two models.

Let me explain why. In the Drude model, all the free electrons are affected by the external electric field, whatever their velocity. In the Pauli model we have to take account of the occupancies or otherwise of the available quantum states. A filled quantum state can accommodate no further electrons. The only electrons that can be affected by the applied field are those occupying states near the Fermi surface. Some of the electrons will change their wave-vector in order to restore the distribution to a symmetrical one. Some of the electrons will change their wavevectors in order to make the distribution more unsymmetrial.

19.4   THE BAND THEORY

The free electron theory proved quite successful for describing the properties of metallic conductors, but failed when it came to dielectrics and to semiconductors. This should not come as a surprise, since dielectrics are substances that definitely do not conduct electric current (insulators), and semiconductors are somewhere in between conductors and dielectrics.

19.5   THE KRONIG–PENNEY MODEL (1930)

The Kronig–Penney model goes one step further than the free electron model and takes account of the variation of the potential due to the presence of the stationary lattice cations.

To take a one-dimensional case, the potential has the form sketched in Figure 19.5.

The highest potential is half-way between the cations (the black circles) and the potential tends to minus infinity as the position of the ion is approached. The potential function is complicated, and a mathematical analysis is outside the scope of this text, but the most important conclusion is that the electron may possess energies only within certain bands but not outside them. We say that

There are allowed and forbidden bands of energy in a solid

I can easily illustrate band behavior in rather more modern molecular orbital language; consider the Hückel π-electron treatment (Chapter 17) of a linear polyene with n carbon

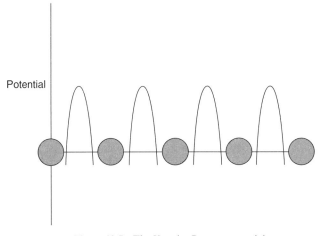

Potential

Figure 19.5   The Kronig–Penney potential

atoms in the conjugated system. The eigenvalues of the Hückel Hamiltonian matrix can be found analytically to be

$$\varepsilon_i = \alpha_C + \beta_{CC} \cos\left(\frac{i\pi}{n+1}\right) \text{ where } i = 1, 2, \ldots, n. \qquad (19.19)$$

The orbital energies of maximum possible deviation from α_C in the Hückel π-electron treatment have values $\alpha_C \pm 2\beta_{CC}$ and the minimum possible is α_C. Typical values are given in Table 19.1 in the form $\alpha_C + x_i\beta_{CC}$, and only the bonding orbitals are shown. The corresponding antibonding orbitals are of the form $\alpha\text{-}x_i\beta$ (in other words, the orbital energies are symmetrically disposed about the zero level α).

These can be plotted on an energy level diagram, and what happens is that eventually (for large n) we find a band structure as shown in Figure 19.6.

Table 19.1   Hückel orbital energies for a linear polyene

$n = 2$	4	6	8	10	20
1.000	1.618	1.802	1.879	1.919	1.978
	0.618	1.247	1.532	1.683	1.911
		0.445	1.000	1.310	1.802
			0.347	0.831	1.652
				0.285	1.466
					1.247
					1.000
					0.731
					0.445

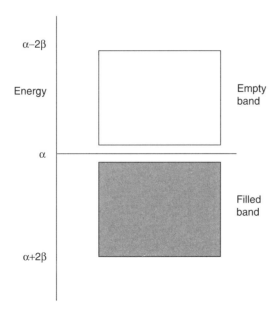

Figure 19.6   Band structure in Hückel theory

The Fermi level corresponds to the highest occupied orbital, and as the size of the conjugated system increases this orbital energy level tends to α_C, the energy of an 'isolated' conjugated carbon atom.

Note the band structure; all the occupied orbital energies are within a band than ranges from $\alpha_C + 2\beta_{CC}$ to (about) 0, and all the unoccupied orbitals lie in the range (about) 0 to $\alpha_C - 2\beta_{CC}$ (you need to recall that β_{CC} is a negative quantity). Notice that the highest occupied level (the Fermi level) in the 'occupied' band is very close to the lowest value of the 'unoccupied' band. According to the Pauli principle, the lowest energy configuration is where all the occupied levels are doubly occupied. It therefore needs very little energy to promote an electron from the top of the occupied band to the bottom of the unoccupied band. The behavior is typical for semiconductors.

This band structure is not an artefact of the Hückel π-electron treatment, nor of the Konig–Penney. The band structure is different for metals and for insulators.

In order to give a rough and ready model of a metal, I have taken a linear array of 10 lithium atoms (Figure 19.7) separated by a constant distance (300 pm, as in the solid). An *Ab Initio* calculation at the HF/STO-3G level of theory gave the orbital energies shown in the extract from the output file in Table 19.2. (I took the ground state to be a singlet spin state, with the lowest 15 orbitals doubly occupied. The conclusions do not depend on this assumption.)

Output from this particular package is in atomic units of energy $(e^2/4\pi\epsilon_0 a_0)$. You can see the 10 inner shell orbitals and the five fully occupied valence orbitals. The interesting thing is that there is no obvious gap between the occupied valence orbitals and the empty ('virtual') ones. There are unfilled orbitals lying very close to the Fermi level, and it requires very little energy to excite the uppermost electrons. This means that the electrons are very mobile, which is reflected in the ability of a metal to conduct electricity. The electrical conductivity of metals is a property characteristic of partially filled bands of orbitals.

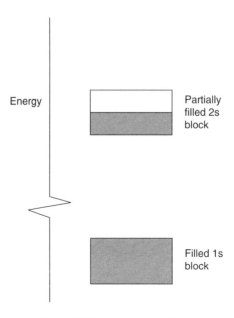

Figure 19.7   Linear array of Li atoms

Table 19.2   *Ab Initio* treatment of a linear array of 10 Li atoms

occupied –	−2.36697	−2.36697	−2.35662	−2.35662	−2.35234
occupied –	−2.35205	−2.34918	−2.34917	−2.34342	−2.34342
occupied –	−0.18229	−0.17480	−0.16144	−0.14272	−0.12037
virtual –	−0.00409	0.02381	0.05387	0.08248	0.10142
virtual –	0.10142	0.10748	0.10748	0.11072	0.11484
virtual –	0.11836	0.11836	0.13349	0.13349	0.15261
virtual –	0.15261	0.17606	0.17606	0.20568	0.20568
virtual –	0.23024	0.23450	0.23450	0.25753	0.26031
virtual –	0.26031	0.27902	0.27902	0.29258	0.33220
virtual –	0.38321	0.42566	0.47280	0.51598	0.54766

Finally I should tell you that insulators have a quite different band structure (Figure 19.8). Because the energy gap between the top of the filled band and the bottom of the empty band is large, an insulator cannot carry any current.

19.6   SEMICONDUCTORS

In classifying a material as a conductor, insulator or a semiconductor, we have to consider the energy gap (ε_g) between the filled band and the empty band and the temperature in addition to the rather more obvious variable, the Fermi energy.

I can illustrate the principles by considering silicon. Silicon is the most widely used semiconductor material, and it has an energy gap of 1.12 eV (Figure 19.8). At 0 K the 'filled band' is completely filled, and the upper band is completely empty. Hence pure silicon is an insulator at 0 K.

At higher temperatures, some electrons are thermally excited from the lower band to the upper band. This leaves an equal number of holes in the lower band. When an external electric field is applied to pure silicon, both the excited electrons and the holes contribute to the current. The electrons flow against the direction of the applied field, and the holes flow in the direction of the applied field. Even though the negatively charged electrons flow in the opposite sense to the electric field, they generate a conventional current in the same direction as the applied field. The total current is just the sum of the current carried by the electrons and the current carried by the holes.

Here is some semiconductor jargon for you. The lower band(s) that are completely filled at 0 K are called the *valence bands*. Chemists would distinguish between inner shell bands and true valence bands. The upper band, which is empty at 0 K is referred to as the *conduction band*.

In order to find the number of electrons in the conduction band, I just take the density of states times the Fermi probability and integrate from the bottom of the conduction band to the top of the band. The integral is a particularly tedious one, and I will just quote the result; the number density n of electrons in the conduction band is given by

$$n_e = A_e T^{3/2} \exp\left[-\frac{(\varepsilon_c - \varepsilon_F)}{k_B T}\right] \tag{19.20}$$

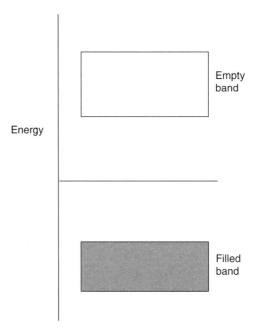

Figure 19.8   Band structure of an insulator

A_e is a constant and ε_C is the energy of the bottom of the conduction band. A similar calculation on the holes leads to the result

$$n_h = A_h T^{3/2} \exp\left[-\frac{(\varepsilon_F - \varepsilon_v)}{k_B T}\right] \tag{19.21}$$

where A_h is a constant and ε_v is the energy of the top of the filled valence band. The number of holes is equal to the number of conduction electrons in a pure semiconductor and

$$n_e n_h = A_e A_h T^3 \exp\left(-\frac{(\varepsilon_C - \varepsilon_v)}{k_B T}\right)$$

so we have finally

$$n_e = n_h = \sqrt{A_e A_h} T^{3/2} \exp\left(-\frac{(\varepsilon_C - \varepsilon_v)}{2 k_B T}\right) \tag{19.22}$$

This expression shows how the number densities of the electrons and the holes vary with temperature.

19.6.1   Doped Semiconductors

Materials such as pure silicon are known in the microelectronics trade as *intrinsic* semiconductors. Manufacturers of semiconductor devices use materials to which small amounts of impurities have been carefully added in a process called *doping*. Doped materials have a number of advantages over pure semiconductors, in particular the conductivity can be sub-

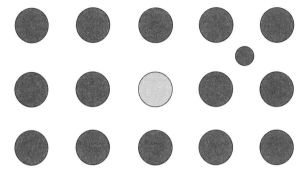

Figure 19.9   An *n*-type semiconductor

stantially increased and made far less temperature dependent. A pure semiconductor carries equal numbers of electrons and holes; a doped semiconductor can be made to contain predominantly either electrons or holes, and it is this property that is important for the operation of integrated circuits.

A semiconductor in which electrons are the major carrier of electric current is called an *n-type semiconductor*, and a material in which the holes are the major carrier is called a *p-type semiconductor*.

If we dope silicon (group IV) with a small amount of a group V element such as arsenic, then each arsenic atom essentially occupies a silicon lattice site in an unperturbed crystal lattice. Compared with Si, each As atom has an extra electron, which might be thought to enter the conduction band and so be free to move through the crystal lattice. This simple picture needs a moment of thought. If the As atom does indeed lose the extra electron to the conduction band, then it becomes positively charged. We can think of the As^+ atom and the extra electron as forming a pseudo hydrogen atom where the extra electron is loosely associated with the As^+ cation (Figure 19.9).

Figure 19.9 shows the usual array of silicon atoms, doped with a single arsenic atom. I have omitted all the conduction electrons due to the silicon atoms, but have indicated the 'extra' electron due to the arsenic atom.

It turns out that the energies of the extra electrons lie just below the conduction band; so close that the majority of these electrons get promoted to the conduction band at ordinary temperatures due to the thermal energy. This has a profound effect on the electrical conductivity.

If instead we now dope silicon with a small percentage of a group III element such as boron or gallium, rather than a group V element, we get a similar effect for a different reason. Each trivalent atom produces a hole in the conduction band, and the hole contributes to the conductivity.

20 Modeling Polymeric Materials

A *polymer* is a large molecule constructed from smaller covalently bonded molecular structural units called *monomers*. The essential requirement that a small molecule should qualify as a monomer is that it should be at least bifunctional; this is necessary in order that the monomers can link together in order to form a chain. A polymer sample will contain chains of very many different lengths, and so it is not possible to assign a specific formula or molar mass to such a species. If you want some idea of size, fix on a molar mass figure of $500\,000\,\text{g mol}^{-1}$.

I   II

The following molecules I and II typically qualify as monomers
Hydroxyethanoic acid I can condense with other (identical) hydroxy acid molecules through the –OH and –COOH groups to form a linear polymer, and the polymerization reaction in this case consists of a series of simple organic condensation reactions such as

The reaction can now be repeated a large number of times to give an example of a *homopolymer*, where all the monomer units are the same.

The double bond in the vinyl chloride II is also bifunctional and activation by a free radical leads to polymer formation

$$R\cdot \quad + \quad \underset{H}{\overset{H}{\diagdown}}C = C \underset{Cl}{\overset{H}{\diagup}} \quad \longrightarrow \quad RH_2C - CHCl\cdot \quad \longrightarrow$$

$$H_2CR - \underset{HCl}{C} - H_2C - CHCl\cdot$$

Polymers are usually named after their source, but a wide variety of trade names are in common use. The prefix *poly* is attached to the name of a monomer in addition polymers, and so polyethene and polystyrene denote polymers prepared from these single monomers. When the monomer has a multi-worded name then this is enclosed in parentheses and prefixed with poly, for example poly(vinyl chloride).

If the chains are composed of two different types of monomer then we talk about a *copolymer*; for example 6:6 nylon is made by the condensation reaction between adipic acid (III) and hexamethylene diamine (IV)

$$\underset{HO}{\overset{O}{\diagdown}}C - \underset{H_2}{C} - \underset{H_2}{C} - \underset{H_2}{C} - \underset{H_2}{C} - \underset{OH}{\overset{O}{\diagup}}C \quad + \quad H_2N - \underset{H_2}{C} - \underset{H_2}{C} - \underset{H_2}{C} - \underset{H_2}{C} - \underset{H_2}{C} - \underset{H_2}{C} - NH_2$$

III   IV

$$\longrightarrow$$

$$\underset{HO}{\overset{O}{\diagdown}}C - \underset{H_2}{C} - \underset{H_2}{C} - \underset{H_2}{C} - \underset{H_2}{C} - \overset{O}{\diagup}\quad HN - \underset{H_2}{C} - \underset{H_2}{C} - \underset{H_2}{C} - \underset{H_2}{C} - \underset{H_2}{C} - \underset{H_2}{C} - NH_2$$

Copolymers prepared from bifunctional monomers (as distinct from monomers with a larger number of functional groups) can be divided into four main categories

(i) Statistical copolymers, where the distribution of the two monomers A and B in the chain is random

...AABABBAABBBABBBBAABAAAAB...

(ii) Alternating copolymers where there is a regular placement along the chain

...ABABABABABABABABABABABAB...

(iii) Block copolymers where substantial blocks of the monomer A or of the monomer B are present

...AAAAABBAAAAABBBBBABBABBA....

(iv) Graft copolymers in which blocks of one monomer are grafted onto the backbone of another.

$$
\begin{array}{cccc}
A & A & A & A \\
A & A & A & A \\
A & A & A & A
\end{array}
$$

BBBBBBBBBBBBBBBBBBBBBBB...

If we link together two polymer chains by a chemical reaction, then we arrive at a *cross-linked* polymer. I will give you a specific example (rubber) in a later section.

20.1 SIZE AND SHAPE

One of the most important features that distinguishes a polymer from a monomer is that it is not possible to assign an exact molar mass to a polymer sample. A given polymer chain obviously has a chemical formula and molar mass, but the length of a polymer chain is determined by random events, and the output from a polymerization reaction is a mixture of chains of different lengths.

Colligative properties, such as the osmotic pressure of a solution, determine the number average molar mass defined as

$$
\langle M \rangle_n = \frac{\sum N_i M_i}{\sum N_i} \tag{20.1}
$$

where N_i is the number of species with molar mass M_i.

Light scattering measurements give a measure of the size rather than the number of molecules, and we obtain what is known as the weight average

$$
\langle M \rangle_w = \frac{\sum N_i M_i^2}{\sum N_i M_i} \tag{20.2}
$$

The difference between these two is a measure of the spread ('dispersion') of the molar masses.

Some measure of the size of a polymer chain can be inferred from the molar mass, but what is the shape of the chain and its average length at 298 K?

The simplest place to start the discussion is with an alkane chain. If we take arbitrarily an alkane chain of 10 000 carbon atoms (which implies a molar mass of about 1.4×10^5), most chemists would guess that the lowest energy conformation of the chain would correspond to the all-*trans* conformation. So they would draw decane $C_{10}H_{22}$ as

Every group of CH_2CH_2 atoms in the chain has a choice of three possible local energy minima (interchanged by rotation about the C–C bond) and so a total of 3^{10} shapes are available for this chain of 10 carbon atoms, and all of these possible shapes will be very close to each other in energy. So in spite of the fact that the lowest energy state might well indeed be the all-*trans* conformation, there will be a spread amongst these 3^{10} coiled chains depending on the temperature (through the Boltzmann factor). The energy difference between a *gauche* and a *trans* state is about $3.3\,\text{kJ}\,\text{mol}^{-1}$ for polyethene and the ratio of such *trans* to *gauche* states is given from the Boltzmann formula as

$$\frac{n_{\text{g}}}{n_{\text{t}}} = 2\exp\left(-\frac{3.3\,\text{kJ}\,\text{mol}^{-1}}{RT}\right) \tag{20.3}$$

The factor of 2 arises because there are twice as many *gauche* conformations as there are *trans*. So we need to examine critically the possible conformations of a simple polymer chain, and then consider their Boltzmann probabilities.

20.2 EARLY MODELS FOR POLYMER CHAINS

The spatial conformation of a long chain polymer is usually discussed in terms of the skeleton of heavy atoms, and the vectors of the bonds joining these successive heavy atoms. This is shown in Figure 20.1. I have followed convention and numbered the atoms sequentially A_0, $A_1\ldots A_n$, and I have taken atom A_0 as the coordinate origin. The chain will not generally be planar.

Vector $\mathbf{r}_1$ points from atom A_0 to A_1, vector $\mathbf{r}_2$ points from atom A_1 to A_2 and so on. It is clear from elementary vector analysis that the position vector of the end atom can be written

$$\mathbf{r} = \sum_{i=1}^{n} \mathbf{r}_i \tag{20.4}$$

The scalar distance between A_0 and A_n is called the *end-to-end distance* and is given by

$$r^2 = \mathbf{r}\cdot\mathbf{r} = \sum_{i=1}^{n}\sum_{j=1}^{n} \mathbf{r}_i\cdot\mathbf{r}_j \tag{20.5}$$

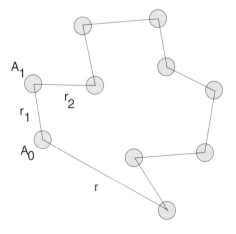

Figure 20.1   A long chain polymer

This sum may be expressed in a more useful form in terms of the bond lengths and the scalar products between the vectors pointing along the bonds as

$$r^2 = \sum_{i=1}^{n} r_i^2 + 2 \sum_{i=1}^{n-1} \sum_{j=i+1}^{n} \mathbf{r}_i \cdot \mathbf{r}_j \tag{20.6}$$

The factor of 2 arises because I have written the sums to run over distinct pairs of atoms.

The number of conformations available to a long chain molecule is very large, and it is futile to attempt to study any one of them in isolation. Instead we adopt the methods of statistical thermodynamics and look for appropriate averages over an ensemble of molecules. We focus attention on the distance between the end (heavy) atoms in the chain, and enquire about the probability that atom A_n will occupy a volume element $d\tau$ (which we would write $dx\, dy\, dz$ in rectangular Cartesian coordinates) at vector position $\mathbf{r}$. I will write this probability $W(\mathbf{r})\, d\tau$ and it can be thought of as either a time average for a given polymer chain or an ensemble average for many identical chains subject to identical conditions. These two interpretations are equivalent, according to the ergodic hypothesis.

For a very long chain, we might expect that the movement of the final atom A_n in the chain relative to the atom A_0 at the origin would resemble the velocity distribution among the molecules of a gas, as discussed in an earlier chapter. If this is the case, then the components of the velocity vector of the atom A_n, in the x, y and z directions should be independent of each other, and the total probability should depend only on the magnitude of the velocity vector and not on its direction in space. With this assumption, it can be shown that $W(\mathbf{r})$, the distribution function, depends only on the scalar distance r and not on the absolute position in space given by the vector $\mathbf{r}$ and

$$W(r) = \left(\frac{3}{2\pi\langle r^2 \rangle}\right)^{3/2} \exp\left(-\frac{3}{2\langle r^2 \rangle} r^2\right) \tag{20.7}$$

where $\langle r^2 \rangle$ is the square of the magnitude of the end-to-end distance, averaged over all configurations of the chain. The pre-exponential term is a normalizing factor to ensure that $W(r)$ has the following properties expected of such a probability.

$$\int_0^\infty W(r) 4\pi r^2 \, dr = 1$$

$$\int_0^\infty r^2 W(r) 4\pi r^2 \, dr = \langle r^2 \rangle$$

The derivation of this expression for $W(r)$ assumes a long chain, typically (say) 10^{20} bonds. Distribution functions for short chains (say 10^4 bonds) are not easily expressible in concise mathematical form.

The statistical average of the end-to-end distance is often referred to as the second moment

$$\langle r^2 \rangle = \sum_{i=1}^{n} \langle r_i^2 \rangle + 2 \sum_{i=1}^{n-1} \sum_{j=i+1}^{n} \langle \mathbf{r}_i \cdot \mathbf{r}_j \rangle \tag{20.8}$$

If we assume that all n of the heavy atom bonds are of equal length l then we write

$$\langle r^2 \rangle = nl^2 + 2 \sum_{i=1}^{n} \sum_{j=i+1}^{n} \langle \mathbf{r}_i \cdot \mathbf{r}_j \rangle \tag{20.9}$$

The second term in the double summation is the scalar product of two fixed length bond vectors averaged over all available conformations, and I will describe a couple of simple but significant models that have been used to evaluate the scalar products.

20.2.1 The Freely Jointed Chain

The freely jointed chain consists of n bonds of fixed length with the angles at the bond junctions free to assume all values with equal probability. Rotations about bonds are therefore completely free, and the directions of neighboring bonds are not correlated. Averaged over the ensemble,

$$\langle \mathbf{r}_i \cdot \mathbf{r}_j \rangle = 0 \tag{20.10}$$

and so

$$\langle r^2 \rangle = nl^2$$

The *characteristic ratio*, defined by

$$C_n = \frac{\langle r^2 \rangle}{nl^2} \tag{20.11}$$

is often encountered in such studies. For the freely jointed chain, C_n turns out to be unity for any length n of the chain.

20.2.2 The Freely Rotating Chain

In this model we take n bonds of fixed length joined at fixed bond angles θ as defined in Figure 20.2. Free rotation is permitted about each bond, so that the dihedral angles can take any value. Not only that, the barrier to rotation is set as zero so that every possible conformation produced during the dihedral angle rotation is equally likely.

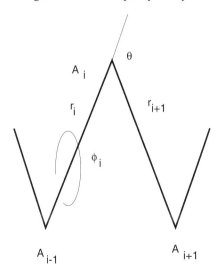

Figure 20.2   Fragment of chain showing dihedral and bond angles

The projection of bond $i + 1$ on bond i is $l \cos \theta$. The projection of this bond in a direction perpendicular to bond i averages to zero under the assumption of free rotation. This gives us immediately

$$\langle \mathbf{r}_{i+1} \cdot \mathbf{r}_i \rangle = l^2 \cos \theta \qquad (20.12)$$

The projection of bond $i + 2$ on bond i is $l \cos^2 \theta$, the projection of bond $i + k$ on bond i is $l \cos^k \theta$ and so the formula for the second moment becomes

$$\langle r^2 \rangle = l^2 \left[n + 2 \sum_{i=1}^{n-1} \sum_{j=i+1}^{n} (\cos \theta)^{j-i} \right] \qquad (20.13)$$

This can be summed to give a characteristic ratio

$$C_n = \frac{1 + \cos \theta}{1 - \cos \theta} - \frac{2 \cos \theta}{n} \frac{1 - \cos^n \theta}{(1 - \cos \theta)^2} \qquad (20.14)$$

and so C_n for a freely rotating chain varies roughly as $1/n$. For an infinite chain of tetra-hedrally bonded atoms, $n = \infty$, $\cos \theta = 1/3$ and so $C_\infty = 2$. In fact this characteristic ratio typically turns out to be about 7 and so these simple models leave much to be desired. The problem is that they do not take account of the internal rotation potential.

20.3   BARRIERS TO INTERNAL ROTATION

I mentioned the importance of dihedral bending in our discussion of molecular mechanics. In the mid nineteenth century, it was believed that free rotation could occur about any single bond, but that rotation about multiple bonds was restricted. As time passed by, the concept of free rotation became more and more suspect, and there is now to be found an impressive literature of experimentally determined potential barriers hindering rotation about single bonds, usually C−C bonds. The often-quoted value for ethane is $12.25 \pm 0.11 \, \text{kJ mol}^{-1}$.

Most experimental data have been determined from microwave spectroscopy. According to the normal selection rules ethane should not absorb in a microwave experiment since it does not have a permanent electric dipole moment. However, under conditions of long path length, transitions that should formally have zero intensity become weakly allowed and one fits the observed energy level differences to a potential of the form

$$U = U_0 + U_3(1 - \cos 3\theta) + U_6(1 - \cos 6\theta) + \cdots \qquad (20.15)$$

It is usually adequate to take just the U_3 term.

The experiment is not an easy one as the absorption of radiation is weak. The barrier is easily calculated by *Ab Initio* methods; what we do is to calculate the energy of a staggered and eclipsed form of ethane and subtract. This gives the barrier to rotation. U_3 is half of this and U_0 is a disposable constant.

Table 20.1   Barriers to rotation about a single CC bond

	R_{CC} (pm)	R_{CH} (pm)	CCH ($^{\circ}$)	$\varepsilon(E_h)$	Barrier (kJ mol^{-1})
HF/6-311G**					12.86
Staggered	152.7	108.6	111.2	-79.251708	
Eclipsed	154.1	108.5	111.7	-79.246810	
MP2/6-311 G**					13.38
Staggered	152.9	109.4	111.1	-79.570888	
Eclipsed	154.2	109.3	111.6	-79.565792	

Workers in the field emphasize the importance of geometry optimization; one generally finds that the eclipsed structure is a little more 'open' than the staggered one. Hartree–Fock calculations are perfectly acceptable for studies of this kind (Table 20.1).

Molecular mechanics force fields generally parameterize the barrier.

20.4   MOLECULAR DYNAMICS SIMULATIONS

For the sake of argument, let us consider a poly(methylene) chain of n atoms. What we need to do is to examine the dynamics of the chain for typical values of n and at typical temperatures.

To get started, consider the alkane decane $C_{10}H_{22}$

Decane is of course a very short chain compared to a real life polymer. I ran a molecular

dynamics calculation on decane at 300 K, starting from the straight chain conformation above.

Figure 20.3 shows how the end-to-end distance (Å) varied through the experiment.

The first part of the calculation takes a molecular mechanics minimum energy structure at 0 K for decane and gradually heats it to the final temperature (300 K in this case). Ignore the first 10% of Figure 20.3 for this reason. The end-to-end distance does indeed decrease significantly through the calculation and the final conformation is somewhat crumpled but the decane chain is not sufficiently long to allow for any spectacular visual effects (Figure 20.4).

A similar calculation with a chain of 100 atoms $(C_{100}H_{202})$ gave me a corresponding molecular dynamics snapshot (Figure 20.5).

20.5   THE ELASTICITY OF RUBBER

Natural rubber (V) is essentially a hydrocarbon whose constitution was established by Faraday to be $(C_5H_8)_n$. It is a polymer of isoprene, with a perfectly regular chain. Every

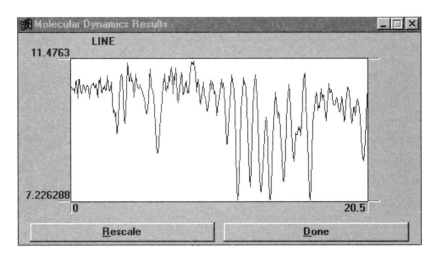

Figure 20.3 End-to-end distance of decane at 300 K

Figure 20.4 Decane molecular dynamics snapshot at 300 K

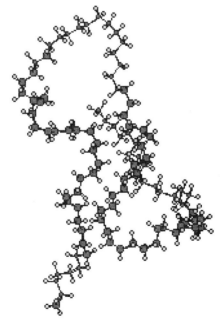

Figure 20.5 Snapshot of a chain with 100 atoms ($C_{100}H_{202}$) at 300 K

fourth carbon atom in the chain carries the methyl side group. The presence of a double bond in each monomer unit determines the chemical reactivity of the compound and its ability to react with sulfur in the vulcanization process. The vulcanization process involves forming cross-chains by reaction of sulfur across the double bonds of different chains. This cross linking prevents flow in the material. The double bond is also responsible for the susceptibility of rubber to oxidation.

The structure of gutta-percha (VI), the other natural polymer of isoprene, differs slightly but significantly from that of natural rubber; in natural rubber the C−C single bonds form the *cis* conformation whilst in gutta-percha they form the *trans* conformation. In both V and VI, the two single bonds adjacent to the double bond remain permanently fixed in a single plane but there is more or less free rotation about the other bonds.

The most obvious physical characteristic of rubber is the high degree of deformability exhibited under small stresses. Samples can be extended by 500–1000% but Hooke's Law does not apply except as a limiting law for very small extensions. A simple explanation of rubber elasticity is as follows. The polymeric molecule exists in a crumpled ball state at (say) 300 K because of the availability of very many conformational states due to rotations about the CC bonds, all of which are similar in energy. According to the Boltzmann formula, all are roughly equally probable and so we will observe an ensemble average where the polymer is crumpled. The polymer can be easily stretched because of the ease of untangling the crumpled ball.

In addition to these familiar properties though, rubber also exhibits the so-called Grough–Joule effects

- Rubber held in a stretched state and under a constant load contracts on heating.
- Rubber gives out heat when stretched.

These two properties are not peculiar to natural rubber, but are characteristic of a wide range of synthetic polymers.

When a rubber band of length l is stretched, it exerts a restoring force f which is a function of the length and of the temperature. If we consider a fixed temperature then the work done on the band in an extension dl is

$$dw = f \, dl \qquad (20.16)$$

and this is in addition to any pV work done in the expansion. The first law of thermodynamics therefore becomes

$$dU = dq - p \, dV + f \, dl \qquad (20.17)$$

Usually the $p\,dV$ term is 10^3 times smaller than the $f\,dl$ term, and so we ignore it for discussions of rubber elasticity. We therefore have

$$dU = dq + f\,dl \tag{20.18}$$

which is

$$dU = T\,dS + f\,dl$$

for a reversible change. Hence, the Helmholtz energy expression can be derived as

$$dA = f\,dl - S\,dT \tag{20.19}$$

Using the simple properties of exact differentials discussed in the Appendix we have

$$dA = \left(\frac{\partial A}{\partial l}\right)_T dl + \left(\frac{\partial A}{\partial T}\right)_i dT \tag{20.20}$$

and so

$$f = \left(\frac{\partial A}{\partial l}\right)_T \tag{20.21}$$

and

$$S = -\left(\frac{\partial A}{\partial T}\right)_l$$

The Helmholtz energy A is defined as $U\text{-}TS$ and so we can also write

$$f = \left(\frac{\partial U}{\partial l}\right)_T - T\left(\frac{\partial S}{\partial l}\right)_T \tag{20.22}$$

Experimental data show that the second term (the entropy term) is far more important than the first term (the internal energy term) for discussions of rubber elasticity. To evaluate the second term we need to find a way to measure $(\partial S/\partial l)_T$ in terms of easily accessible physical quantities. We start again from

$$dA = f\,dl - S\,dT$$

and

$$dA = \left(\frac{\partial A}{\partial l}\right)_T dl + \left(\frac{\partial A}{\partial T}\right)_l dT$$

Now the second mixed differentials of A with respect to l and to T are equal

$$\frac{\partial^2 A}{\partial l\,T\partial l} = \frac{\partial^2 A}{\partial l\,\partial T}$$

and so

$$\left(\frac{\partial S}{\partial l}\right)_T = \left(\frac{\partial f}{\partial T}\right)_L \tag{20.23}$$

Thus $(\partial S/\partial l)_T$ may be obtained from measurements of force *versus* temperature at constant length.

In order to make progress with the discussion, I need to introduce an equation of state for rubber. From our qualitative description of rubber elasticity given above, we can say that the force exerted by an extended piece of rubber arises from the fact that the chains can adopt very many configurations of similar energy. This is an entropy-type effect, and the entropy is expected to decrease as the chains are extended. A simple statistical treatment along the lines given earlier in the Chapter suggests that the entropy is given by

$$S = A - B\left(\frac{2}{\lambda} + \lambda^2\right) \tag{20.24}$$

where A and B are positive constants and $\lambda = l/l_0$ is the fractional chain extension. Substitution of $(\partial S/\partial l)_T$ into the equation

$$-\left(\frac{\partial S}{\partial l}\right)_T = \left(\frac{\partial f}{\partial T}\right)_t \tag{20.25}$$

and subsequent integration yields an expression for the force f

$$f = CT\left(\lambda - \frac{1}{\lambda^2}\right)$$

which is an equation of state for rubber. C is a constant, given by $2B/l_0$. The model predicts correctly that rubber should contract when heated under the action of a constant force.

21　Modeling Liquids

Of the three states of matter, gases are said to be the easiest to model because the constituent atoms or molecules are so far apart that we can ignore the intermolecular forces apart from during their brief collisions. This makes it reasonable to regard an ideal gas as a good first approximation to a real gas. Thus the kinetic theory of gases was brought to an advanced state of development by the end of the nineteenth century. The atoms or molecules in a crystalline solid are arranged in a regular geometric pattern and it is therefore quite reasonable to regard a regular crystalline solid as the starting point for a discussion of a real solid. Once this regular pattern was thoroughly understood at the beginning of the twentieth century, the theory of the solid state made rapid progress.

It is a fact that liquids are much harder to study than solids or gases, and that is why I have left them to the end of this text. We often say that there is neither complete order nor complete disorder in liquids; perhaps you would like to know the experimental evidence for this statement?

21.1　THE RADIAL DISTRIBUTION FUNCTION

For most of this chapter, I will assume that the particles making up a pure liquid are very simple ones and I will draw them as spheres. These particles are allowed to interact with each other through a suitable pair potential, but for most of the chapter I will assume that this pair potential depends only on the separation between the particles and has no angular dependency.

Since there are typically 10^{23} particles in a macroscopic liquid sample, it would be foolish to try and specify the exact position of each and every particle at any given instant of time. What we do instead is to specify the structure in an average sense by means of the *radial distribution function $g(r)$*. Consider Figure 21.1 as a snapshot of the particles in a simple liquid.

What I do is to pick a typical particle as the coordinate origin, the grey particle in Figure 21.1, and then draw two spheres of radii r and $r + \mathrm{d}r$ with that particle as centre. I count the

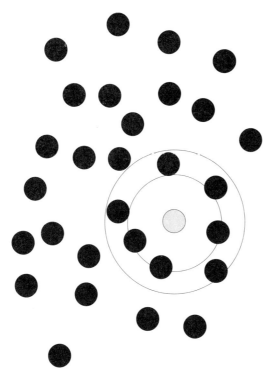

Figure 21.1   Snapshot of s simple liquid

number of particles whose centers lie within the two spheres, and repeat the experiment for a large number of particles each taken as the origin. If the result of experiment i is $g_i(r)dr$ and if I repeat the measurement N times then the radial distribution function $g(r)$ is defined by the average of these measurements

$$g(r)\,dr = \frac{1}{N} \sum_{i=1}^{N} g_i(r)\,dr \qquad (21.1)$$

This process has to be carried out for many complete shells over the range of values of r thought to be significant, and provided that the number of experiments is sufficiently large then $g(r)$ will be independent of which particle we choose as the center and should be a characteristic of the liquid itself.

It turns out that we also have to take account of the size of the sample; a particle at the boundary experiences quite different forces to a particle in the body of the sample. But for the minute, assume that the sample is essentially infinite.

Radial distribution functions for liquids can be deduced experimentally from diffraction studies. These experiments were first performed in the 1940s.

To fix our ideas about the difference between gases, liquids and solids, consider first the case of the simple cubic solid shown in Figure 21.2. If I call the nearest neighbor distance a, then each atom is surrounded by 6 nearest neighbors at a distance a, 12 next-nearest neighbors at a distance $\sqrt{2}\,a$, 8 next-next nearest neighbor neighbors at a distance of $\sqrt{3}\,a$ and so on. If we ignore the vibrational energy of each atom for the minute, then we would expect to find a radial distribution function similar to the one sketched in Figure 21.3.

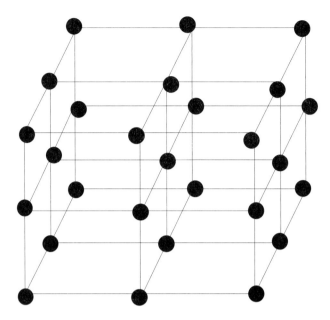

Figure 21.2 Simple cubic lattice

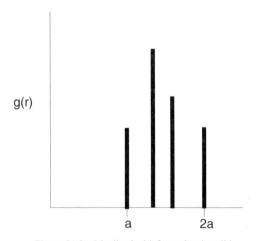

Figure 21.3 Idealized $g(r)$ for a simple solid

In fact the atoms do have vibrational kinetic energy, and as a consequence the peaks in $g(r)$ become smeared out. The area under each peak is proportional to the number of neighboring atoms.

For an ideal gas, the motion is completely random. The number of ideal gas particles contained between the two spheres of radii r and $r + dr$ is proportional to the enclosed volume and so $g(r) = 4\pi r^2$.

The interesting thing is that the $g(r)$ curve for a liquid at low temperature resembles that for a solid, and at high temperature it resembles that for an ideal gas. At intermediate temperatures the $g(r)$ curve appears to be a superposition of these two. Figure 21.4 shows $g(r)$ for a

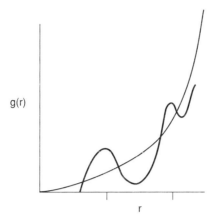

Figure 21.4 Experimental radial distribution function for liquid argon

liquid at an intermediate temperature, and the two features can be clearly seen. The thick curve is the experimental result for liquid argon, and I have included for comparison the curve that you would expect for an ideal gas.

21.2 THE INTERNAL ENERGY OF A LIQUID

All of the equilibrium properties of a liquid can be deduced from the internal energy U and the equation of state. The internal energy of a given amount n of substance may be written as a function of any two of the variables p, V and T.

According to the equipartition of energy principle, the average translational kinetic energy of any particle within a substance is

$$\langle \varepsilon_{\text{trans}} \rangle = \frac{3}{2} k_{\text{B}} T \tag{21.2}$$

and this holds for any substance in equilibrium whether it is a gas, solid or liquid. For simple liquids, which we can imagine as featureless particles with no internal vibrational motion, this expression gives the translational kinetic energy. The total kinetic energy for N particles is therefore N times this amount.

The internal potential energy can be deduced with the help of Figure 21.1 which I used in my discussion of the radial distribution function $g(r)$. First of all let me write the mutual pair potential energy between two particles distance r as $U^{\text{pair}}(r)$. I have left the expression general; it could be a Lennard–Jones expression, it could be a potential determined from a highly accurate quantum mechanical calculation or whatever.

If I label the particle at the center of the sphere in Figure 21.1 number 1, and take for the sake of argument one of the particles in the shell between the two spheres to be particle number 2, then their contribution to the mutual potential energy is $U^{\text{pair}}_{12}(r)$.

The number of particles between the two spheres of radii r and $r + dr$ is given by $g(r)\,dr$ and so the total contribution to the mutual potential energy is

$$\varepsilon_{\text{pot}} = \varepsilon^{\text{pair}}(r) g(r)\,dr \tag{21.3}$$

To find the mutual potential energy of particle 1 and all the others in the liquid we have to integrate from $r = 0$ to infinity

$$\varepsilon_{\text{pot}} = \int_0^\infty \varepsilon^{\text{pair}}(r)g(r)\,\mathrm{d}r \qquad (21.4)$$

and finally, in order to calculate the total mutual potential energy of the liquid we have to add together the contributions from all of the N particles present. In line with our discussion in previous chapters, this gives

$$U_{\text{pot}} = \frac{1}{2}N\int_0^\infty \varepsilon^{\text{pair}}(r)g(r)\,\mathrm{d}r \qquad (21.5)$$

which is the potential energy contribution to the internal energy of a simple liquid. The total internal energy is therefore

$$U = \frac{3}{2}Nk_{\mathrm{B}}T + \frac{1}{2}N\int_0^\infty \varepsilon^{\text{pair}}(r)g(r)\,\mathrm{d}r \qquad (21.6)$$

21.3   THE VIRIAL EQUATION OF STATE

In Chapter 1 I introduced a number of equations of state, and mentioned that they have to be capable of describing the properties of a substance

- over a range of temperatures, pressures and molar volumes, and
- in the gaseous, solid and liquid phases.

Equations of state such as the Van der Waals equation

$$\left(p + \frac{a}{V_{\mathrm{m}}^2}\right)(V_{\mathrm{m}} - b) = RT$$

can be said to have some basis in theory, whilst equations of state such as the Beattie–Bridgeman equation

$$pV_{\mathrm{m}}^2 = RT\left[V_{\mathrm{m}} + B_0\left(1 - \frac{b}{V_{\mathrm{m}}}\right)\right]\left(1 - \frac{c}{V_{\mathrm{m}}T^3}\right) - A_0\left(1 - \frac{a}{V_{\mathrm{m}}}\right)$$

contain very many constants that have to be determined by fitting experimental pVT data. Neither type give very much insight into the molecular processes responsible for the pVT behavior. The equilibrium properties of a liquid are determined by the forces between the particles that make up the liquid, and the virial equation of state gives a more direct link between the microscopic and macroscopic properties.

The physical significance of the coefficients in the virial equation of state for a liquid can be demonstrated in a number of ways. Let me get us started by considering N spherical particles each of mass m, contained in a cubic box of side L.

The size of the box is taken to be large in comparison to the size of a single liquid particle. Later on in the proof, I will have to consider the forces between the particles and the walls of the container.

Let me focus attention on particle i moving in the box. As this particle moves it will be subject to some varying force that I will write $\mathbf{F}_i$ and so from Newton's second law

$$\mathbf{F}_i = m\frac{d\mathbf{v}_i}{dt} \tag{21.7}$$

If I now take the scalar product of both side of this equation with $\mathbf{r}_i$ I get

$$\mathbf{r}_i \cdot \mathbf{F}_i = m\mathbf{r}_i \cdot \left(\frac{d\mathbf{v}}{dt}\right) \tag{21.8}$$

Consider now the vector identity

$$\frac{d}{dt}(\mathbf{r}_i \cdot \mathbf{v}_i) = \mathbf{r}_i \cdot \frac{d\mathbf{v}_i}{dt} + \frac{d\mathbf{r}_i}{dt} \cdot \mathbf{v}_i$$

which can also be written

$$\frac{d}{dt}(\mathbf{r}_i \cdot \mathbf{v}_i) = \mathbf{r}_i \cdot \frac{d\mathbf{v}_i}{dt} + v_i^2$$

On comparison of the above equations, I have

$$\mathbf{r}_i \cdot \mathbf{F}_i = m\left[\frac{d}{dt}(\mathbf{r}_i \cdot \mathbf{v}_i) - v_i^2\right]$$

or

$$-\frac{1}{2}\mathbf{r}_i \cdot \mathbf{F}_i = -\frac{1}{2}m\frac{d}{dt}\mathbf{r}_i \cdot \mathbf{v}_i + \frac{1}{2}mv_i^2 \tag{21.9}$$

The next step in the derivation is to sum corresponding terms on both sides of the equation for each particle in the box. For N particles each of mass m, this gives

$$-\frac{1}{2}\sum_{i=1}^{N}\mathbf{r}_i \cdot \mathbf{F}_i = \frac{1}{2}m\frac{d}{dt}\sum_{i=1}^{N}\mathbf{r}_i \cdot \mathbf{v}_i + \frac{1}{2}m\sum_{i=1}^{N}v_i^2 \tag{21.10}$$

Finally, we take an average over all the particles in the box. For an equilibrium distribution of particles, this can be regarded as either an average for a single box over time or an average, at a fixed time, over a large number of identical boxes. If I denote the average as $\langle\cdots\rangle$ then we have

$$-\frac{1}{2}\left\langle\sum_{i=1}^{N}\mathbf{r}_i \cdot \mathbf{F}_i\right\rangle = -\frac{m}{2}\frac{d}{dt}\left\langle\sum_{i=1}^{N}\mathbf{r}_i \cdot \mathbf{v}_i\right\rangle + \frac{1}{2}m\left\langle\sum_{i=1}^{N}v_i^2\right\rangle \tag{21.11}$$

The second term on the right hand side is obviously the mean kinetic energy of all the particles in the box. This must be $3/2\ Nk_BT$, according to the equipartition of energy principle.

Whatever the value of the first average quantity in brackets on the right hand side it cannot vary with time because we are dealing with an equilibrium state and so the first time derivative must vanish.

$$-\frac{m}{2}\frac{d}{dt}\left\langle\sum_{i=1}^{N}\mathbf{r}_i \cdot \mathbf{v}_i\right\rangle = 0$$

and so we have

$$-\frac{1}{2}\left\langle \sum_{i=1}^{N} \mathbf{r}_i \cdot \mathbf{F}_i \right\rangle = \frac{1}{2}\, m \left\langle \sum_{i=1}^{N} v_i^2 \right\rangle \qquad (21.12)$$

The summation term on the left-hand side $-\frac{1}{2}\langle \sum_{i=1}^{N} \mathbf{r}_i \cdot \mathbf{F}_i \rangle$ involving the forces and coordinates is often referred to as the *virial of Clausius*.

In order to derive a virial equation of state, we note that the force on each fluid particle i can be split into two parts; the internal force due to the other fluid particles and the force due to interaction with the walls of the container. The average value of the virial of the wall forces can be simply related to the pressure, which is only significant when the particle comes close to the walls.

The derivation of the equation of state depends on the fact that the virial of Clausius can be shown to be written in two different ways (although I have not proved this);

$$-\frac{1}{2}\left\langle \sum_{i=1}^{N} \mathbf{r}_i \cdot \mathbf{F} \right\rangle = \frac{3}{2}\left(Nk_\mathrm{B}T - pV\right) \qquad (21.13)$$

$$-\frac{1}{2}\left\langle \sum_{i=1}^{N} \mathbf{r}_i \cdot \mathbf{F} \right\rangle = \frac{N}{4}\int_{0}^{\infty} \frac{\mathrm{d}\varepsilon^{\mathrm{pair}}(r)}{\mathrm{d}r}\, rg(r)\,\mathrm{d}r \qquad (21.14)$$

If these two expressions are true for any fluid composed of particles whose pair potential is spherically symmetrical, then equating the right hand sides of the two equations gives the result

$$pV = Nk_\mathrm{B}T - \frac{N}{6}\int_{0}^{\infty} \frac{\mathrm{d}U^{\mathrm{pair}}(r)}{\mathrm{d}r}\, rg(r)\,\mathrm{d}r$$

In order to apply the virial equation of state to a liquid in which the particles interact according to a known pair potential, it is necessary to know $g(r)$ very accurately indeed.

21.4 COMPUTER SIMULATION

The first computer simulation of a liquid was carried out in 1953 at the Los Alamos National Laboratories. The power of the mainframe MANIAC computer used was far less than that of the PC that I used to write this text. The early work by Metropolis laid the foundations for the modern Monte Carlo technique. The original models were the highly idealized representations of molecules that I will discuss in the next few sections but within a few years simulations were carried out using the Leonard–Jones pair potential. This made it possible to compare data from experiment (for example, on liquid argon).

The first Molecular Dymamics (MD) calculations on liquids were carried out in the 1950s, but it was not until the 1960s that MD calculations with realistic pair potentials first appeared in the literature.

One liquid differs from another because of the interactions between the particles, so I had better now turn to the topic of the pair potential. I want you to distinguish between the intermolecular potential (that between two complete molecular species) and the intramolecular potential (which relates to the constituent atoms in a given molecule).

21.4.1 The Pair Potential

First of all, let me remind you of a well-established intermolecular potential discussed in Chapter 6, the Lennard–Jones 12-6 potential. This relates at its simplest to a pair of identical interacting (but not bonding) atoms, separated by distance r. Over the years, the ideas behind the Lennard–Jones potential have been extended to include simple interacting molecular systems of high symmetry such as dinitrogen (a linear molecule), methane (a tetrahedral molecule), carbon dioxide (a linear molecule) and benzene (a planar regular hexagonal molecule).

Such models should be regarded with caution because they ignore any angular dependence of the molecular interaction. I wrote the Lennard–Jones potential in two equivalent ways

$$U_{\mathrm{LJ}}(r) = \frac{A}{r^{12}} - \frac{B}{r^6}$$

and

$$U_{\mathrm{LJ}}(r) = 4\varepsilon\left[\left(\frac{\sigma}{r}\right)^{12} - \left(\frac{\sigma}{r}\right)^6\right]$$

where the 'well depth' ε corresponds to the minimum value of U_{LJ} and the 'distance of closest approach' σ is the value of r for which U is zero. The coefficients A and B can be deduced in a number of different ways, from example by studying the scattering of atoms or simple molecules in molecular beams. If the two atoms or simple molecules are identical then the following values of ε and σ shown in Table 21.1 are often used.

The well depth ε is often reported as $\varepsilon/k_{\mathrm{B}}$, which has the dimensions of temperature.

Such a potential as written applies to the interaction of a pair of atoms or simple molecules in the gas phase (Figure 21.5). The first thing to say is that interaction between unlike pairs can be approximated using the Lorentz–Berthelot mixing rules; for species A and B we would use

$$\sigma_{\mathrm{AB}} = \frac{1}{2}(\sigma_{\mathrm{AA}} + \sigma_{\mathrm{BB}})$$

$$\varepsilon_{\mathrm{AB}} = \sqrt{\varepsilon_{\mathrm{AA}}\varepsilon_{\mathrm{BB}}} \tag{21.16}$$

What we do in a MD computer simulation is to choose a suitable box for the N particles of mass m in order to reproduce the density of interest. N might typically be several thousands. We would then normally fix the temperature, and ensure that the starting configuration of the

Table 21.1 Lennard–Jones 12-6 parameters

	$(\varepsilon/k_{\mathrm{B}})(K)$	$\sigma\,(\mathrm{pm})$
He	10.22	258
Ne	35.7	279
Ar	124	342
Xe	229	406
N_2	91.5	368
CO_2	190	400
CH_4	137	382
C_6H_6	440	527

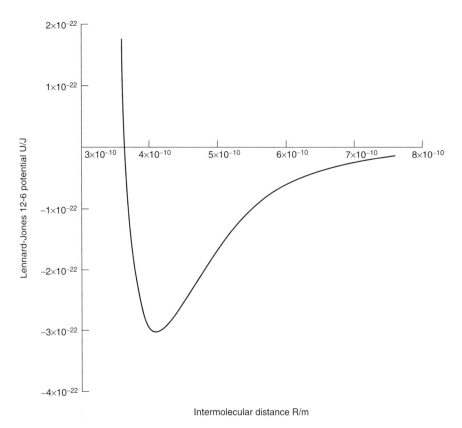

Figure 21.5 Lennard–Jones potential for dinitrogen ... dinitrogen

particles had the correct random distribution of energies. In order to allow for the effect of the finite size of the box compared to the number of particles, we allow for an infinite repeat box as discussed earlier.

Finally, we solve the equations of motion by the Verlet algorithm or any other suitable algorithm. As we progress along the timescale of the (so-called) experiment we keep track of (for example) the temperature and the pressure. These can be adjusted if necessary but finally we arrive at physical quantities that can be compared with experiment. We then go back and modify the Lennard–Jones parameters until we get a good fit between simulation and experiment.

The thing to note is that the gas-phase pair potential may not be exactly suitable for modeling a fluid, because the density of particles in a fluid is much higher than that in a gas, and what we do instead is to seek an effective pair potential. This is one with the same form as above but with modified parameters that take account of the fact that each extra particle can modify the potential of the particles already present.

21.4.2 Pairwise Additivity

Consider the two interacting particles A and B fixed at certain positions in space, as shown in Figure 21.6. Assume for the minute that C is not present and I will write their mutual potential

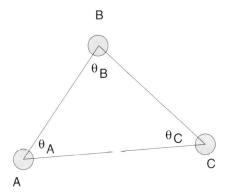

Figure 21.6 Three interacting bodies A, B and C

energy as $U(A, B)$. If A and B are point charges then their mutual potential energy is given exactly by

$$U(A, B) = \frac{Q_A Q_B}{4\pi\epsilon_0 R_{AB}}$$

If we now think of B and C (in the absence of A) then the mutual potential energy is $U(B, C)$. Finally, application of these considerations to A and C gives a mutual potential energy of $U(A, C)$.

If A, B and C are point charges and the potential is an electrostatic one then the mutual potential energy of the three particles is just the sum of these three terms

$$U(A, B, C) = U(A, B) + U(A, C) + U(B, C)$$

and we say that the potential is pair-wise additive. If A, B and C are now polarizable particles such as molecules, addition of particle C to the pair A, B can change their mutual potential energy (for example, by inducing a dipole moment in the two molecules A and B) and so on. The difference

$$U(A, B, C) - U(A, B) - U(A, C) - U(B, C)$$

is not necessarily zero, and it is called the three-body contribution to the mutual potential energy.

Electrostatic mutual potential energies between point charges are exactly pairwise additive; the mutual electrostatic potential for three point charges is exactly equal to the sum of the pair terms. The gravitational potential energy is also exactly pairwise additive.

It can be shown that the dispersion energy is pairwise additive in the R^{-6} term, with non-additive contributions appearing only in the higher powers of $1/R$.

The dipole induction term is not pairwise additive. This can easily be seen by considering a special case of Figure 21.6 but with the three particles (atoms, or molecules) arranged in a straight line A...B...C, with the distance between A and B equal to the distance between B and C. If A and C are the same chemical species then the dipole induced by A in B will be exactly balanced by the dipole induced by C in B, yet A by itself will induce a non-zero dipole in B, as will C. The net induction potential energy is not equal to the sum of the induction energies due to each pair.

Axilrod and Teller investigated this effect in the very specific case of the dispersion contribution to the lattice energy of crystalline argon, and they found a three-body correction term of some 10% which can be written

$$U^{(3)} = K \frac{\cos\vartheta_A \cos\vartheta_B \cos\vartheta_C}{(r_{AB}r_{BC}r_{AC})^3} \qquad (21.17)$$

We do not normally include three-body effects in liquid modeling. This is to do with pragmatism rather than a belief that such effects are negligible (which they certainly are not). The 'inner loop' of a liquid modeling program invariably involves the double sum over the number of particles present, in order to evaluate the forces. If there are N particles present, this evaluation is an N^2 process. To add a third inner loop would increase the time taken by a factor of N. What we do is to seek an 'effective' two-body potential where the three body terms are compensated for by a careful choice of the two-body parameters.

21.4.3   Model Pair Potentials

The construction of intermolecular potentia.l energy functions is an important one, and a great deal of research has gone into this activity. Over the years, several idealized pair potentials have appeared in the literature. I will record them here for reference. In the hard sphere potential we have

$$U^{HS}(r) = \begin{cases} \infty & \text{if } r < \sigma \\ 0 & \text{if } r \geq \sigma \end{cases} \qquad (21.18)$$

In the square well potential

$$U^{SW}(r) = \begin{cases} \infty & \text{if } r < \sigma_1 \\ -\varepsilon & \text{if } \sigma_1 \leq r \leq \sigma_2 \\ 0 & \text{if } r > \sigma_2 \end{cases} \qquad (21.19)$$

In the soft-sphere potential

$$U^{SS}(r) = \varepsilon \left(\frac{\sigma}{r}\right)^{\nu} \qquad (21.20)$$

where ν is a parameter often chosen to be an integer. Soft sphere potentials contain no attractive parts.

In addition, it is common to divide potentials into attractive and repulsive parts. For the Lennard–Jones 12-6 potential

$$U^{LJR}(r) = \begin{cases} U^{LJ}(r) + \varepsilon & \text{if } r \leq r_{min} \\ 0 & \text{if } r > r_{min} \end{cases} \qquad (21.21)$$

and

$$U^{LJA}(r) = \begin{cases} -\varepsilon & \text{if } r < r_{min} \\ U^{LJ}(r) & \text{if } r > r_{min} \end{cases} \qquad (21.22)$$

21.4.4 Site–site Intermolecular Potentials

If we knew exactly the charge distribution in molecule A and molecule B, then we would be able to calculate the interaction using classical electrostatics. As I explained earlier, a charge distribution is described by its multipoles and therefore a knowledge of the non-zero multipoles in A and B is required. In the spirit of the multipole expansion, it is hoped that only a few terms will be needed in the expansion.

Electric dipole moments can be routinely measured in the gas phase from the effect of an applied electric field on a pure rotational spectrum (the microwave Stark effect). Prior to 1970 the only direct routes to molecular electric quadrupole moments were the Kerr and the Cotton–Mouton effects. They can now be obtained to fair accuracy from the effect of a magnetic field on a pure rotational spectrum (molecular microwave Zeeman spectroscopy). Higher moments are much more elusive animals, and very few are known with any degree of accuracy.

What is done in point electrostatic models is to assign point charges to positions within a molecule in such a way that the point charges reproduce a given electric moment. The interaction between molecules is then accounted for from the electrostatic potential energy between pairs of point charges on different molecules.

For example, HF has an electric dipole moment of 6.093×10^{-30} C m and a bond length of 91.7 pm. This dipole moment can be reproduced by placing equal and opposite charges of 6.65×10^{-20} C at the nuclei. This corresponds to 0.415 of an electron transferred from H to F.

Dinitrogen has a zero dipole moment but a non-zero quadrupole moment and this can be represented as a charge of -0.5 e at each nuclear position and $+1$e midway between the nuclei. In order to account for the electrostatic interaction between HF and N_2 we would sum over contributions from the charges on either atom. That is to say, we would calculate the interactions between $+Q_A$ and both of the $-Q_B$s, $+Q_A$ with $2Q_B$, $-Q_A$ with both of the $-Q_B$s and finally $-Q_A$ with the $+2Q_B$ at the center of the dinitrogen molecule (Figure 21.7). Intermolecular potentials that are based on the interaction of specific sites

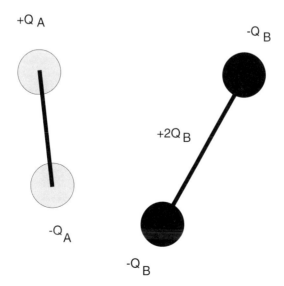

Figure 21.7 Lennard–Jones 12-6 parameters

within the molecules are called *site–site interactions*. The sites need not correspond to nuclear positions but they often do.

Alternatively, we can model the interaction of A and B in terms of Lennard–Jones interactions between the sites (almost always chosen to be the nuclei) in molecule A with the sites in molecule B.

21.5   CLASSICAL AND QUANTUM LIQUIDS

If we examine the four liquids helium, benzene, water and aluminum then we see that they have very different simple properties (Table 21.2)

Liquid helium has unique properties. It cannot be solidified at atmospheric pressure however low the temperature, and it is the only pure substance for which two liquid phases can coexist in equilibrium. These two liquid phases are referred to as He(I) and He(II). He(I) behaves pretty much like any other liquid but He(II) has very many fascinating properties such as superfluidity.

According to the equipartition of energy principle, the translational energy of an atom at temperature T is $(3/2)\,k_B T$. Calculation of the de Broglie wavelength shows that

$$\lambda_{dB} = \frac{h}{\sqrt{3 k_B T m}}$$

and even at $4\,\mathrm{K}$ the de Broglie wavelength of a helium atom is $6.3 \times 10^{-10}\,\mathrm{m}$, that is to say it is of the order of magnitude of molecular dimensions.

[4]He is often referred to as a 'quantum' liquid and we know from a previous chapter that we should therefore look for a quantum mechanical treatment of the liquid. Not only that, its zero point energy is greater than $k_B T$ and so the solid tends to shake itself to pieces.

Luckily, quantum liquids are rare. Most liquids can be studied by the methods of classical mechanics, and they are usually referred to as classical liquids.

Table 21.2   Physical properties of some simple liquids

	Helium	Benzene	Water	Aluminum
T_m (K)	0	279	273	933
T_b (K)	4.21	353	373	2740
T_c (K)	520	562	647	7740
$\rho\,(\mathrm{g\,cm})^{-3}$ at T_m	0.125	0.899	1.000	2.380

Appendix: A Mathematical Toolkit

I want to start this Appendix by quoting an important theorem that enables functions of a single variable $f(x)$ to be expanded in a power series of x about some known point.

A.1 TAYLOR'S THEOREM

If $f(x)$ is a continuous and differentiable function then

$$f(x) = f(a) + \frac{(x-a)}{1!}\left(\frac{df}{dx}\right)_{x=a} + \frac{(x-a)^2}{2!}\left(\frac{d^2f}{dx^2}\right)_{x=a} + \cdots \tag{A.1}$$

where the notation $(df/dx)_{x=a}$ means that we have to evaluate the first derivative at the point $x = a$.

A.2 PARTIAL DIFFERENTIATION

Consider a function of two variables x and y defined by

$$f(x, y) = x^2 - y^2$$

The value of $f(x, y)$ is defined for every pair of numbers x and y and we can represent $f(x, y)$ as a contour diagram as in Figure 10.10.

A.2.1 First Partial Derivatives

The first partial differential of $f(x, y)$ with respect to x is defined as the limit

$$\left(\frac{\partial f}{\partial x}\right)_y = \text{Lim}\left[\frac{f(x + \delta x, y) - f(x, y)}{\delta x}\right] \tag{A.2}$$

Similarly the first partial derivative of $f(x, y)$ with respect to y is defined as

$$\left(\frac{\partial f}{\partial y}\right)_x = \text{Lim} \left[\frac{f(x, y + \delta y) - f(x, y)}{\delta y}\right] \tag{A.3}$$

A.2.2 Total Derivatives

Consider now the change in the value of the function $f(x, y)$ due to changes δx in x and δy in y. Then we have

$$\delta f = f(x + \delta x, y + \delta y) - f(x, y)$$

$$= f(x + \delta x, y + \delta y) - f(x, y + \delta y) + f(x, y + \delta y) - f(x, y)$$

and so we see that

$$\partial f \approx \frac{\partial f}{\partial x} \delta x + \frac{\partial f}{\partial y} \delta y$$

In the limit we have

$$df = \frac{\partial f}{\partial x} dx + \frac{\partial f}{\partial y} dy \tag{A.4}$$

Suppose now that both x and y are differentiable functions of another variable t. Then

$$\frac{df}{dt} = \frac{\partial f}{\partial x} \frac{dx}{dt} + \frac{\partial f}{\partial y} \frac{dy}{dt} \tag{A.5}$$

A.3 SCALARS AND VECTORS

A scalar s is a quantity such as mass, temperature, time that can be represented by a real number. The modulus of s, denoted $|s|$, is the value of s irrespective of sign and is always positive.

A vector quantity $\mathbf{v}$ is a quantity such as displacement, force, velocity that has both a magnitude and a direction in space.

The magnitude or length or modulus of vector $\mathbf{v}$ is denoted by v or $|\mathbf{v}|$. Magnitudes are always positive.

A unit vector is one that has unit modulus. A unit vector in the direction of vector $\mathbf{v}$ is $\mathbf{v}/v$. I will generally write such unit vectors as $\hat{\mathbf{v}}$.

A.3.1 Vector Addition and Scalar Multiplication

The vector $k\,\mathbf{v}$ (where k is a positive scalar) is a vector in the same direction as $\mathbf{v}$ but with k times the length. The direction of $k\,\mathbf{v}$ is in the direction of $\mathbf{v}$ if k is positive, but in the opposite direction if k is negative (Figure A.1).

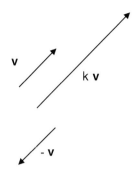

Figure A.1   Scalar multiplication of a vector

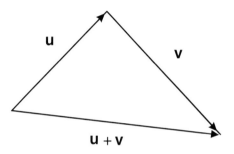

Figure A.2   Addition of two vectors

Vectors **u** and **v** are added together according to the parallelogram rule; we draw **u** and then add **v** onto **u** according to Figure A.2.

A.3.2   Coordinate Systems

There are two basic coordinate systems that we shall use

(i) Cartesian coordinates
    A point P has Cartesian coordinates (x, y, z) (Figure A.3). The Cartesian unit vectors are $\mathbf{e}_x$, $\mathbf{e}_y$ and $\mathbf{e}_z$ directed parallel to the x, y and z axes, respectively.
    The position vector **r** of point P can be expressed as

$$\mathbf{r} = x\mathbf{e}_x + y\mathbf{e}_y + z\mathbf{e}_z \tag{A.6}$$

(ii) Spherical polar coordinates
    Point P has spherical polar coordinates (r, θ, ϕ). The r coordinate is the distance of the point P from the origin O. The angle θ is the angle that the line OP makes with the positive z axis system. The angle ϕ is the azimuthal angle measured in the x-y plane from the positive x axis.
    The unit vectors in this system are $\mathbf{e}_r$, $\mathbf{e}_\theta$ and $\mathbf{e}_\phi$. The unit vector $\mathbf{e}_r$ is directed outwards from the coordinate origin O to point P, the unit vector $\mathbf{e}_\theta$ is normal to the line OP in the plane

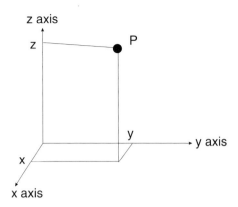

Figure A.3 Cartesian components of a vector

containing the z axis and OP, and in the direction of increasing θ. The unit vector $\mathbf{e}_\phi$ is tangential to the circle shown in Figure A.4 and points in the direction of increasing ϕ.

The coordinates (r, θ, ϕ) are related to Cartesian coordinates by

$$
\begin{aligned}
x &= r \sin \theta \cos \phi \\
y &= r \sin \theta \sin \phi \\
z &= r \cos \theta
\end{aligned}
\tag{A.7}
$$

The values of r, θ and ϕ are restricted as follows

$$
\begin{aligned}
&r \geq 0 \\
&0 \leq \theta \leq \pi \\
&0 \leq \phi < 2\pi
\end{aligned}
$$

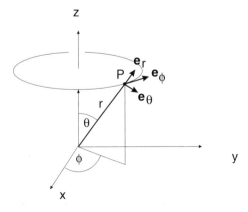

Figure A.4 Spherical polar coordinates

A.3.3 Cartesian Components of a Vector

In view of the laws for vector addition and scalar multiplication, vectors $\mathbf{u}$ and $\mathbf{v}$ can be specified by their Cartesian components

$$\mathbf{u} = u_x\mathbf{e}_x + u_y\mathbf{e}_y + u_z\mathbf{e}_z$$

$$\mathbf{v} = v_x\mathbf{e}_x + v_y\mathbf{e}_y + v_z\mathbf{e}_z$$

The vector $k\mathbf{u}$ has components

$$k\mathbf{u} = ku_x\mathbf{e}_x + ku_y\mathbf{e}_y + ku_z\mathbf{e}_z$$

whilst the vector sum $\mathbf{u} + \mathbf{v}$ has components

$$\mathbf{u} + \mathbf{v} = (u_x + v_x)\mathbf{e}_x + (u_y + v_y)\mathbf{e}_y + (u_z + v_z)\mathbf{e}_z \tag{A.8}$$

A.3.4 The Dot (or Scalar) Product

The dot product of two vectors $\mathbf{u}$ and $\mathbf{v}$ is

$$\mathbf{u}\cdot\mathbf{v} = |\mathbf{u}||\mathbf{v}|\cos\theta \tag{A.9}$$

where θ is the angle between $\mathbf{u}$ and $\mathbf{v}$, and $|\mathbf{u}|$, $|\mathbf{v}|$ are the magnitudes of the vectors. If $\mathbf{u.v} = 0$ and neither $\mathbf{u}$ nor $\mathbf{v}$ is a zero vector, then we say that $\mathbf{u}$ and $\mathbf{v}$ are *orthogonal*.

Dot products obey the rules

$$\mathbf{u}\cdot\mathbf{v} = \mathbf{v}\cdot\mathbf{u}$$

$$\mathbf{u}\cdot(\mathbf{v} + \mathbf{w}) = \mathbf{u}\cdot\mathbf{v} + \mathbf{u}\cdot\mathbf{w} \tag{A.10}$$

and the Cartesian unit vectors satisfy

$$\mathbf{e}_x\cdot\mathbf{e}_x = \mathbf{e}_y\cdot\mathbf{e}_y = \mathbf{e}_z\cdot\mathbf{e}_z = 1$$

$$\mathbf{e}_x\cdot\mathbf{e}_y = \mathbf{e}_x\cdot\mathbf{e}_z = \mathbf{e}_y\cdot\mathbf{e}_z = 0 \tag{A.11}$$

It follows that the dot product of $\mathbf{u}$ and $\mathbf{v}$ can be written

$$\mathbf{u}\cdot\mathbf{v} = u_xv_x + u_yv_y + u_zv_z \tag{A.12}$$

and the modulus of vector $\mathbf{v}$ is

$$|\mathbf{v}| = (\mathbf{v}\cdot\mathbf{v})^{1/2} = (v_x^2 + v_y^2 + v_x^2)^{1/2} \tag{A.13}$$

It also follows that the angle between the vectors $\mathbf{u}$ and $\mathbf{v}$ is given by

$$\cos\theta = \frac{u_xv_x + u_yv_y + u_zv_z}{(u_x^2 + u_y^2 + u_z^2)^{1/2}(v_x^2 + v_y^2 + v_z^2)^{1/2}} \tag{A.14}$$

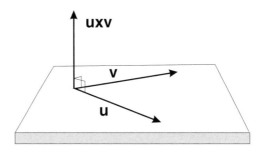

Figure A.5   The cross (or vector) product

A.3.5   The Cross (or Vector) Product

The cross product of two vectors **u** and **v** is

$$\mathbf{u} \times \mathbf{v} = |\mathbf{u}||\mathbf{v}| \sin \theta \mathbf{n} \tag{A.15}$$

where θ is the angle between **u** and **v** and **n** a unit vector normal to the plane containing the vectors **u** and **v**. The direction of this unit normal is given by the direction in which a screw would advance if rotated from the direction of **u** to the direction of **v** (Figure A.5).

The cross product obeys the laws

$$\mathbf{u} \times \mathbf{v} = -\mathbf{v} \times \mathbf{u}$$
$$\mathbf{u} \times (\mathbf{v} + \mathbf{w}) = \mathbf{u} \times \mathbf{v} + \mathbf{u} \times \mathbf{w} \tag{A.16}$$

The Cartesian unit vectors satisfy

$$\mathbf{e}_x \times \mathbf{e}_x = \mathbf{e}_y \times \mathbf{e}_y = \mathbf{e}_z \times \mathbf{e}_z = 0$$
$$\mathbf{e}_x \times \mathbf{e}_y = \mathbf{e}_z, \quad \mathbf{e}_y \times \mathbf{e}_z = \mathbf{e}_x, \quad \mathbf{e}_z \times \mathbf{e}_x = \mathbf{e}_y \tag{A.17}$$

and in terms of cartesian components

$$\mathbf{u} \times \mathbf{v} = (u_y v_z - u_z v_y)\mathbf{e}_x + (u_z v_x - u_x v_z)\mathbf{e}_y + (u_x v_y - u_y v_x)\mathbf{e}_z. \tag{A.18}$$

If $\mathbf{u} \times \mathbf{v} = \mathbf{0}$ and neither **u** nor **v** is a zero vector, then **u** and **v** are either parallel or antiparallel.

A.4   SCALAR AND VECTOR FIELDS

Mathematically, a field is a function that describes a physical property at points in space. In a *scalar field*, this physical property is completely described by a single value for each point (e.g., temperature, density, electrostatic potential). For *vector fields*, both a direction and a magnitude are required for each point (e.g. gravitation, electrostatic field intensity).

A.5   VECTOR CALCULUS

A.5.1   Differentiation of Fields

Suppose that the vector field $\mathbf{u}(t)$ is a continuous function of the scalar variable t. As t varies, so does $\mathbf{u}$ and if $\mathbf{u}$ denotes the position vector of a point P, then P moves along a continuous curve in space as t varies (Figure A.6)

For most of this book we will identify the variable t as time and so we will be interested in the trajectory of particles along curves in space.

By analogy with ordinary differential calculus, the ratio $\mathrm{d}\mathbf{u}/\mathrm{d}t$ is defined as the limit of the ratio $\delta\mathbf{u}/\delta t$ as the interval δt becomes progressively smaller.

$$
\begin{aligned}
\frac{\mathrm{d}\mathbf{u}}{\mathrm{d}t} &= \mathrm{Lim}\,\frac{\delta\mathbf{u}}{\delta t} \\
&= \mathrm{Lim}\left(\frac{\delta u_x}{\delta t}\,\mathbf{e}_x + \frac{\delta u_y}{\delta t}\,\mathbf{e}_y + \frac{\delta u_z}{\delta t}\,\mathbf{e}_z\right) \\
&= \frac{\mathrm{d}u_x}{\mathrm{d}t}\,\mathbf{e}_x + \frac{\mathrm{d}u_y}{\mathrm{d}t}\,\mathbf{e}_y + \frac{\mathrm{d}u_z}{\mathrm{d}t}\,\mathbf{e}_z
\end{aligned}
\tag{A.19}
$$

The derivative of a vector is the vector sum of the derivatives of its components. The usual rules for differentiation apply.

$$
\begin{aligned}
\frac{\mathrm{d}}{\mathrm{d}t}(\mathbf{u}+\mathbf{v}) &= \frac{\mathrm{d}\mathbf{u}}{\mathrm{d}t} + \frac{\mathrm{d}\mathbf{v}}{\mathrm{d}t} \\
\frac{\mathrm{d}}{\mathrm{d}t}(k\mathbf{u}) &= k\,\frac{\mathrm{d}\mathbf{u}}{\mathrm{d}t} \\
\frac{\mathrm{d}}{\mathrm{d}t}(f\mathbf{u}) &= \frac{\mathrm{d}f}{\mathrm{d}t}\,\mathbf{u} + f\,\frac{\mathrm{d}\mathbf{u}}{\mathrm{d}t}
\end{aligned}
\tag{A.20}
$$

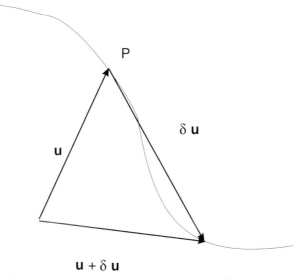

Figure A.6   Construct needed to discuss vector differentiation

where k is a scalar and f a scalar field. The following two rules relate to the scalar and vector products;

$$\frac{d}{dt}\mathbf{u}\cdot\mathbf{v} = \frac{d\mathbf{u}}{dt}\cdot\mathbf{v} + \mathbf{u}\cdot\frac{d\mathbf{v}}{dt}$$

$$\frac{d}{dt}\mathbf{u}\times\mathbf{v} = \frac{d\mathbf{u}}{dt}\times\mathbf{v} + \mathbf{u}\times\frac{d\mathbf{v}}{dt}$$

(A.21)

A.5.2 The Gradient

Suppose that $f(x,y,z)$ is a scalar field, and we wish to investigate how f changes between the points $\mathbf{r}$ and $\mathbf{r} + d\mathbf{r}$. Here

$$d\mathbf{r} = \mathbf{e}_x dx + \mathbf{e}_y dy + \mathbf{e}_z dz$$

We know from Section A.2 that

$$df = \left(\frac{\partial f}{\partial x}\right)dx + \left(\frac{\partial f}{\partial y}\right)dy + \left(\frac{\partial f}{dz}\right)dz$$

and so we write df as a scalar product

$$df = \left(\frac{\partial f}{\partial x}\mathbf{e}_x + \frac{\partial f}{\partial y}\mathbf{e}_y + \frac{\partial f}{\partial z}\mathbf{e}_z\right)\cdot(\mathbf{e}_x\,dx + \mathbf{e}_y\,dy + \mathbf{e}_z\,dz)$$

The first vector on the right-hand side is called the *gradient of f*, and it is written *grad f* in this text.

$$\mathrm{grad}\, f = \frac{\partial f}{\partial x}\mathbf{e}_x + \frac{\partial f}{\partial y}\mathbf{e}_y + \frac{\partial f}{dz}\mathbf{e}_z$$

(A.22)

An alternative notation involves the use of the so-called gradient operator ∇ (pronounced *del*)

$$\nabla = \frac{\partial}{\partial x}\mathbf{e}_x + \frac{\partial}{\partial y}\mathbf{e}_y + \frac{\partial}{\partial z}\mathbf{e}_z$$

(A.23)

and so the gradient of f is ∇f.

In spherical polar coordinates, the corresponding expression for grad f is

$$\mathrm{grad}\, f = \frac{\partial f}{\partial r}\mathbf{e}_r + \frac{1}{r}\frac{\partial f}{\partial \theta}\mathbf{e}_\theta + \frac{1}{r\sin\theta}\frac{\partial f}{\partial \phi}\mathbf{e}_\phi$$

(A.24)

grad f is a vector field whose direction at any point is the direction in which f is increasing most rapidly and whose magnitude is the rate of change of f in that direction. The spatial rate of change of the scalar field f in the direction of an arbitrary unit vector $\mathbf{e}$ is given by $\mathbf{e}\cdot\mathrm{grad}\, f$.

A.5.3 The Laplacian

It is worth noting at this point that the *Laplacian operator* $\nabla^2 = \nabla\cdot\nabla$ plays an important role in this text. In Cartesian coordinates,

$$\nabla^2 f = (\nabla\cdot\nabla)f = \frac{\partial^2 f}{\partial x^2} + \frac{\partial^2 f}{\partial y^2} + \frac{\partial^2 f}{\partial z^2}$$

(A.25)

and in spherical polar coordinates

$$\nabla^2 f = \frac{1}{r^2}\frac{\partial}{\partial r}\left(r^2\frac{\partial f}{\partial r}\right) + \frac{1}{r^2\sin\theta}\frac{\partial}{\partial\theta}\left(\sin\theta\frac{\partial f}{\partial\theta}\right) + \frac{1}{r^2\sin^2\theta}\frac{\partial^2 f}{\partial\phi^2} \tag{A.26}$$

A.6   VOLUME INTEGRALS OF SCALAR FIELDS

In order to evaluate quantities such as the mass or electric charge contained within a region of space, it is necessary to evaluate *volume integrals*.

For example, suppose that the electric charge density ρ inside a cubic box whose faces are the planes $x = 0$, $x = 1$; $y = 0$, $y = 1$ and $z = 0$, $z = 1$ is given by

$$\rho = \rho_0(x + y + z)$$

where ρ_0 is a constant (Figure A.7).

We divide each axis into differential elements, dxs along the x axis, dys along the y axis and dzs along the z axis, giving a number of infinitesimal differential volume elements each of volume dx dy dz. The charge enclosed by each differential volume element dx dy dz is ρ dx dy dz and so the total Q enclosed by the box is

$$Q = \int \rho\,\mathrm{d}\tau = \iiint \rho_0(x + y + z)\,\mathrm{d}x\,\mathrm{d}y\,\mathrm{d}z$$

To evaluate this volume integral we form three single integrals. First of all, we draw within the region a column having cross section dz, and constant x and y (Figure A.8).

In order to add these contributions to Q, we integrate between the limits of $z = 0$ and $z = 1$ to give

$$\int_0^1 \rho_0(x + y + z)\,\mathrm{d}z = \rho_0\left(x + y + \frac{1}{2}\right)$$

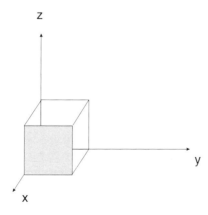

Figure A.7   A cubic volume

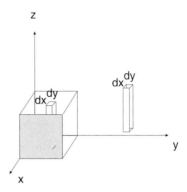

Figure A.8 The first step in integration

Next we form the y integral by drawing a slice parallel to the $y - z$ plane and including the column. The limits of the y integration are $y = 0$ and $y = 1$.

$$\int_0^1 \rho_0 \left(x + y + \frac{1}{2} \right) \mathrm{d}y = \rho_0(x + 1)$$

Finally we complete the volume by adding together all the slabs, and the limits of the x integration are $x = 0$ to $x = 1$.

$$\int_0^1 \rho_0(x + 1)\, \mathrm{d}x = \frac{3}{2}\, \rho_0$$

Sometimes the symmetry of a problem will lend itself to spherical polar coordinates. It can be shown that the volume element is

$$\mathrm{d}\tau = r^2 \sin\theta \mathrm{d}\,\theta \mathrm{d}\,\phi\, \mathrm{d}r \tag{A.27}$$

In the special case where the problem has no angular dependence, we can integrate over θ and ϕ to give

$$\mathrm{d}\tau = 4\pi r^2\, \mathrm{d}r \tag{A.28}$$

I have used the symbol $\mathrm{d}\tau$ to denote a volume element; other commonly used symbols are $\mathrm{d}V$, $\mathrm{d}v$, $\mathrm{d}\mathbf{r}$.

A.7 LINE INTEGRALS

Figure A.9 refers to a particle at point P moving along a curve in space under the influence of a force $\mathbf{F}$. The work done w in moving the short distance $\delta\mathbf{r}$ is $\mathbf{F} \cdot \delta\mathbf{r}$. In order to calculate the total work done as the particle moves from point A to point B, we divide the curve into N small segments with position vectors $\mathbf{r}_1, \mathbf{r}_2, \ldots \mathbf{r}_N$. We then have

$$w \approx \sum_{i=1}^{N} \mathbf{F}(\mathbf{r}_i) \cdot \delta\mathbf{r}_i \tag{A.29}$$

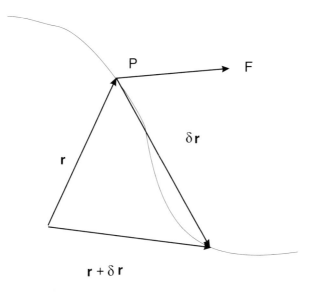

Figure A.9 Work done under the influence of a force

We let the number of points become large, and the summation approaches a limit called the *line integral*.

$$\int_C \mathbf{F} \cdot d\mathbf{r} = \mathrm{Lim} \sum_{i=1}^{N} \mathbf{F}(\mathbf{r}_i) \cdot \delta\mathbf{r}_i \qquad (A.30)$$

In this case it gives the work done by the force $\mathbf{F}$ as the body moves from point A to point B in space.

If the points A and B coincide, then the line integral refers to a closed curve and we denote the line integral

$$\oint_C \mathbf{F} \cdot d\mathbf{r} \qquad (A.31)$$

If the line integral of a certain vector field $\mathbf{A}$ is zero around any arbitrary closed path, then the vector field is called a *conservative field*. It can be shown that every conservative field can be written as the gradient of a suitable scalar field, related by

$$\mathbf{A} = -\mathrm{grad}\,\phi \qquad (A.32)$$

The importance of fields such as ϕ is that changes in ϕ depend only on the starting point and finishing point, and not on the path chosen to get between these points.

It can also be shown that the differential

$$M\,dx + N\,dy$$

has these properties provided that

$$\frac{\partial M}{\partial y} = \frac{\partial N}{\partial x}$$

In which case we call $M\,dx + N\,dy$ an *exact differential*.

A.8 MATRICES AND DETERMINANTS

A.8.1 Determinants

The set of simultaneous linear equations

$$a_{11}x_1 + a_{12}x_2 = b_1$$

$$a_{21}x_1 + a_{22}x_2 = b_2$$

has solution

$$x_1 = \frac{b_1 a_{22} - b_2 a_{12}}{a_{11}a_{22} - a_{12}a_{21}}$$

$$x_2 = \frac{b_2 a_{11} - b_2 a_{21}}{a_{11}a_{22} - a_{12}a_{21}}$$

If we define the symbol

$$D_2 = \begin{vmatrix} a_{11} & a_{12} \\ a_{21} & a_{22} \end{vmatrix} \equiv (a_{11}a_{22} - a_{12}a_{21}) \tag{A.33}$$

then these solutions may be written as

$$x_1 = \frac{\begin{vmatrix} b_1 & a_{12} \\ b_2 & a_{22} \end{vmatrix}}{D_2} \tag{A.34}$$

$$x_2 = \frac{\begin{vmatrix} a_{11} & b_1 \\ a_{21} & b_b \end{vmatrix}}{D_2} \tag{A.35}$$

It is usual to speak of the symbol defined by $A \dots$ as a *determinant of order 2*, and we speak of the elements, rows, columns and diagonal in an obvious manner.

Consideration of the set of three simultaneous linear equations

$$a_{11}x_1 + a_{12}x_2 + a_{13}x_3 = b_1$$

$$a_{21}x_1 + a_{22}x_2 + a_{23}x_3 = b_2$$

$$a_{31}x_1 + a_{32}x_2 + a_{33}x_3 = b_3$$

shows that the solution involves a denominator

$$D_3 = a_{11}a_{22}a_{33} - a_{11}a_{23}a_{32} + a_{21}a_{32}a_{13} - a_{21}a_{12}a_{33} + a_{31}a_{12}a_{23} - a_{39}a_{22}a_{13}$$

We define

$$D_3 = \begin{vmatrix} a_{11} & a_{12} & a_{13} \\ a_{21} & a_{22} & a_{23} \\ a_{31} & a_{32} & a_{33} \end{vmatrix}$$

which is called a determinant of order 3. By inspection, it can be written as

$$a_{11}\begin{vmatrix} a_{22} & a_{23} \\ a_{32} & a_{33} \end{vmatrix} - a_{21}\begin{vmatrix} a_{12} & a_{13} \\ a_{32} & a_{33} \end{vmatrix} + a_{31}\begin{vmatrix} a_{12} & a_{13} \\ a_{22} & a_{23} \end{vmatrix}$$

The three determinants of order 2 are called the minors of a_{11}, a_{21} and a_{31}, respectively; they are the determinants produced by striking out the first column and successive rows of D_3. A little analysis shows that D_3 may be expanded down any of its columns or along any of its rows by suitably combining products of elements and their minors.

A.8.2 Properties of Determinants

The value of a determinant is unchanged by interchanging the elements of all corresponding rows and columns. So for example

$$
\begin{vmatrix} a_{11} & a_{12} & a_{13} \\ a_{21} & a_{22} & a_{23} \\ a_{31} & a_{32} & a_{33} \end{vmatrix} = \begin{vmatrix} a_{11} & a_{21} & a_{31} \\ a_{12} & a_{22} & a_{32} \\ a_{13} & a_{23} & a_{33} \end{vmatrix}
$$

The sign of a determinant is reversed by interchanging any two of its rows or columns. So for example

$$
\begin{vmatrix} a_{11} & a_{12} & a_{13} \\ a_{21} & a_{22} & a_{23} \\ a_{31} & a_{32} & a_{33} \end{vmatrix} = - \begin{vmatrix} a_{11} & a_{13} & a_{12} \\ a_{21} & a_{23} & a_{22} \\ a_{31} & a_{33} & a_{32} \end{vmatrix}
$$

The value of a determinant is zero if any two of its rows (or columns) are identical. So for example

$$
\begin{vmatrix} a_{11} & a_{12} & a_{12} \\ a_{21} & a_{22} & a_{22} \\ a_{31} & a_{32} & a_{32} \end{vmatrix} = 0
$$

The value of a determinant is unchanged if equal multiples of any row (or column) are added to the corresponding elements of any other rows or columns. So for example

$$
\begin{vmatrix} a_{11} & a_{12} & a_{13} \\ a_{21} & a_{22} & a_{23} \\ a_{31} & a_{32} & a_{33} \end{vmatrix} = \begin{vmatrix} a_{11} & a_{12} + ka_{11} & a_{13} \\ a_{21} & a_{22} + ka_{21} & a_{23} \\ a_{31} & a_{32} + ka_{31} & a_{33} \end{vmatrix}
$$

A.8.3 Evaluation of Determinants

The inductive definition given above quickly becomes impractical for large determinants. Practical methods involve repeated use of the last property above in order to transform a given determinant to one such as

$$
\begin{vmatrix} a_{11} & a_{12} & a_{13} \\ 0 & a_{22} & a_{23} \\ 0 & 0 & a_{33} \end{vmatrix} = a_{11}a_{22}a_{33}
$$

We say that this determinant is in *triangular form*

A.8.4 Matrices

A matrix is a set of mn quantities arranged in a rectangular array of m rows and n columns, for example

$$\mathbf{A} = \begin{pmatrix} a_{11} & a_{12} & \cdots & a_{1n} \\ a_{21} & a_{22} & \cdots & a_{2n} \\ \vdots & \vdots & \cdots & \vdots \\ a_{m1} & a_{m2} & \cdots & a_{mn} \end{pmatrix} \tag{A.36}$$

Throughout this text, I will denote matrices by bold letters, just like vectors. The matrix above is said to be of *order m by n* denoted $(m \times n)$. If $m = n$, the matrix is said to be square of order n. The null matrix of order n has all $a_{ij} = 0$, and a square matrix whose only non-zero elements are the diagonal elements is said to be a *diagonal matrix*. Thus

$$A = \begin{pmatrix} a_{11} & 0 & 0 \\ 0 & a_{22} & 0 \\ 0 & 0 & a_{33} \end{pmatrix} \tag{A.37}$$

is a diagonal matrix. The unit matrix of order n is a diagonal matrix whose diagonal elements are all equal to 1.

A *row vector* is a matrix of order $1 \times n$ and a *column vector* is a matrix of order $n \times 1$. For example $(a_1, a_2, \ldots, a_n)$ is a row vector and

$$\begin{pmatrix} a_1 \\ a_2 \\ \vdots \\ a_n \end{pmatrix}$$

is a column vector.

A.8.5 The Transpose of a Matrix

Interchanging rows and columns of a determinant leaves its value unchanged, but interchanging the rows and columns of a matrix A produces a new matrix called the *transpose* $\mathbf{A}^{\mathrm{T}}$. Thus for example if

$$\mathbf{A} = \begin{pmatrix} a_{11} & a_{12} \\ a_{21} & a_{22} \\ a_{31} & a_{32} \end{pmatrix} \text{ then } \mathbf{A}^{\mathrm{T}} = \begin{pmatrix} a_{11} & a_{21} & a_{31} \\ a_{12} & a_{22} & a_{32} \end{pmatrix}$$

In the case that $\mathbf{A} = \mathbf{A}^{\mathrm{T}}$ then we say that the matrix $\mathbf{A}$ is *symmetric*. A symmetric matrix is square and has $a_{ij} = a_{ji}$.

A.8.6 The Hermitian Transpose of a Matrix

If $\mathbf{A}$ is a matrix with complex elements a_{ij}, then the *complex conjugate* $\mathbf{A}^*$ is the matrix with elements a_{ij}^*. That is to say, we take the complex conjugate of all the elements. For example, if

$$\mathbf{A} = \begin{pmatrix} 1+j & j \\ 2j+1 & 2 \\ -j & 3 \end{pmatrix} \text{ then } \mathbf{A}^* = \begin{pmatrix} 1-j & -j \\ -2j+1 & 2 \\ j & 3 \end{pmatrix}$$

(I have used the symbol j for the square root of -1; the symbol i is also used in other texts.)

The *hermitian transpose* $\mathbf{A}^+$ of matrix $\mathbf{A}$ is defined as the transpose of the complex conjugate. This is the same as the complex conjugate of the transpose, so if

$$\mathbf{A} = \begin{pmatrix} 1+j & j \\ 2j+1 & 2 \\ -j & 3 \end{pmatrix} \text{ then } \mathbf{A}^+ = \begin{pmatrix} 1-j & -2j+1 & j \\ -j & 2 & 3 \end{pmatrix}$$

A matrix that is equal to its own hermitian transpose is called a *hermitian* matrix. So for example

$$\mathbf{A} = \begin{pmatrix} 5 & 1+j \\ 1-j & 6 \end{pmatrix}$$

is a hermitian matrix. The diagonal elements of a hermitian matrix are always real.

A.8.7 Algebra of Matrices

If $\mathbf{A}$ and $\mathbf{B}$ are two matrices of the same order with elements a_{ij} and b_{ij}, then their sum $\mathbf{C} = \mathbf{A} + \mathbf{B}$ is defined as the matrix whose elements $c_{ij} = a_{ij} + b_{ij}$.

Two matrices $\mathbf{A}$ and $\mathbf{B}$ with elements a_{ij} and b_{ij} are equal only if they are of the same order, and all their corresponding elements are equal $a_{ij} = b_{ij}$.

The result of multiplying a matrix $\mathbf{A}$ whose elements are a_{ij} by a scalar k is a matrix whose elements are ka_{ij}.

The definition of *matrix multiplication* is such that two matrices $\mathbf{A}$ and $\mathbf{B}$ can only be multiplied together to form their product $\mathbf{AB}$ when the number of columns of $\mathbf{A}$ is equal to the number of rows of $\mathbf{B}$. Suppose $\mathbf{A}$ is a matrix of order $(m \times p)$, and $\mathbf{B}$ is a matrix of order $(p \times n)$. Their product $\mathbf{C} = \mathbf{AB}$ is a matrix of order $(m \times n)$ with elements

$$c_{ij} = \sum_{k=1}^{p} a_{ik} b_{kj}$$

Thus for example if

$$\mathbf{A} = \begin{pmatrix} 1 & 3 \\ 2 & 4 \\ 3 & 6 \end{pmatrix} \quad \mathbf{B} = \begin{pmatrix} 1 & 2 & 3 \\ 4 & 5 & 6 \end{pmatrix}$$

then

$$\mathbf{AB} = \begin{pmatrix} 13 & 17 & 21 \\ 18 & 24 & 30 \\ 27 & 36 & 45 \end{pmatrix} \qquad \mathbf{BA} = \begin{pmatrix} 14 & 29 \\ 32 & 68 \end{pmatrix}$$

so we see from this simple example that $\mathbf{AB} \neq \mathbf{BA}$. This non-commutative property of matrices appears even when $\mathbf{A}$ and $\mathbf{B}$ are such that their products have the same order. If $\mathbf{AB} = \mathbf{BA}$ then we say that the matrices *commute*.

A.8.8 The Inverse Matrix

Let $\mathbf{A}$ be the square matrix

$$\mathbf{A} = \begin{pmatrix} a_{11} & a_{12} & \cdots & a_{1n} \\ a_{21} & a_{22} & \cdots & a_{2n} \\ \vdots & \vdots & \cdots & \vdots \\ a_{n1} & a_{n2} & \cdots & a_{nn} \end{pmatrix}$$

If a suitable matrix $\mathbf{X}$ can be found such that $\mathbf{AX} = \mathbf{1}$ then we refer to $\mathbf{X}$ as the *inverse* of $\mathbf{A}$, and write it $\mathbf{A}^{-1}$. We say that $\mathbf{A}$ is *invertible* or *non-singular*. Not all matrices are invertible; for example,

$$\mathbf{A} = \begin{pmatrix} 1 & 1 \\ 1 & 1 \end{pmatrix}$$

is not invertible.

The statement above gives no clue as to the calculation of an inverse. Let me derive for you a formal relationship which in principle could be used to calculate the inverse of a matrix.

The *determinant of a matrix* $\mathbf{A}$ is only defined when the matrix is square, and it is then the determinant of the matrix elements. It is often written det $\mathbf{A}$.

The *adjoint* of $\mathbf{A}$ is defined as the transpose of the matrix of its cofactors. In other words, if A_{rs} is the cofactor formed by striking out the row and column in which a_{rs} occurs, then

$$\text{adj } \mathbf{A} = \begin{pmatrix} A_{11} & A_{21} & \cdots & A_{n1} \\ A_{12} & A_{22} & \cdots & A_{n2} \\ \vdots & \vdots & \cdots & \vdots \\ A_{1n} & A_{2n} & \cdots & A_{nn} \end{pmatrix} \tag{A.38}$$

Consider now the product $\mathbf{A}$ adj $\mathbf{A}$

$$\mathbf{A} \text{ adj } \mathbf{A} = \begin{pmatrix} a_{11} & a_{12} & \cdots & a_{1n} \\ a_{21} & a_{22} & \cdots & A_{n2} \\ \vdots & \vdots & \cdots & \vdots \\ A_{1n} & A_{2n} & \cdots & a_{nn} \end{pmatrix} \begin{pmatrix} A_{11} & A_{21} & \cdots & A_{n1} \\ A_{12} & A_{22} & \cdots & A_{n2} \\ \vdots & \vdots & \cdots & \vdots \\ A_{1n} & A_{2n} & \cdots & A_{nn} \end{pmatrix}$$

which comes to

$$\mathbf{A} \text{ adj } \mathbf{A} = \begin{pmatrix} \det A & 0 & \cdots & 0 \\ 0 & \det A & \cdots & 0 \\ \vdots & \vdots & \cdots & \vdots \\ 0 & 0 & \cdots & \det A \end{pmatrix}$$

So provided that det $\mathbf{A}$ is non-zero, we see that

$$\mathbf{A} \frac{\text{adj } \mathbf{A}}{\det \mathbf{A}} = 1 \qquad \text{(A.39)}$$

which gives a formal method for the calculation of an inverse

$$\mathbf{A}^{-1} = \frac{\text{adj } \mathbf{A}}{\det \mathbf{A}} \qquad \text{(A.40)}$$

If det $\mathbf{A} = 0$ then the matrix is singular.

A.8.9 Matrix Eigenvalues and Eigenvectors

Consider once again the $(n \times n)$ matrix $\mathbf{A}$

$$\mathbf{A} = \begin{pmatrix} a_{11} & a_{12} & \cdots & a_{1n} \\ a_{21} & a_{22} & \cdots & a_{2n} \\ \vdots & \vdots & \cdots & \vdots \\ a_{n1} & a_{n2} & \cdots & a_{nn} \end{pmatrix}$$

If we form the product of $\mathbf{A}$ with an arbitrary column vector $\mathbf{u}$ then sometimes it happens that the product $\mathbf{Au}$ is a linear multiple λ of $\mathbf{u}$.

$$\mathbf{Au} = \lambda \mathbf{u} \qquad \text{(A.41)}$$

In this case we say that $\mathbf{u}$ is an *eigenvector* of $\mathbf{A}$ with *eigenvalue* λ.

Thus for example if $n = 3$ and

$$\mathbf{A} = \begin{pmatrix} 0 & 1 & 0 \\ 1 & 0 & 1 \\ 0 & 1 & 0 \end{pmatrix}$$

then there are exactly three eigenvalues and associated eigenvectors.

$$\lambda_1 = -\sqrt{2} \quad \text{and} \quad \mathbf{u}_1^T = \left(-\frac{1}{2} \quad \frac{\sqrt{2}}{2} \quad \frac{1}{2} \right)$$

$$\lambda_2 = 0 \quad \text{and} \quad \mathbf{u}_2^T = \left(-\frac{1}{2} \quad \frac{\sqrt{2}}{2} \quad \frac{1}{2} \right)$$

$$\lambda_3 = -\sqrt{2} \quad \text{and} \quad \mathbf{u}_3^T = \left(-\frac{1}{2} \quad \frac{\sqrt{2}}{2} \quad \frac{1}{2} \right)$$

Suggestions for Further Reading

Burkert, U. and Allinger, N. L. (1982) *Molecular Mechanics*, American Chemical Society, Washington, DC.

Eyring, H., Walter, J. and Kimball, G. E. (1944) *Quantum Chemistry*, John Wiley & Sons, New York.

Fischer, C. F. (1977) *The Hartree–Fock Method for Atoms: A Numerical Approach*, John Wiley & Sons, New York.

French, A. P. and Taylor, E. F. (1978) *An Introduction to Quantum Physics*, van Nostrand, Reinhold (UK), Cambridge.

Grant, I. S. and Phillips, W. R. (1990) *Electromagnetism*, John Wiley & Sons, Chichester.

Hansen, J. P. and McDonald, I. R. (1986) *Theory of Simple Liquids*, Academic Press, London.

Hartree, D. R. (1957) *The Calculation of Atomic Structures*, John Wiley & Sons, New York.

Heyes, D. M. (1998) *The Liquid State. Applications of Molecular Simulations*, John Wiley & Sons, Chichester.

Herzberg, G. and Huber, K. P. (1979) *Constants of Diatomic Molecules*, van Nostrand, Princeton, NJ.

Hinchliffe, A. (1996) *Modelling Molecular Structures*, John Wiley & Sons, Chichester.

Kittell, C. and Kroemer, H. (1980) *Thermal Physics*, W. H. Freeman, San Francisco.

LeFèvre, R. J. W. (1938) *Dipole Moments*, Methuen, London.

Maczek, A. (1998) *Statistical Thermodynamics*, Oxford University Press, Oxford.

March, N. H. (1992) *Electron Density Theory of Atoms and Molecules*, Academic Press, London.

McWeeny, R. and Sutcliffe, B. T. (1969) *Methods of Molecular Quantum Mechanics*, Academic Press, London.

Murrell, J. N. and Jenkins, A. D. (1994) *Properties of Liquids and Solutions*, John Wiley & Sons, Chichester.

Open University Course S272 (1986) *The Physics of Matter, A Second Level Course*, Open University Press, Milton Keynes.

Parr, R. G. and Yang, W. (1989) *Density Functional Theory of Atoms and Molecules*, Oxford University Press, Oxford.

Poirier, R., Kari, R. and Csizmadia, I. G. (1985) *Handbook of Gaussian Basis Sets*, Elsevier, Amsterdam.

Pryde, J. A. (1969) *The Liquid State*, Hutchinson & Co., London.

Streitwieser, A. (1961) *Molecular Orbital Theory for Organic Chemists*, John Wiley & Sons, New York.

Wilson Jr, E. B., Decius, J. C. and Cross, P. C. (1955) *Molecular Vibrations. The Theory of Infrared and Raman Vibrational Spectra*, McGraw-Hill, New York.

Index